Introduction to Genomic Signal Processing with Control

Introduction to Genomic Signal Processing with Control

Aniruddha Datta
Edward R. Dougherty

CRC Press
Taylor & Francis Group
Boca Raton London New York

CRC Press is an imprint of the
Taylor & Francis Group, an informa business

CRC Press
Taylor & Francis Group
6000 Broken Sound Parkway NW, Suite 300
Boca Raton, FL 33487-2742

Printed in the United States of America on acid-free paper
10 9 8 7 6 5 4 3 2 1

International Standard Book Number-10: 0-8493-7198-8 (Hardcover)
International Standard Book Number-13: 978-0-8493-7198-1 (Hardcover)

Library of Congress Cataloging-in-Publication Data

Datta, Aniruddha, 1963-
Introduction to genomic signal processing with control / Aniruddha Datta and Edward R. Dougherty.
p. ; cm.
Includes bibliographical references and index.
ISBN-13: 978-0-8493-7198-1 (alk. paper)
ISBN-10: 0-8493-7198-8 (alk. paper)
1. Cellular signal transduction. 2. Genetic regulation. 3. Control theory. I. Dougherty, Edward R. II. Title.
[DNLM: 1. Gene Expression Regulation--physiology. 2. Genomics. 3. Models, Theoretical. 4. Signal Transduction--genetics.]
QU 475 D234i 2007]

QP517.C45D383 2007
571.7'4--dc22 2006024030

Dedication

TO MICHAEL L. BITTNER

We dedicate this book to Michael L. Bittner. A little more than a decade ago, he sat in a bagel shop somewhere in Maryland and explained to Edward Dougherty his belief that engineering methods might be used to intervene in and control genetic regulatory activity with the goal of mitigating the likelihood of disease or its progression. Much of our work in genomic signal processing, in particular the last chapter of this book, has been guided by the vision and active participation of Michael Bittner.

Preface

Recently, the Human Genome Project announced one of the most stunning achievements in the history of science: the generation of a reference sequence of the human genome. This has brought unprecedented interest to the area of genomics, which concerns the study of large sets of genes with the goal of understanding collective function, rather than that of individual genes. Such a study is important since cellular control and its failure in disease result from multivariate activity among cohorts of genes. Very recent research indicates that engineering approaches for prediction, signal processing and control are quite well suited for studying this kind of multivariate interaction. The aim of this book is to provide the readers with an introduction to genomic signal processing, including a state-of-the-art account of the use of control theory to obtain optimal intervention strategies for gene regulatory networks, and to make readers aware of some of the open research challenges.

Successful realization of the full potential of the area of genomics will require the collective skill and creativity of a diverse set of researchers such as biologists (including medical practitioners), statisticians and engineers. This can be possible only when each of these groups is willing and able to venture out of its immediate area of expertise, a task that is not always easy to undertake. We consider ourselves to be mathematically trained engineers who have had the opportunity to learn some basic biology while pursuing engineering research motivated by medical applications. Our main motivation for writing this book is to enable other interested readers with a similar mathematical background to get a reasonably good working knowledge of the genomics-related engineering research while having to expend only a small fraction of the time and effort that we ourselves had to originally invest.

Since the book is targeted primarily at readers whose biology background is practically non-existent, we felt that the book could be made self-contained only by including a substantial amount of material on molecular biology. However, it should be pointed out at the very outset that our molecular biology presentation, which is to a substantial extent gleaned from the classic, *Essential Cell Biology*, by Alberts et. al. [1], is *molecular biology seen through the eyes of engineers* and is the minimum needed for appropriately motivating the genomics-related engineering research presented here. As such, it is our sincere hope that the serious reader, after gaining the skeletal molecular biology knowledge from this text, will be motivated to consult other books such as the one by Alberts et. al. for additional details.

The book provides a tutorial introduction to the current engineering research in genomics. The necessary molecular biology background is presented, and techniques from signal processing and control are used to (i) unearth inter-gene relationships, (ii) carry out gene based classification of disease, (iii) model genetic regulatory networks and (iv) alter (i.e. control) their dynamic behavior. The book can be divided into two parts. In the first eleven chapters, the focus is on building up the necessary molecular biology background. No prior exposure to molecular biology is assumed. In the last six chapters, the focus is on discussing the application of engineering approaches for attacking some of the challenging research problems that arise in genomics-related research.

The book begins with a basic review of organic chemistry leading to the introduction of DNA, RNA and proteins. This is followed by a description of the processes of transcription and translation and the genetic code that is used to carry out the latter. Control of gene expression is also discussed. Genetic engineering tools such as microarrays, PCR, etc., are introduced and cell cycle control and tissue renewal in multi-cellular organisms is discussed. Cancer is introduced as the breakdown of normal cell cycle control. Having covered the basics of genomics, the book proceeds to engineering applications.

The engineering techniques of classification and clustering are shown to be appropriate for carrying out gene-based disease classification. Classification then leads naturally to expression prediction which in turn leads to genetic regulatory networks. Finally, the book concludes with a discussion of control approaches that can be used to alter the behavior of such networks in the hope that this alteration will move the network from undesirable (diseased) states to more desirable (disease-free) ones.

Several people contributed to the writing of this book in many different ways. First and foremost, we would like to thank the authors of *Essential Cell Biology* and Garland Publishers, Inc. for allowing us to use a large number of figures from that book. These figures, which have been identified by proper citations at appropriate places in the text, are so informative that it is hard to imagine how we could have effectively described some of the related material without being able to refer to them. Several of our doctoral students helped at different stages with the preparation of the book. Yufei Xiao helped during the initial stages with the preparation of the first set of class notes that ultimately became this book, Ashish Choudhary was a major player in helping us get the book to its final form, and Ranadip Pal assisted us during the intermediate stages on an as-needed basis.

We would like to thank R. Kishan Baheti, Director of the Power, Control and Adaptive Networks Program and John Cozzens, Director of the Theoretical Foundations Cluster Program at the National Science Foundation for supporting our research(Grant Nos. ECS-0355227 and CCF-0514644). Financial support from the National Cancer Institute (Grant No. CA90301), the National Human Genome Research Institute, the Translational Genomics Research Institute, the University of Texas M. D. Anderson Cancer Center, and the Texas A&M Department of Electrical and Computer Engineering is also thankfully acknowledged. Last, but not the least, we would like to thank our families without whose understanding and patience, this project could not have been completed.

Aniruddha Datta and Edward R. Dougherty
College Station, Texas
June 2006.

Contents

1

Introduction

In this introductory chapter, we provide a very brief introduction to cells and point out some general characteristics and associated terminology. A cell is the basic unit of life and all living creatures are made of cells. Nothing smaller than a cell can be truly called living, e.g., viruses, which are essentially genetic material encapsulated in a protein coat, have no ability to replicate by themselves, and are therefore nonliving. The only way in which they can replicate is by hijacking the replication machinery of a living cell, and this is what usually happens in a viral infection.

Each cell is typically about $5 - 20\mu m$ in diameter and this small size means that a cell can only be viewed under a microscope. The word *cell* is due to Robert Hooke, who in 1665 reported examining a piece of cork under a microscope and found it to be composed of a large number of small chambers, which he called *cells*. Fig. 1.1 shows the schematic diagram for a typical plant cell. Enclosing the entire cell is a limiting boundary called the *plasma membrane*. Inside the cell there are a number of compartments or *organelles*. Of these, the most prominent one is the *nucleus*, which serves as the store house of most of the genetic information. The *endoplasmic reticulum* is a continuation of the nuclear envelope and is the site at which many of the cell components are synthesized. The *Golgi apparatus* is the site where components synthesized in the endoplasmic reticulum undergo appropriate modifications before being passed on to their destinations. There are many other smaller organelles in a cell and some of these are shown in Fig. 1.1. The *mitochondria* are the sites at which most of the energy generation for the cell takes place; the *chloroplasts* for a plant cell are the sites where photosynthesis takes place; and *peroxisomes* are the organelles used to compartmentalize cellular reactions that release the dangerous chemical hydrogen peroxide. If one removes all the organelles from the cell, then what remains is called the *cytosol*.

Cells can be grouped into two broad categories depending on whether they contain a nucleus or not. *Procaryotes* (or procaryotic cells) do not contain a nucleus and other organelles. All bacteria are procaryotic cells. *Eucaryotes* (or eucaryotic cells) on the other hand are characterized by the presence of a nucleus. All multicellular organisms, including humans, are made up of eucaryotic cells, while a yeast would be an example of a unicellular eucaryote.

The cytosol in a cell is a concentrated aqueous (watery) gel of large and small molecules. The plasma membrane and the membranes of the other organelles, on the other hand, are made of *lipids*. These molecules have the

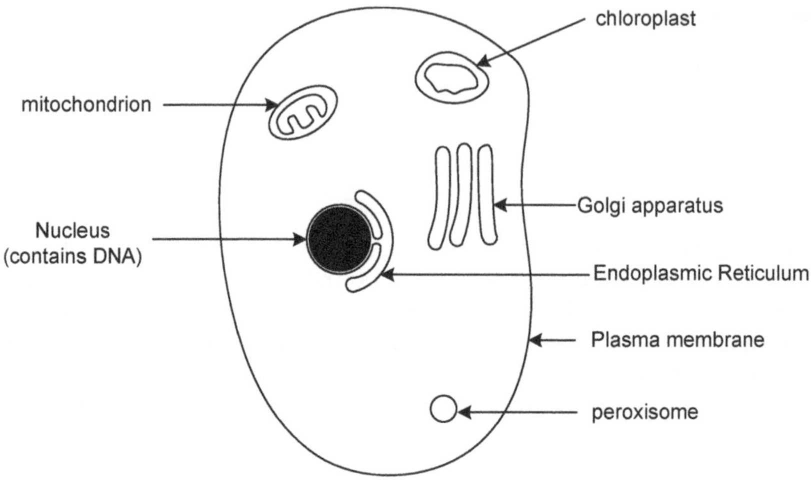

FIGURE 1.1 A typical cell

property that they have a *hydrophilic* (water loving) part and a *hydrophobic* (water hating) part, as shown in Fig. 1.2. So when they come in contact with water, they align appropriately, as shown in the figure, to form a *lipid bilayer*, which is the basis of all cell membranes. The lipid bilayer also facilitates the transportation of hydrophilic materials from the outside of a cell to its cytosol and vice versa, or for that matter from one membrane-bounded organelle to another. All that is required is that the hydrophilic material get encapsulated in a lipid bilayer, forming what is called a *vesicle* which can then be transported across the aqueous environment inside the cell to its appropriate destination organelle. Once the destination is reached, the lipid bilayer of the vesicle "merges" with that of the destination organelle with the result that the aqueous contents of the vesicle are transferred to the interior of the organelle. This vesicular transport is schematically illustrated in Fig. 1.2.

The cytosol of a eucaryotic cell contains what is called a *cytoskeleton*. As the name suggests, this provides structure to the interior of the cell and is made up of three types of proteins: (i) *actin filaments*; (ii) *microtubules*; and (iii) *intermediate filaments*. The cytoskeleton performs important functions such as (i) providing mechanical strength to the cell and its neighbors (intermediate filaments); (ii) generating contractile forces, for instance, in muscle cells (actin filaments); (iii) providing tracks along which different cell components can be moved (microtubules); and (iv) distributing the DNA properly between the daughter cells at cell division (microtubules).

Although cells from different organisms and even from the same organism vary enormously in size and function, they all have the same underlying chem-

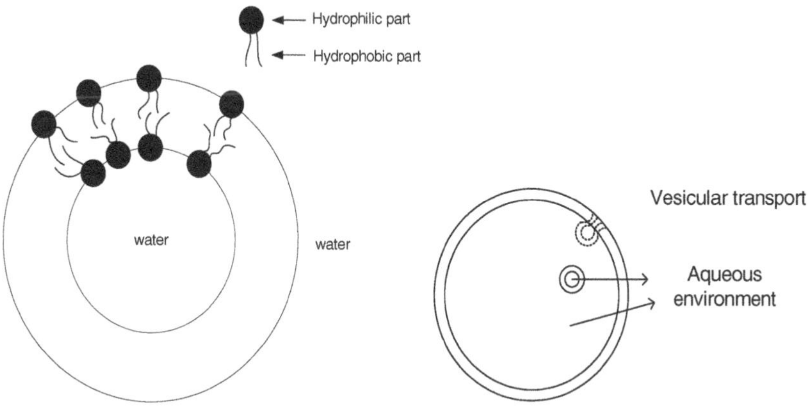

FIGURE 1.2 Lipid bilayer and vesicular transport

istry. As we will see later in this book, the genetic information in the DNA is coded using the same blocks in all organisms from humans all the way down to the simplest unicellular ones. Furthermore, this information is interpreted using essentially the same machinery to produce proteins and different proteins are made up of the same 20 blocks put together in different ways. This greatly simplifies the study of many aspects of molecular biology since they are not organism specific. In order to appreciate Genomic Signal Processing, one must have some basic understanding about genes, proteins and their interactions. Consequently, about one half of this book (the next 10 chapters) is devoted to buiding up this necessary background in sufficient detail. The details about the role of the cytoskeleton and the lipid membranes, although important in their own right, are not crucial to the theme of this book. Consequently, for these topics, only a superficial discussion has been included in this introductory chapter. For a more detailed treatment, the reader is referred to [1].

We conclude this chapter with a mention of *model organisms* that are used for molecular biology studies. These model organisms, which are used for carrying out experiments, must be simple and capable of quickly replicating themselves. For procaryotes, the *E. coli* bacteria, which contain a few genes and reproduce every 20 minutes, is, by far, the most widely used model of choice; for flowering plants, it is the plant *Arabidopsis* which produces thousands of offsprings in 8 to 10 weeks; for insects, it is the *Drosophila* or fruit fly; for mammals, it is rats and mice; and for unicellular eucaryotes, it is the yeast.

2

Review of Organic Chemistry

In this chapter, we provide a brief introduction to organic chemistry. The chemical properties of biological molecules play a crucial role in making all known life possible and so any discussion of molecular biology would have to necessarily include some discussion of organic chemistry. Although the discussion here is far from exhaustive, it is essentially self-contained and should provide a good introduction to anyone who has had some exposure to basic chemistry in the past. For a more detailed treatment, the reader is referred to [2].

Matter is made up of combinations of *elements* — substances such as hydrogen or carbon that cannot be broken down or converted into other substances by chemical means. An *atom* is the smallest particle of an element that still retains its distinctive chemical properties. *Molecules* are formed by two or more atoms of the same element or of different elements combining together in a chemical fashion. An atom is made up of three kinds of subatomic particles: *protons* (mass $= 1$, charge $= +1$); *neutrons* (mass $= 1$, charge $= 0$); and *electrons* (mass $\simeq 0$, charge $= -1$). Protons and neutrons reside in the nucleus of the atom while electrons revolve around the nucleus in certain orbits.

Each element has a fixed number of protons in the nucleus of each of its atoms and this number is referred to as the *atomic number.* We next list a few elements that occur over and over again in organic molecules, along with their atomic numbers: hydrogen (H) has an atomic number of 1; oxygen (O) has an atomic number of 8; carbon (C) has an atomic number of 6; nitrogen (N) has an atomic number of 7; phosphorous (P) has an atomic number of 15; sodium (Na) has an atomic number of 11; and calcium (Ca) has an atomic number of 20.

The total number of protons and neutrons in the nucleus of an atom of a particular element is referred to as its *atomic weight.* The hydrogen atom has only one proton and one electron and so its atomic weight is 1. Thus the atomic weight of an element is the mass of an atom of that element relative to that of a hydrogen atom. The electrons have negligible mass and do not contribute to the atomic weight. However, since the atom as a whole is electrically neutral, the number of protons must be equal to the number of electrons.

The number of neutrons in an atom of a particular element can vary. For instance, carbon usually occurs as atoms possessing 6 protons and 6 neutrons so that its usual atomic weight is 12. However, sometimes carbon atoms can

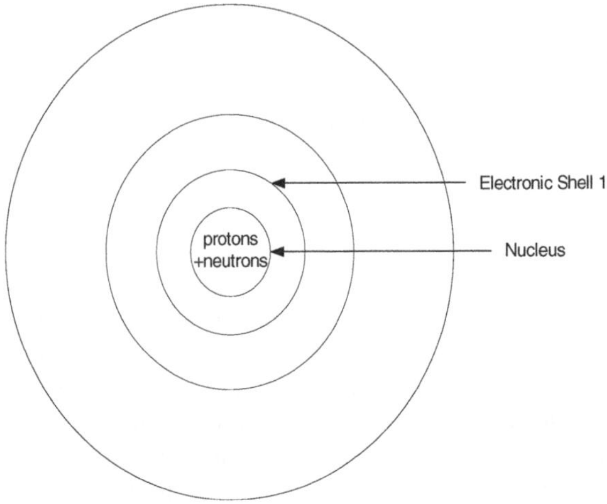

FIGURE 2.1 Schematic diagram of an atom.

have 8 neutrons resulting in a carbon atom with atomic weight 14. These two types of carbon atoms with varying numbers of neutrons are referred to as *isotopes*. Usually, one isotope of an element is the most stable one and the other isotopes radioactively decay towards it with time.

Fig. 2.1 shows the schematic diagram of a typical atom. The protons and neutrons are held together tightly in the nucleus while the electrons revolve around the nucleus in certain discrete orbits called shells. There is a limit to the number of electrons that can be held in each shell. The first shell (the one closest to the nucleus) can accommodate up to 2 electrons, the next one can accommodate up to 8 electrons, the next one up to 8 electrons, and so on. If an atom has its outermost electronic shell completely occupied, then it is unreactive. Helium (He, atomic number = 2), neon (Ne, atomic number = 10), argon (Ar, atomic number = 18) are all unreactive and are, therefore, called inert gases.

2.1 Electrovalent and Covalent Bonds

Atoms that do not have completely filled outer shells are capable of reacting with each other to form compounds. The natural tendency is to acquire completely filled outer shells by donating or accepting electrons (electrovalent or ionic bonds) or by sharing pairs of electrons (covalent bonds). For instance, sodium (Na) has an atomic number of 11 so that its electronic distribution is

$$2 + 8 + 1.$$

Chlorine (Cl), on the other hand, has an atomic number of 17, so that its electronic distribution is

$$2 + 8 + 7.$$

So, the sodium atom has a strong tendency to get rid of the single electron in its outermost shell while chlorine has a strong tendency to acquire one more electron so that it can have a complete outermost shell. Because of this, the sodium atom donates an electron to the chlorine atom to form the compound sodium chloride (or common salt). In the process, the sodium atom becomes a sodium ion with one positive charge (Na^+) while the chlorine atom becomes a chloride ion with one negative charge (Cl^-). The type of bond achieved by this transfer of electrons is called an *electrovalent bond* or *ionic bond.* Such bonds are extremely strong and are held together in place by the forces of electrostatic attraction between the two oppositely charged ions.

Another way in which an atom can achieve a completely filled outermost shell is by sharing pairs of electrons with other atoms. Consider the following three instances:

(1) Hydrogen with an atomic number of 1 has only one electron in its outermost shell. Two hydrogen atoms can share their electrons with each other so that each hydrogen atom can have an outer shell with two electrons. The result is the formation of a hydrogen molecule H_2.

(2) Carbon with an atomic number of 6 has the electronic distribution 2+4. Thus it can share its four outermost electrons with the electrons of four different hydrogen atoms. The result is the formation of the gas methane, CH_4.

(3) Oxygen with an atomic number of 8 has the electronic distribution 2+6. Thus, it can share two of its outermost electrons with the outermost electrons of two hydrogen atoms. The result is the formation of water, H_2O.

The bonds in the above examples, achieved by the sharing of pairs of electrons between atoms, are called *covalent bonds*. The number of electrons that an atom of an element must donate/accept/share to form a complete outermost electronic shell is referred to as its *valence*. Thus, from the above examples, hydrogen has a valence of 1, carbon has a valence of 4, oxygen has a valence of 2, sodium and chlorine each has a valence of 1. Since carbon has a valence of 4, each carbon atom can share up to four pairs of electrons with other atoms to achieve a complete outermost shell. Depending on the number of electron pairs that are shared, there can be different types of covalent bonds. A *single bond* is a covalent bond where only one pair of electrons is shared between the participating atoms. An example is ethane (C_2H_6), shown below, where all the participating carbon and hydrogen atoms are linked together by single bonds indicated by the single dashes. Each dash denotes that a pair of electrons is shared between the participating atoms. A *double bond* is formed when two pairs of electrons are shared between the participating atoms. An example is the gas ethene (C_2H_4), shown below, where the two dashes between the two carbon atoms indicate that two pairs of electrons are being shared.

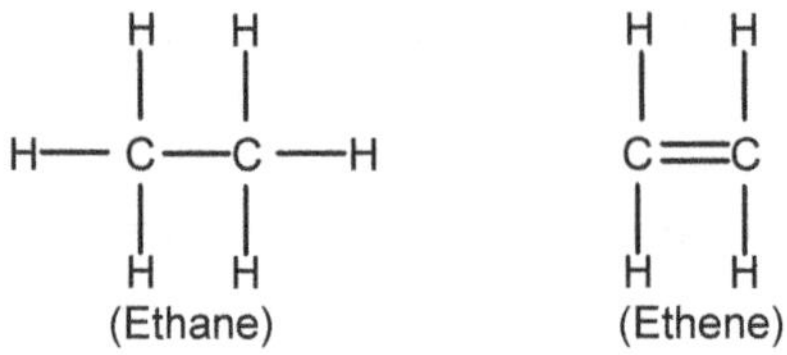

Double bonds are inherently more rigid. For instance, the single carbon-to-carbon bond in ethane allows free rotation about it for the right and left halves of the molecule. On the other hand, the two halves of the ethene molecule cannot be rotated freely about the carbon-to-carbon double bond without twisting the bond itself. This is referred to as *stearic hindrance* and can have profound consequences in determining the structure of a long biological molecule.

When the atoms joined by a single covalent bond belong to different elements, the two atoms usually attract the shared electrons to different degrees. For example, oxygen and nitrogen atoms attract electrons quite strongly while hydrogen attracts electrons quite weakly. Consequently, in a hydrogen-oxygen covalent bond, the hydrogen atom will tend to acquire a positive charge while the oxygen atom will tend to acquire a negative charge. This makes the overall molecule have an uneven charge distribution and such molecules are called *polar*. For instance, the covalent bond between oxygen and hydrogen $-\mathrm{O}-\mathrm{H}$ is polar. The covalent bond between nitrogen and hydrogen $-\mathrm{N}-\mathrm{H}$ is polar. However, the bond between carbon and hydrogen $-\mathrm{C}-\mathrm{H}$ is non-polar since carbon and hydrogen atoms both attract electrons more or less equally. Polar covalent bonds are extremely important in biology because they allow molecules to interact through electrical forces. Consider a molecule of water, $\mathrm{H_2O}$:

$$\mathrm{H}-\mathrm{O}-\mathrm{H}.$$

Here each hydrogen atom is positively charged while the oxygen atom is negatively charged. Thus, each hydrogen atom in a water molecule could form a weak ionic bond with the oxygen atom of another water molecule. Such a weak ionic bond is called a *hydrogen bond* and the hydrogen bonding between the water molecules is known to be the reason for water being a liquid at room temperatures.

Polar molecules readily dissolve in water because of their electrical interactions with the charges on the water molecules. Such substances are called *hydrophilic* (water loving). On the other hand, non-polar molecules such as the hydrocarbons do not dissolve in water. Such substances are called *hydrophobic* (water hating). A molecule that has both hydrophilic and hydrophobic parts is called *amphipathic*. From the point of view of molecular biology, it is important to note that water is the most abundant substance inside cells, and amphipathic molecules are crucial building blocks for all cell membranes.

2.2 Some Chemical Bonds and Groups Commonly Encountered in Biological Molecules

C-H Compounds

Compounds containing only carbon and hydrogen are called *hydrocarbons*. Examples are methane (CH_4) and ethane (C_2H_6), which have the structures shown below.

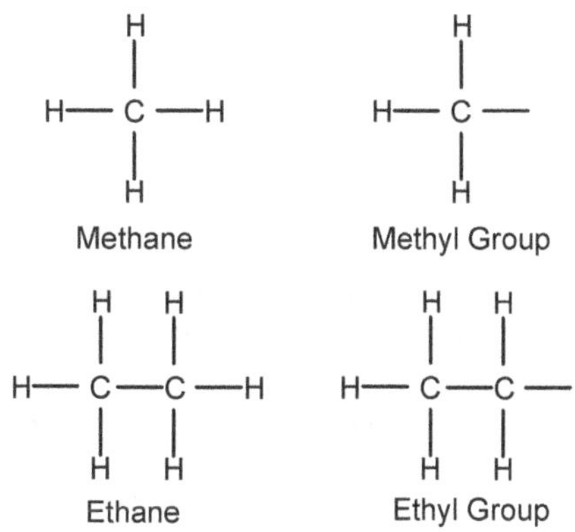

By removing one of the hydrogen atoms from methane, we obtain a highly reactive group called the *methyl group*. The *ethyl group* is similarly derived from ethane.

C–O Compounds

We next consider organic compounds that contain carbon and oxygen in addition to other elements. There are different classes of compounds that fall into this category.

Alcohols are characterized by the presence of the hydroxyl (OH) group and have the general formula R – OH. Here R could be any organic group. When R = CH_3, i.e. the methyl group, the corresponding alcohol is called *methyl alcohol* or *methanol*. When R = C_2H_5 i.e. the ethyl group, the corresponding alcohol is called *ethyl alcohol* or *ethanol*.

Aldehydes are characterized by the general formula

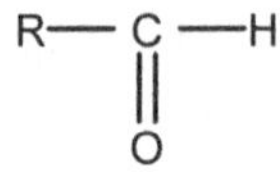

where R is any organic group. When $R = H$, the corresponding aldehyde is called *formaldehyde* and when $R = CH_3$, the corresponding aldehyde is called *acetaldehyde.*

Ketones are characterized by the general formula

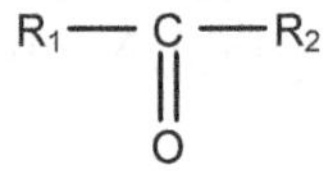

where R_1 and R_2 can be any two organic groups. The C = O (C double bonded with O) is called the *carbonyl* group and it is present in both aldehydes and ketones.

Carboxylic Acids are characterized by the general formula

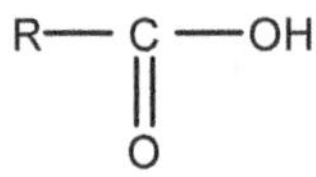

where R is any organic group. When $R = H$, the corresponding acid is called *formic Acid* while when $R = CH_3$, the corresponding acid is called *acetic Acid.* The COOH group present in all carboxylic acids is called a *carboxyl* group.

Recall from inorganic chemistry that substances that release hydrogen ions into solution are called *acids.*, e.g.,

$$\underset{\text{hydrochloric acid}}{\mathrm{HCl}} \longrightarrow \mathrm{H^+} + \mathrm{Cl^-}$$

i.e. hydrochloric acid dissociates in water to yield hydrogen ions and chloride ions. Carboxylic acids are called acids because they *partially* dissociate in solution to yield hydrogen ions:

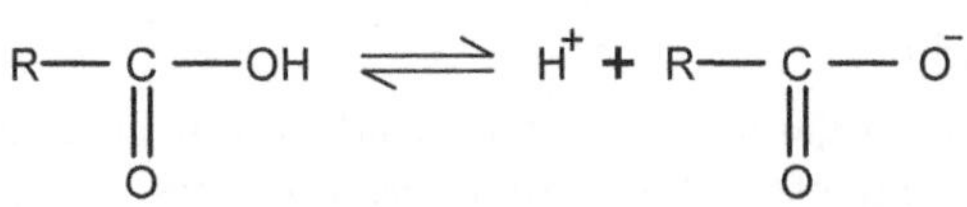

Because of this partial dissociation, organic acids are called *weak acids* as opposed to say hydrochloric acid, which is called a *strong acid.*

The hydrogen ion concentration of a solution is usually measured according to its pH value, which is defined by pH $= -\log_{10}[\mathrm{H}^+]$, where $[H^+]$ is the hydrogen ion concentration in moles per liter. For pure water, which is neither acidic nor basic, $[\mathrm{H}^+] = 10^{-7}$ so that the pH of pure water is 7. For acids $pH < 7$ since acids have a higher concentration of hydrogen ions than pure water. Bases, on the other hand, have a lower concentration of hydrogen ions than pure water and so for bases the $pH > 7$.

In inorganic chemistry, *bases* are defined as substances that reduce the number of hydrogen ions in aqueous solution. For instance, sodium hydroxide (NaOH) dissociates in solution to yield sodium ions and hydroxyl ions:

$$\mathrm{NaOH} \longrightarrow \mathrm{Na^+} + \mathrm{OH^-}.$$

The hydroxyl ions then react with some of the hydrogen ions in the water, thereby reducing the number of hydrogen ions. Thus, NaOH is a base as is the gas ammonia (NH_3) which reacts with a hydrogen ion in solution to produce the ammonium ion (NH_4^+):

$$\mathrm{NH_3} + \mathrm{H^+} \longrightarrow \mathrm{NH_4^+}.$$

Since many bases in inorganic chemistry do contain a hydroxyl group, one could say that alcohols in organic chemistry share some similarity with many of the bases in inorganic chemistry. Indeed, an important fact from inorganic chemistry is that an acid reacts with a base to produce a salt and water. For instance, sodium hydroxide reacts with hydrochloric acid to produce the common salt sodium chloride plus water:

$$\mathrm{NaOH} + \mathrm{HCl} \longrightarrow \mathrm{NaCl} + \mathrm{H_2O}.$$

In organic chemistry, when a carboxylic acid reacts with an alcohol, the reaction produces water and a compound called an *ester*. The chemical equation showing how the bonds get rearranged is given below:

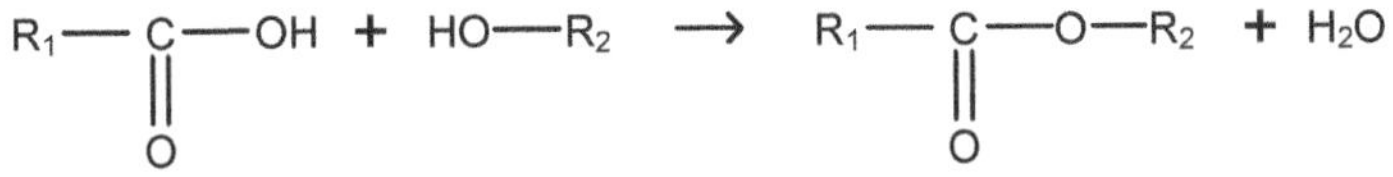

Here R_1 and R_2 are the paticular groups associated with the carboxylic acid and alcohol respectively.

C-N Compounds

There are many organic compounds that contain carbon and nitrogen, in addition to other elements. In fact, nitrogen occurs in several ring compounds, including important constituents of nucleic acids. Because of their fundamental importance to genomics, these ring compounds will be discussed in some detail in Section 2.3. For the present, we focus on *amines* which are organic compounds characterized by the presence of the amino (NH_2) group. A typical amine will have the general formula

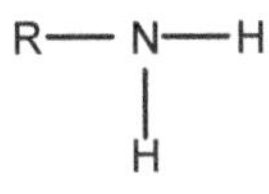

where R is any organic group. An amine can reversibly accept a hydrogen ion from water and, therefore, qualifies as a *weak base*:

$$R - NH_2 \ + \ H^+ \ \rightleftharpoons \ R - NH_3^+.$$

Furthermore, an amine can react with a carboxylic acid to produce an *amide* and water. The general reaction is given below:

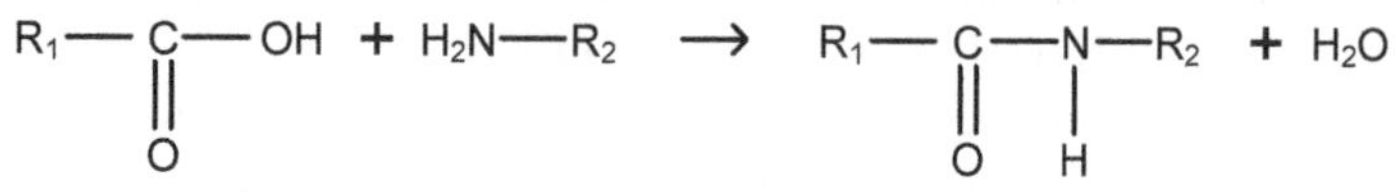

Phosphates

Phosphates are characterized by the presence of the inorganic phosphate ion, which is derived from phosphoric acid (H_3PO_4) by the loss of two hydrogen ions:

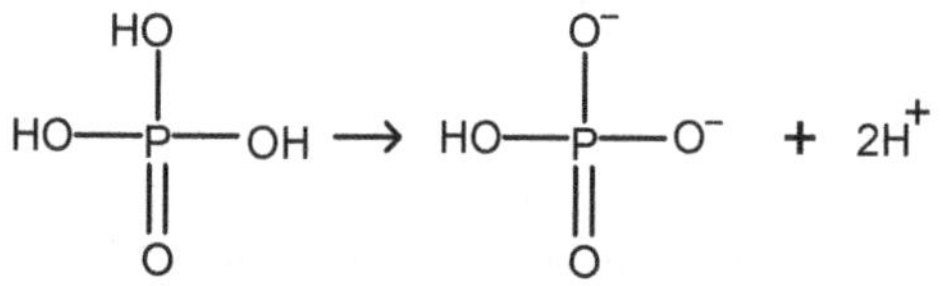

The phosphate ion can take part in many reactions that play an important role in molecular biology. A few instances are listed below.
(i) The inorganic phosphate ion can react with an alcohol to produce a *phosphate ester*:

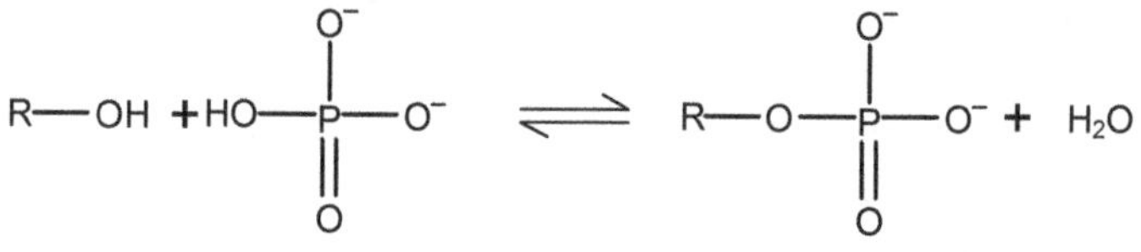

Phosphate ester bonds as shown above are important in the formation of nucleic acids such as DNA and RNA.
(ii) The inorganic phosphate ion can react with the carboxyl group of an organic acid to produce a *carboxylic-phosphoric acid anhydride*.

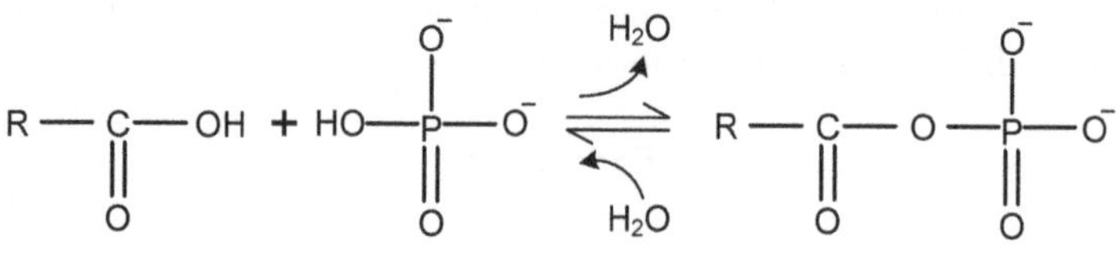

The compound formed above is called an *anhydride*, since it results from the removal of a water molecule.
(iii) Two inorganic phosphate ions can react together to produce a phosphoanhydride bond:

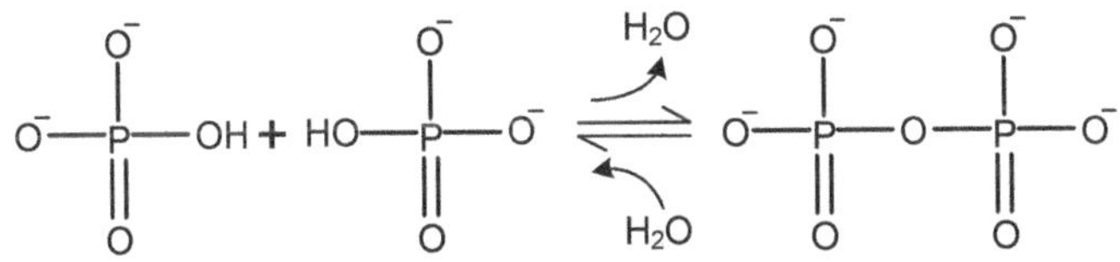

Phosphoanhydride bonds are high energy bonds and the reversibility of the above reaction makes such bonds suitable for energy storage and transfer. In this connection, we note that each of the above three reactions can be reversed by the addition of a water molecule. In organic chemistry, one encounters many reactions of this type.

2.3 Building Blocks for Common Organic Molecules

If we disregard water, almost all the molecules in a cell are based on *carbon*. Furthermore, most of the molecules are giant *macromolecules*. The small organic molecules constitute a very small fraction of the total cell mass. However, fortunately, even the giant macromolecules are made up of small organic molecules that are bonded together by covalent bonds. Cells contain four major families of small organic molecules, or modular units, that are combined together to form the large macromolecules:

1. *Sugars*, which are the building blocks for more complex sugars and carbohydrates;

2. *Fatty acids*, which are the building blocks for fats, lipids and all cell membranes;

3. *Amino acids*, which are the building blocks for proteins; and

4. *Nucleotides*, which are the building blocks for nucleic acids such as DNA and RNA.

Let us focus on these building blocks one by one.

2.3.1 Sugars

The simplest sugars (monosaccharides) are compounds with the general formula $(CH_2O)_n$ where n is usually $3, 4, 5, 6$ or 7. Sugars are also called *carbohydrates* since their general formula suggests that they are somehow built up from carbon and water. Monosaccharides usually occur as aldehydes or ketones. Five-carbon sugars, called *pentoses* and six-carbon sugars, called *hexoses* are very important in molecular biology. Both the aldehyde and ketone versions of these sugars are shown in Figs. 2.2 and 2.3.

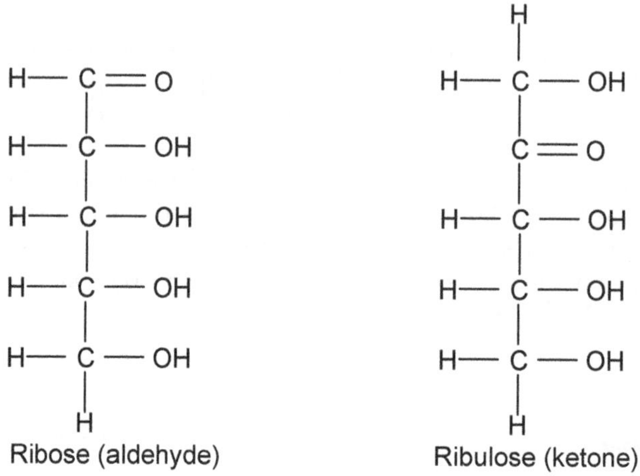

FIGURE 2.2 5-carbon sugars

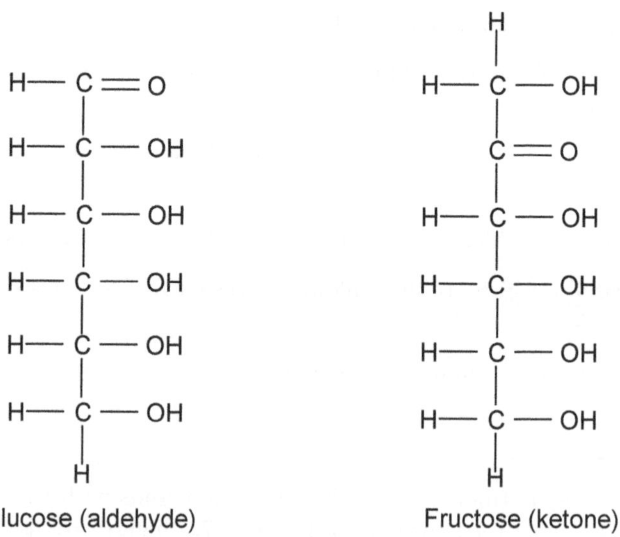

FIGURE 2.3 6-carbon sugars

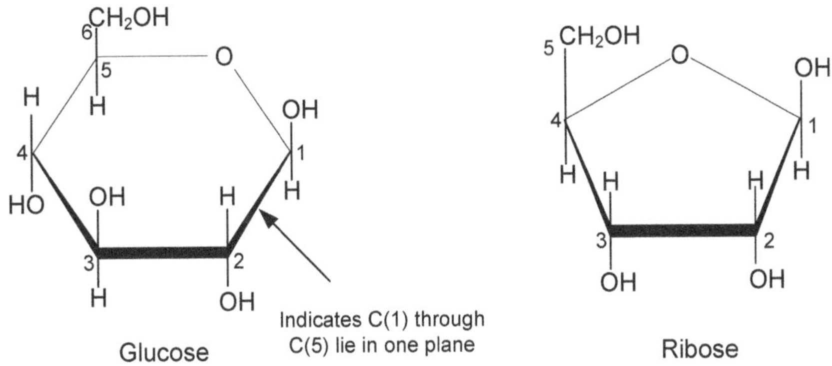

FIGURE 2.4 Ring formation in glucose and ribose

The 5-carbon sugar *ribose* and its derivative *deoxyribose* are important constituents of nucleic acids such as DNA and RNA while the 6-carbon sugar *glucose* serves as an important source of energy. For aldehyde sugars, in the structural representation, the carbon atoms are numbered consecutively with the carbon atom containing the aldehyde group being numbered 1.

In aqueous solution, the aldehyde or ketone group of a sugar molecule tends to react with a hydroxyl group of thc same molecule, thereby closing the molecule into a ring. The ring structures for the sugars glucose and ribose are shown in Fig. 2.4. Since the environment inside a cell is aqueous, it is this ring structure that we will be encountering over and over again in this text and, indeed, in any book on molecular biology. If one were to replace the hydroxyl group at carbon number 2 in ribose with hydrogen then one obtains *deoxyribose*, which is the sugar present in DNA.

In inorganic chemistry, the chemical formula of a compound usually determines a unique structural formula and associated properties. In organic chemistry, on the other hand, compounds having the same chemical formula may have totally different structural formulae and properties. For instance, if we interchange the hydrogen and the hydroxyl group at carbon No.4 in glucose we obtain a different sugar called *galactose*. In another instance, if we interchange the hydrogen and the hydroxyl groups at carbon No.2 in glucose, we obtain the sugar *mannose*. Compounds which have the same chemical formula but different structural formulae are called *isomers* of each other. Organic chemistry is filled with instances of isomers whose different properties make them especially well suited for particular biological functions.

The monosaccharide building blocks can be combined together to yield more complex sugars. Disaccharides are made up of two monosaccharides that are linked together in a *condensation* reaction. This reaction, which is accompanied by the removal of a water molecule, is schematically shown in Fig. 2.5. Since there are many hydroxyl groups on each monosaccharide, two monosaccharides can link together in many different ways. Furthermore, since there are additional hydroxyl groups available on the disaccharide, more monosaccharides can get linked to it, producing chains and branches of various lengths. Short chains are called *oligosaccharides*, while long chains are called *polysaccharides*. An example of a polysaccharide is *glycogen*, which is made up entirely of glucose units linked together. Glycogen serves as an energy storage in animals.

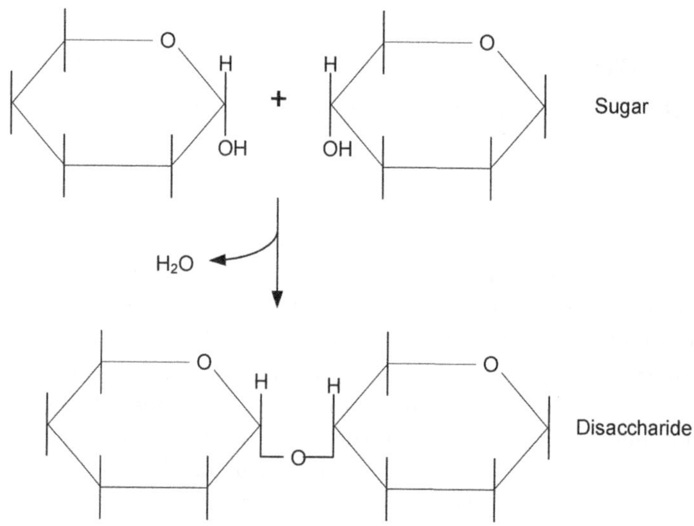

FIGURE 2.5 Condensation reaction showing the formation of a disaccharide.

Condensation reactions are not unique to sugars. In fact, as we will see later in this chapter, they are used to form proteins from amino acids and nucleic acids from nucleotides. The reverse reaction of condensation is called *hydrolysis* (splitting with water) and plays an important role in our digestion of carbohydrates and other foods. We conclude our discussion of sugars by providing some specific examples of disaccharides that can result from linking together two monosaccharide units:

glucose + glucose = maltose;
glucose + galactose = lactose (the sugar found in milk); and
glucose + fructose = sucrose.

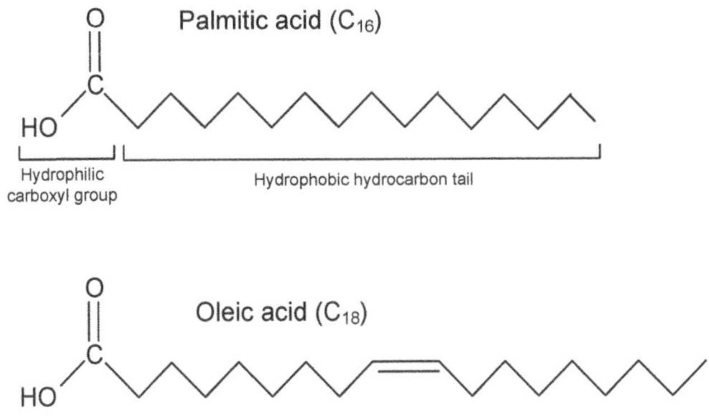

FIGURE 2.6 Fatty acids

2.3.2 Fatty Acids

A fatty acid molecule has two distinct regions: (i) a long hydrocarbon chain, which is hydrophobic and (ii) a carboxyl group, which behaves as a carboxylic acid, is ionized in solution and is extremely hydrophilic. Two examples of fatty acids are shown in Fig. 2.6. Palmitic acid (with 16 carbon atoms) is a *saturated* fatty acid. Saturated fatty acids are characterized by the absence of double bonds between the carbon atoms. Oleic acid, on the other hand, has 18 carbon atoms and is *unsaturated* since it has a double bond between a pair of carbon atoms. The double bond creates stearic hindrance, which is an important factor in determining the final shape of the molecule. This shape, in turn, may be crucial in allowing other molecules in the cell to recognize a molecule of oleic acid uniquely. Fatty acids serve as *concentrated* food reserves in cells, as they can be broken down to produce about six times as much usable energy, weight for weight, as glucose. They are usually stored in cells as triacylglycerol molecules (triglycerides), which consist of three fatty acids joined to a glycerol backbone. The structure of a typical triglyceride is shown in Fig. 2.7 where the three hydroxyl groups in a glycerol molecule have been replaced by three fatty acids with hydrocarbon chains R_1, R_2 and R_3.

Because of the presence of the carboxyl group, fatty acids can react with alcohols and amines to form esters and amides respectively.

Phospholipids are the major constituents of cell membranes. In phospholipids, two of the hydroxyl groups in glycerol are linked to fatty acids, while the third hydroxyl group is linked to phosphoric acid. The phosphate is further linked to one of a variety of small polar groups. The typical structure of a phospholipid is shown in Fig. 2.8.

Since a fatty acid possesses a hydrophilic part and a hydrophobic part, it

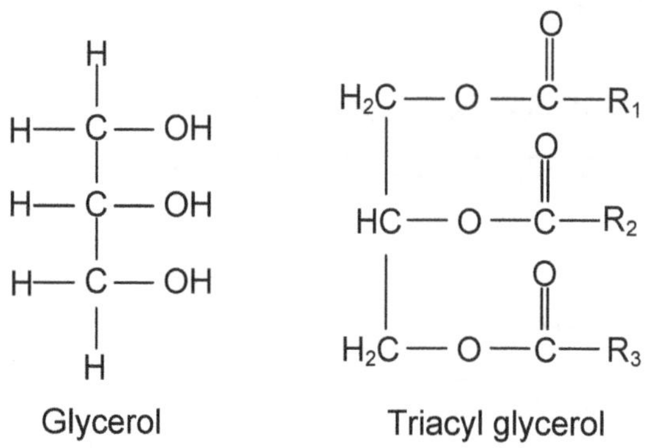

FIGURE 2.7 Structure of triglycerides

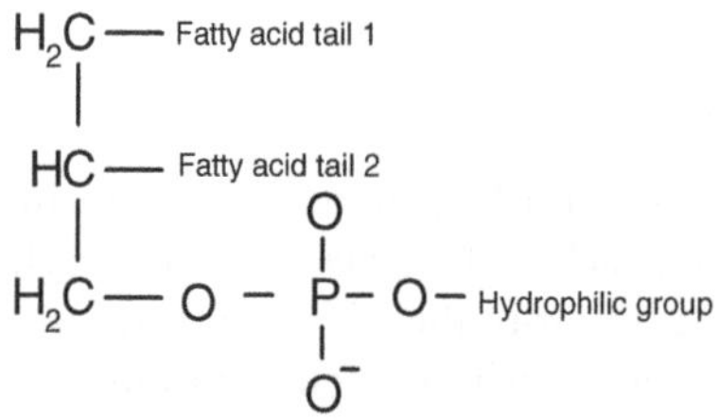

FIGURE 2.8 A typical phospholipid

can form a surface film or small micelles* when it comes in contact with water. Because of similar reasons, phospholipids and glycolipids (lipids made up of two fatty acids for the hydrophobic part and one or more sugar residues for the hydrophilic part) form self-sealing lipid bilayers that are the basis for all cellular membranes. The formation of the surface film and the micelles is illustrated in Fig. 2.9 while the schematic diagram leading to the formation of the lipid bilayer is shown in Fig. 2.10.

2.3.3 Amino Acids

Amino acids are the subunits of proteins — they possess both a carboxylic acid group and an amino group, both linked to a single carbon atom called the *alpha*-carbon. Amino acids differ from each other because of the different side chains that are also attached to the *alpha*-carbon. A typical amino acid has the following structural formula:

*A globular aggregation of molecules.

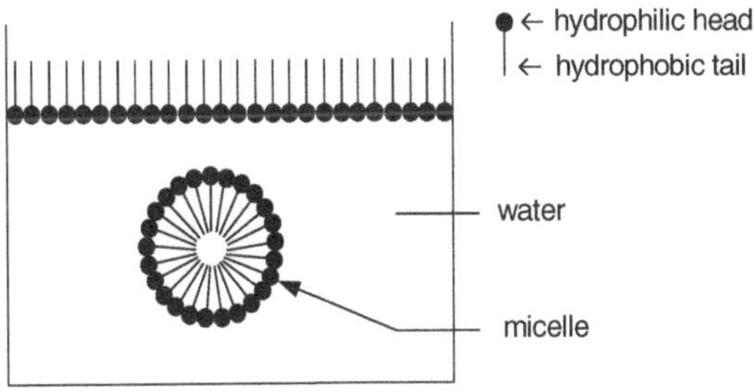

FIGURE 2.9 Formation of a surface film and a micelle.

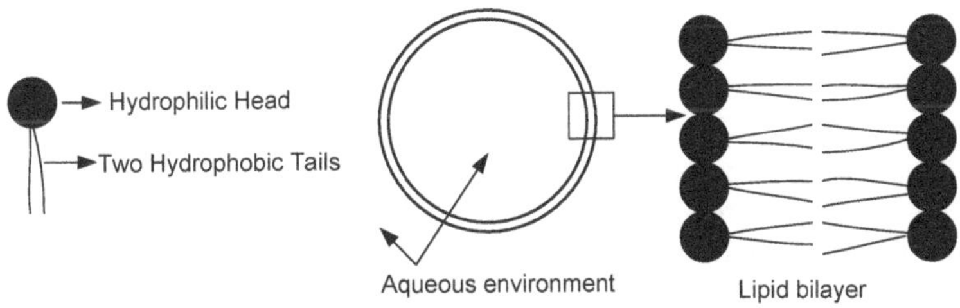

FIGURE 2.10 Formation of the lipid bilayer.

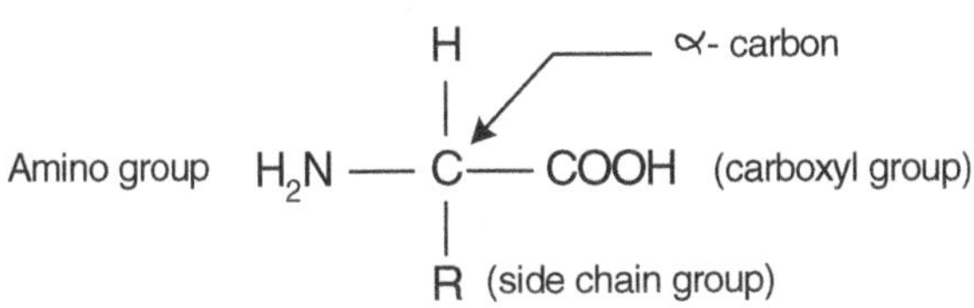

where R is one of 20 different side chains. It is remarkable that the same 20 side chains occur over and over again in nature. The 20 amino acid side chains can be classified under different groups depending on their properties:

1. Basic side chains: These side chains usually contain a nitrogen atom/ amino group which takes up a H^+ ion in solution. Examples of amino acids with basic side chains are lysine, arginine and histidine.

2. Acidic side chains: These amino acids contain a carboxyl group in the side chain. Examples of acidic side chains are R=CH_2COOH corresponding to aspartic acid, and R=C_2H_4COOH corresponding to glutamic acid.

3. Uncharged polar side chains: These amino acids contain a hydrophilic side chain containing an amino group or a hydroxyl group. Examples of amino acids with uncharged polar side chains are asparagine, glutamine, serine, threonine and tyrosine.

4. Non-polar side chains: These side chains contain hydrocarbons and are usually hydrophobic or non-polar. Examples of amino acids with non-polar side chains are alanine, valine, leucine, isoleucine, proline, phenylalanine, methionine, tryptophan, glycine and cysteine.

The structural formula for a representative amino acid (cysteine) is shown below:

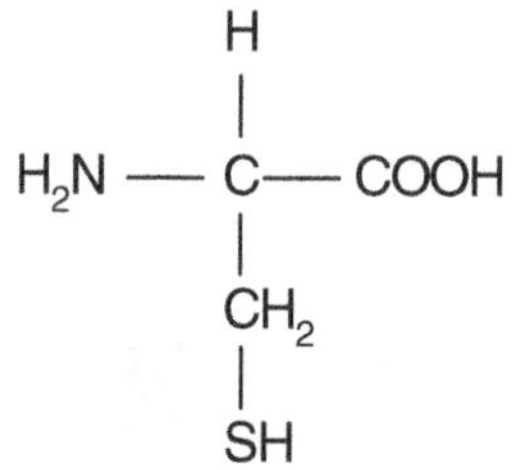

(cysteine)

Disulphide bonds (S-S bonds) can form between two cysteine side chains and can play an important role in determining the 3-dimensional shape of the protein containing these cysteines.

Amino acids can be linked to each other via *peptide bonds* which are essentially amide linkages. The carboxyl group of one amino acid reacts with the amino group of the next amino acid and the two get linked together in a condensation reaction which, as usual, is accompanied by the removal of a water molecule. The condensation reaction linking together two amino acids with side chains R_1 and R_2 is shown in Fig. 2.11. From Fig. 2.11 we note that the resulting molecule has a well-defined directionality: an amino group on one end and a carboxyl group on the other. Hence, it is possible to attach additional amino acids to the existing chain via further condensation reactions. The final chain that results will also have an amino group at one end (called the *amino terminus* or *N-terminus*) and a carboxyl group at the other end (called the *carboxyl terminus* or *C-terminus*). In Fig. 2.11, the four atoms in the box form a rigid planar unit so that there is no rotation about the C-N bond. However, the two bonds linking the α-carbons to the rest of the polypeptide chain allow rotation so that long chains of amino acids are very flexible. Since polypeptide chains can be of any length and at each location we can have 1 out of 20 possible amino acids, there is an enormous variety of proteins that can be synthesized using peptide bonds between amino acids.

2.3.4 Nucleotides

Nucleotides are the subunits of nucleic acids such as DNA and RNA. A nucleotide is a molecule made up of a nitrogen-containing compound linked to a 5-carbon sugar that carries one or more phosphate groups. The nitrogen-containing ring compounds (bases) are of two types: *pyrimidines* and *purines*.

As shown in Fig. 2.12, the pyrimidines are characterized by the presence of a six-member ring, while purines are characterized by the presence of a

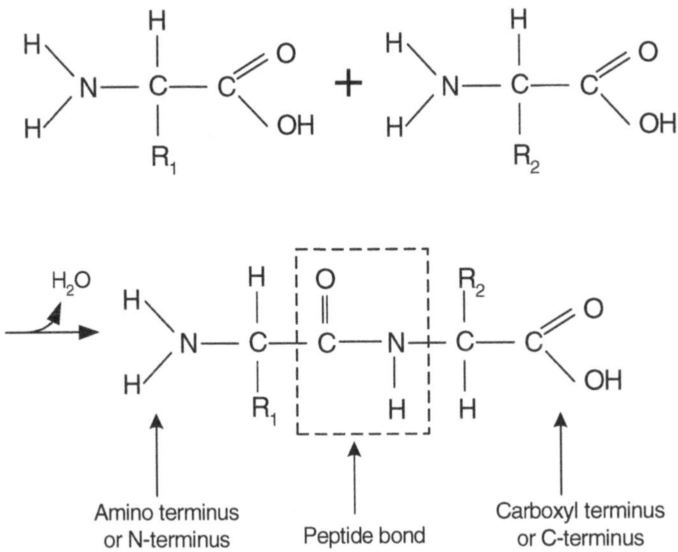

FIGURE 2.11 Condensation reaction linking together two amino acids.

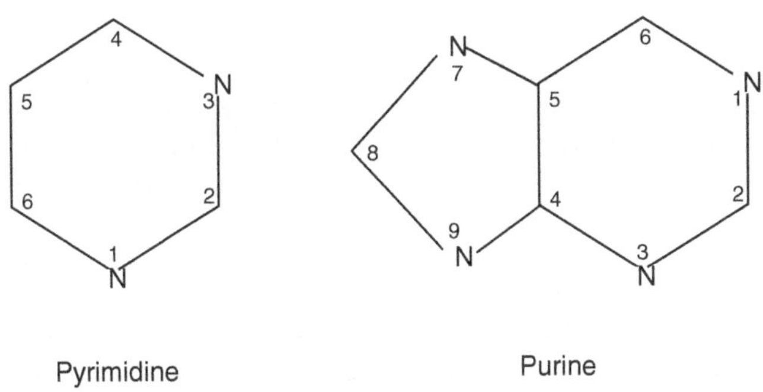

FIGURE 2.12 The two different categories of bases for nucleotides.

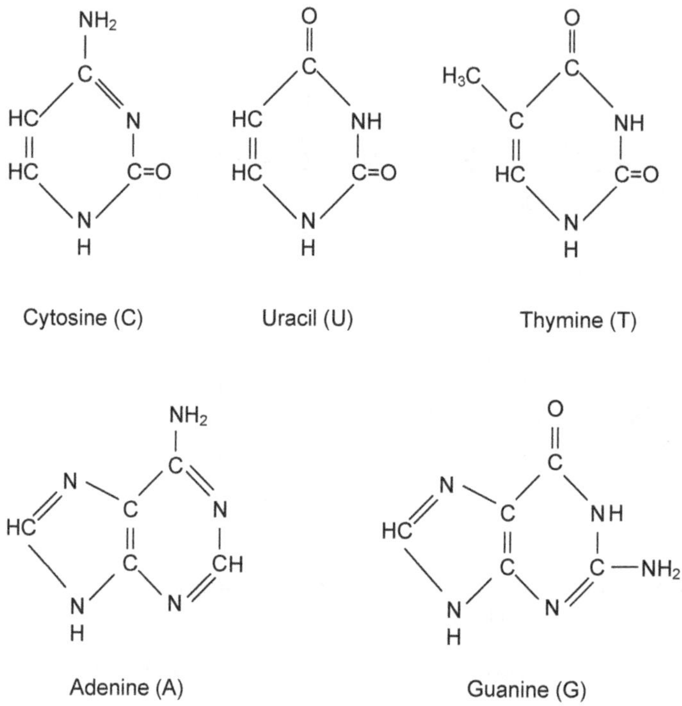

FIGURE 2.13 The different bases for nucleotides.

nine-member ring. The atoms in each case are numbered according to the schemes shown in the figure. Fig. 2.13 shows the structural formulae for the different bases that are found in nucleic acids. There are three pyrimidines namely, *cytosine* (C), *uracil*(U) and *thymine* (T) with thymine differing from uracil in that the hydrogen at position No. 5 in uracil is replaced by a methyl group to obtain thymine. There are two purines which are called *adenine* (A) and *guanine* (G). Although we have included the structural formula for all these five bases here, it should be pointed out that for our purposes, it is not necessary to know the detailed structure of each base. All that needs to be kept in mind is that there are five different bases resulting in five different nucleotides and these provide the variety that is needed to produce a particular nucleic acid by stringing such units together. Let us now describe how a nucleotide is formed starting from a nitrogenous base. A nucleotide is made up of three components: (i) a nitrogenous base, (ii) a 5-carbon sugar (pentose) and (iii) an inorganic phosphate ion. For the nitrogenous base, we could pick any of the five bases shown in Fig. 2.13 while for the pentose, we could have the sugar ribose or its derivative deoxyribose, both of which are shown in Fig. 2.14. The phosphate ion, of course, is the unique one introduced

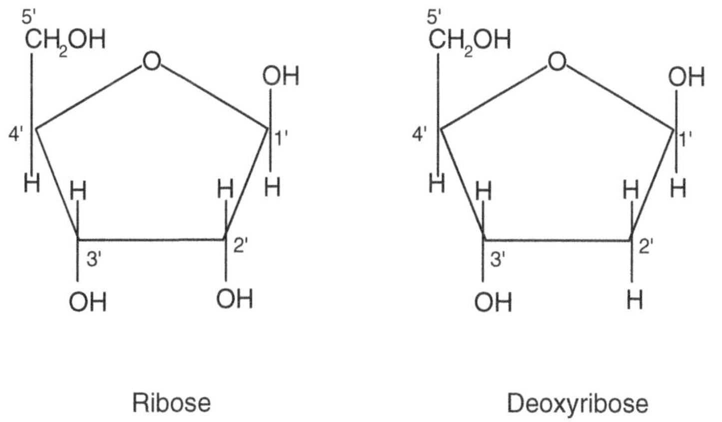

FIGURE 2.14 Pentoses: ribose and deoxyribose in aqueous solution.

earlier in Section 2.2. Suppose, for instance, we have chosen the base cytosine and the sugar ribose. Then, the formation of the corresponding nucleotide is shown in Fig. 2.15. Here the carbon atoms in the sugar are numbered $1'$ through $5'$ since the numbers 1 through 6 for pyrimidines and 1 through 9 for purines have already been taken up for labeling the atoms forming the ring in the nitrogenous base. In Fig. 2.15, the inorganic phosphate is linked to the pentose through a phospho-ester bond at the $5'$ location on the pentose. The combination of the base and the sugar is called a *nucleoside*. If the sugar is ribose, then the bases adenine, guanine, cytosine, thymine and uracil give rise to the nucleosides adenosine, guanosine, cytidine, thymidine and uridine respectively. If the sugar is deoxyribose, the corresponding nucleosides are called deoxyadenosine, deoxyguanosine and so on. As already mentioned, the nitrogenous base, sugar and phosphate ion when combined together as in Fig. 2.15 constitute a nucleotide.

Nucleic acids are formed by stringing nucleotide units together. If the sugar is deoxyribose and the bases are A, G, C, T, we have *deoxyribonucleic acid* or DNA. If the sugar is ribose and the bases are A, G, C, U, then we have *ribonucleic acid* or RNA. In either case, the phosphate group on the $5'$ end of an incoming nucleotide links together with the $3'$ hydroxyl group on the sugar of the previous nucleotide via a *phosphodiester bond* and the elongation of the chain continues by repeating the process with a new incoming nucleotide. The formation of nucleic acids from nucleotides will be treated in more detail in a later chapter.

In addition to serving as the building blocks for DNA and RNA, nucleotides have many other functions: (i) they can serve as store houses of chemical energy; (ii) they combine with other groups to form activated carrier molecules or *coenzymes*; and (iii) they are used as specific signaling molecules in the cell, e.g. cyclic AMP (cAMP). These functions and the appropriate nucleotides

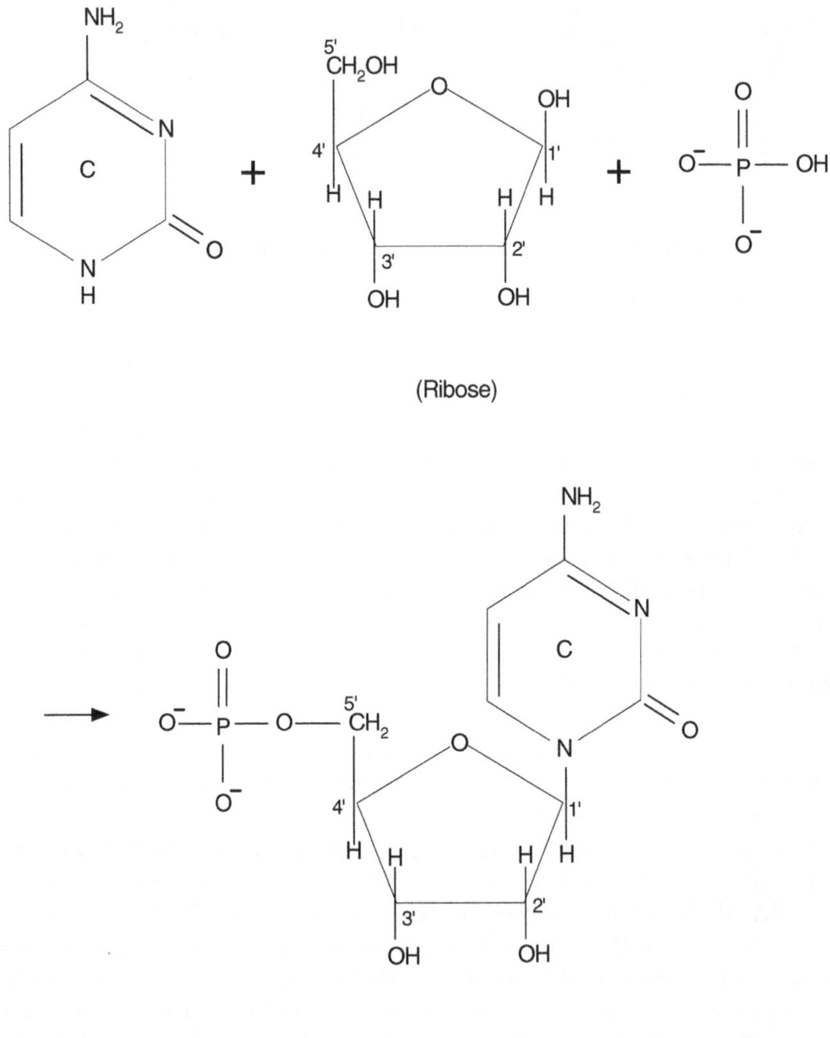

FIGURE 2.15 Formation of the nucleotide cytidine monophosphate.

involved will be discussed in some detail in the next chapter.

We conclude this chapter by noting that (i) macromolecules (polysaccharides, proteins and nucleic acids) found in cells contain a specific sequence of units (sugars, amino acids and nucleotides); these macromolecules are formed by condensation reactions where a new unit is added to a growing chain by the expulsion of a water molecule. (ii) Noncovalent bonds such as hydrogen bonds, hydrophobic forces, etc. make macromolecules assume specific shapes. (iii) These specific shapes as well as the noncovalent bonds allow macromolecules to seek out their appropriate partners and undergo the required reactions with them. These are themes that we will be revisiting in later chapters.

3

Energy Considerations in Biochemical Reactions

In this chapter, we examine the factors that determine whether a particular biochemical reaction is feasible or not and, if it is feasible, what additional factors are needed to get the reaction started. A detailed treatment of these aspects belongs to the subjects of biochemistry and thermodynamics and is beyond the scope of this introductory text. However, there are some general principles that we will be attempting to highlight here. These general principles are not of mere academic interest; indeed, recall from the last chapter that condensation and hydrolysis reactions are really the reverse reactions of each other and consequently, in a given situation, it is important to be able to determine the direction in which the reaction will proceed.

A unique characteristic of living things is that they create and maintain order in a universe that is always tending to greater disorder. This seems somewhat paradoxical since the second law of thermodynamics asserts that the disorder in the universe can only increase, and one begins to wonder if living things obey the laws of thermodynamics. Before proceeding further, let us state two of the laws of thermodynamics:

1st Law Energy can neither be created nor destroyed. It can only be converted from one form to another.

2nd Law The *entropy* of the universe can only increase. The entropy is a measure of disorder.

Living things in general and cells, in particular, obey the basic laws of thermodynamics. When a cell creates and maintains order, the disorder in its environment increases and, for the cell and its environment taken together, the disorder does increase. Thus the second law of thermodynamics is not violated.

To create biological order, the cells in a living organism must perform a never-ending stream of reactions. In fact, each cell is like a tiny chemical factory where thousands of reactions are performed per second. Many of the reactions that a cell performs would not normally occur at biological temperatures. Cells make such reactions possible by using catalysts called *enzymes* and coupling reactions together. There are two main types of chemical reactions in cells. *Catabolic reactions* are the ones that break down foodstuffs into smaller molecules, thereby generating both a useful form of energy for the cell as well as some of the small molecules that the cell needs as building blocks. *Anabolic reactions*, on the other hand, use the energy harnessed by catabolism to drive the synthesis of the many other molecules that form the cell. Catabolic and anabolic reactions together constitute the *metabolism* of the cell.

The second law of thermodynamics is usually stated in terms of entropy which is a measure of disorder. In contrast, the free energy is a measure of the amount of energy that is available to do useful work. A reaction is energetically favorable if it proceeds with a decrease in free energy. Cells generate biological order by coupling the biological order generating reactions (energetically unfavorable) to other heat-generating reactions (which are energetically favorable). Releasing heat to the environment ensures that the entropy of the cell and its surroundings increases, even as biological order is created within the cell itself.

3.1 Some Common Biochemical Reactions

In this section, we briefly discuss some common biochemical reactions that occur inside cells. The particular reactions considered have been chosen to highlight some key aspects that are common to many other biochemical reactions.

3.1.1 Photosynthesis

Animals eat other animals and plants to survive. So plants directly or indirectly provide the primary energy source for all animals. Plants, in turn, trap energy directly from sunlight by a process called *photosynthesis*. Thus, all the energy used by animal cells is ultimately derived from the sun. The overall reaction for photosynthesis, which takes place in the chloroplasts of plant cells, is as follows:

$nCO_2 + nH_2O + \text{sunlight} \longrightarrow \text{(sugars)}\ (CH_2O)_n + nO_2 + \text{heat energy}.$

The actual reactions take place in two stages. In stage 1, the energy from sunlight is captured and transiently stored as chemical bond energy in specialized small molecules that act as carriers of energy and reactive chemical groups. Molecular oxygen derived from the splitting of water by light is released as a waste product. In stage 2, the energy carriers from the first stage are used to drive a process in which sugars are manufactured from carbon dioxide and water. The sugars produced are then used both as a source of chemical bond energy and as source of materials to make the many other small and large organic molecules that are essential to the plant cell. Our simplified discussion of photosynthesis does emphasize something which is common to several cellular reactions: the need for molecules which can store energy and reactive chemical groups and transport them from one location to another.

3.1.2 Cellular Respiration

Cells obtain energy by the oxidation of organic molecules. For carbohydrates, the general reaction is:

$$(CH_2O)\text{ (sugars)} + O_2 \longrightarrow CO_2 + H_2O + \text{energy.}$$

This reaction is referred to as cellular respiration since it is similar to us breathing in oxygen and breathing out carbon dioxide and water vapor. Cells do not carry out this entire reaction in one step because then too much energy would be released and the cell would not be able to store it in a useful fashion. Instead, most of it would be dissipated as heat. The cell carries out this reaction in several incremental steps, each resulting in the release of a small amount of energy which the cell can store in specialized energy carriers. In the process, about 50% of the bond energy in a sugar such as glucose is stored in a retrievable form by the cell. This is highly efficient compared to an automobile engine, say, where the conversion efficiency is only 20%, the rest being dissipated as heat. The general principle illustrated by cellular respiration is that cells have devised mechanisms to carry out reactions at highly controlled rates.

3.1.3 Oxidation and Reduction

Recall from inorganic chemistry that *oxidation* refers to reactions that involve the addition of oxygen or the removal of hydrogen. Similarly, *reduction* refers to reactions that involve the addition of hydrogen or the removal of oxygen. For instance, in the reaction

$$CH_2O\text{(sugar)} + O_2 \longrightarrow CO_2 + H_2O$$

the sugar molecule is oxidized while the oxygen molecule is reduced. The definitions of oxidation and reduction can be extended to reactions where oxygen and/or hydrogen are not involved. This is done by redefining an oxidation reaction as one where a loss of electrons occurs, while a reduction reaction is defined as one where a gain of electrons takes place. For instance, when a ferrous iron ion (Fe^{2+}) gets converted to a ferric iron ion (Fe^{3+}), it loses an electron and is said to be oxidized. On the other hand, when the chlorine atom gains an electron to form a chloride ion (say to form the compound sodium chloride), it is said to be reduced. In a chemical reaction, since the total number of electrons does not change, whenever one compound gets oxidized, another compound gets reduced. Thus oxidation and reduction occur simultaneously in a chemical reaction and such reactions are called *redox* reactions.

The terms oxidation and reduction apply even when there is only a partial shift of electrons between atoms linked by a covalent bond. When a carbon atom becomes covalently bonded to an atom with a strong affinity for elec-

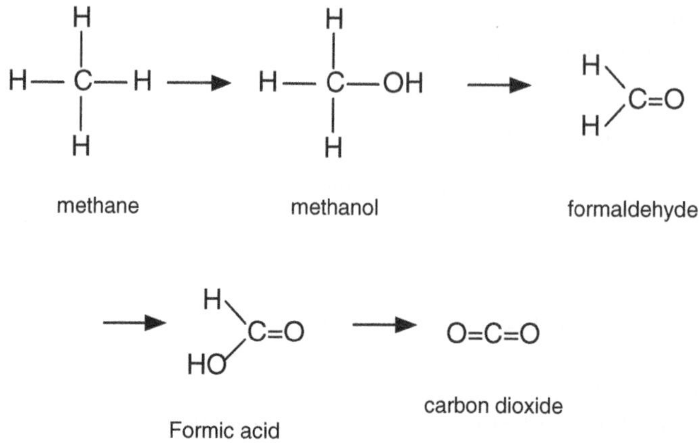

FIGURE 3.1 Steps in the oxidation of methane

trons, such as oxygen, chlorine, or sulphur, for example, it gives up more than its equal share of electrons and forms a *polar* covalent bond and the carbon atom is said to have been *oxidized.* This notion is in agreement with the step-by-step oxidation of methane shown in Fig. 3.1.

3.2 Role of Enzymes

From the second law of thermodynamics, chemical reactions proceed only in a direction that leads to a loss of free energy. Such reactions are said to be *energetically favorable.* Even such reactions do not occur spontaneously — the molecules require activation energy, i.e., a kick over an energy barrier to get the reaction going. For instance, consider the burning of a piece of wood. It is an energetically favorable reaction; the free energy decreases and the piece of wood can never be retrieved from the ashes, smoke, heat, etc. However, the reaction does not occur unless the kick over the energy barrier is provided by a heat source such as a lighted match. The general situation is graphically illustrated in Fig. 3.2 (red line). The reaction X→Y is energetically favorable since $c < b$. However, to make this reaction occur, X must receive an energy kick so that it can go over the hump a.

One way to try to overcome the energy barrier is to supply heat energy so that the molecules of the reactants become more energetic and ultimately go over the energy barrier. However, living cells have to operate within a narrow temperature range and such a mechanism for boosting the energy level may not work. Instead, the necessary boost over the energy barrier is provided by a specialized class of proteins called *enzymes.* Enzymes are among the most

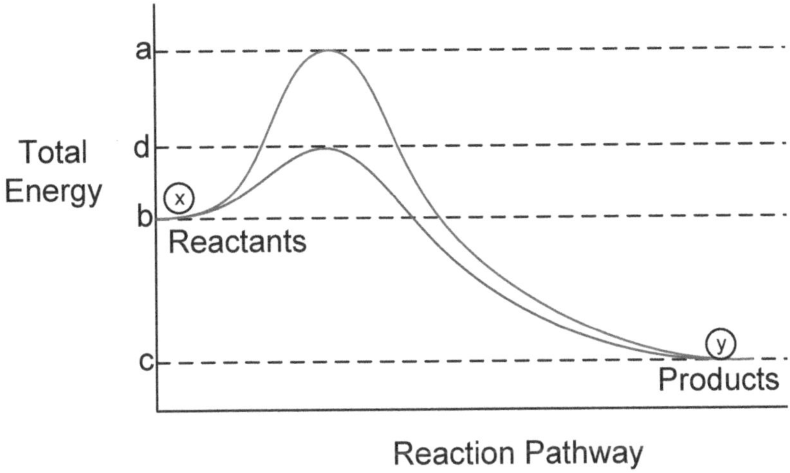

FIGURE 3.2 Activation energy requirement in the absence (red) and presence (blue) of enzymes (See color figure following page 146).

effective catalysts known, often speeding up reactions by a factor of as much as 10^{14}, thereby allowing reactions that would not otherwise occur to proceed rapidly at room temperatures. Each enzyme binds tightly to one or two molecules called *substrates*, and holds them in a way that greatly reduces the activation energy of a particular chemical reaction that the bound substrates can then undergo. The reduction in activation energy is illustrated in Fig. 3.2 (blue line).

Once the substrates have reacted, the enzyme dissociates from the products and is free to bind additional substrate molecules. Enzymes are also highly selective and each enzyme catalyzes only a particular reaction. Thus, by using different enzymes, a cell can regulate a number of reactions in a somewhat decoupled way.

Enzymes find their substrates through rapid diffusion. Inside the cell, the different molecules are randomly colliding with each other. Most of these encounters result in almost immediate dissociation. However, when a substrate encounters the appropriate enzyme, immediate association takes place since the two molecules are "made for each other", as it were. In other words, the substrate molecule fits into the enzyme molecule like a "hand in a glove" and several weak bonds are formed linking the two of them together and facilitating the necessary reaction.

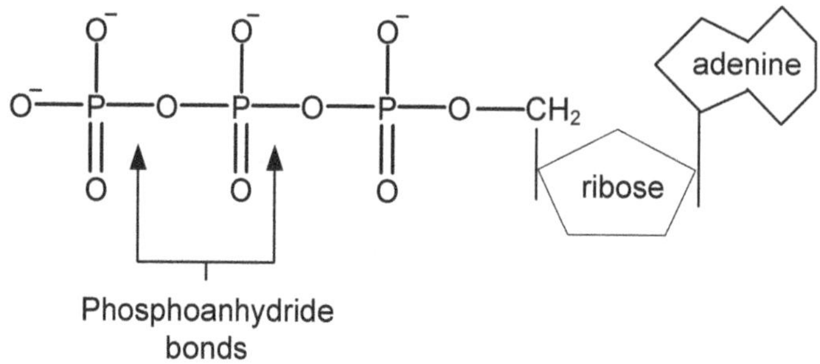

FIGURE 3.3 Structure of adenosine triphosphate (ATP).

3.3 Feasibility of Chemical Reactions

Chemical reactions are energetically favorable if they are accompanied by a decrease in the free energy, i.e., the change in free energy $\Delta G < 0$. Many reactions inside cells, e.g. the synthesis of complex carbohydrates from sugars are *energetically unfavorable*. Such reactions will occur only if they are coupled to an energetically favorable reaction such that the combined free energy change is negative. Before presenting an example of such a *coupled reaction*, let us introduce adenosine tri phosphate (ATP), which is a molecule that is capable of storing a large amount of energy in its phosphoanhydride bonds and can participate in many coupled reactions. The structure of the ATP molecule is shown in Fig. 3.3. From this figure, it is clear that ATP is obtained by attaching two inorganic phosphate ions in cascade to the adenosine monophosphate (AMP) molecule introduced in the last chapter. The two phosphate ions are linked to the AMP molecule by two high energy phosphoanhydride bonds. One or both of these bonds can be readily hydrolyzed to yield adenosine diphosphate (ADP) or adenosine monophosphate (AMP) and both the reactions are energetically favorable. We are now ready to present an example of a coupled reaction.

Example 3.1

Consider the condensation reaction involved in the formation of sucrose:

$$\text{glucose} + \text{fructose} \longrightarrow \text{sucrose}.$$

This reaction has a *standard* free energy change (free energy change when reactants and products have the same concentration), $\Delta G^0 = 5.5$ kcal/mole. Since this number is positive, this reaction will not occur by itself. On the other hand, consider the hydrolysis of ATP, which is an energetically favorable reaction having $\Delta G^0 = -7.3 kcal/mole$:

$$\text{ATP} \longrightarrow \text{ADP} + \text{P}_\text{i}.$$

The two reactions above can be coupled together as follows. The ATP is readily hydrolyzed to ADP and the high energy phosphate ion is incorporated into location 1 on the glucose molecule to yield glucose-1-phosphate:

$$\text{glucose} + \text{ATP} \longrightarrow \text{glucose}-1-\text{P} + \text{ADP}$$

The glucose-1 phosphate then reacts with fructose to produce sucrose and the inorganic phosphate ion:

$$\text{glucose}-1-\text{P} + \text{fructose} \longrightarrow \text{sucrose} + \text{P}_\text{i}$$

Since the free energy change for sequential reactions are additive, the standard free energy change for the overall reaction is $\Delta G^0 = -1.8$kcal/mole and the net result is that sucrose is made in a reaction driven by the hydrolysis of ATP.

□

The free energy change for a reaction depends on the concentration of the reactants. Consider a simple reaction where a reactant A gets converted into the product B:

$$A \longrightarrow B.$$

The free energy change for this reaction is given by [2]

$$\Delta G = \Delta G^0 + 0.616\ln\frac{[\text{B}]}{[\text{A}]}$$

where ΔG^0 is the standard free energy change, $[B]$=concentration of B in moles/liter and $[A]$=concentration of A in moles/liter. From the above relationship, if $[B] = [A]$ then

$$\Delta G = \Delta G^0 = \text{standard free energy change for the reaction.}$$

At equilibrium, the concentration of the reactants and products does not change and hence $\Delta G = 0$. Thus, if $[A_e]$ and $[B_e]$ denote the equilibrium concentrations of $[A]$ and $[B]$ respectively, then

$$\Delta G^0 = -0.616\ln\frac{[\text{B}_\text{e}]}{[\text{A}_\text{e}]}.$$

The quantity $K := \frac{[B_e]}{[A_e]}$ is called the equilibrium constant of the reaction.

Cells can speed up reactions in at least two ways: (i) they rapidly remove the products into a separate compartment (a procedure called compartmentalization); and (ii) they can use the products of one reaction as the reactants of an *immediately* following reaction. In the second case above, cells can cause the energetically unfavorable transition X→Y to occur if an enzyme catalyzing the X→Y reaction is supplemented by a second enzyme that catalyzes the energetically favorable reaction Y→Z. If the overall free energy change for the reaction X→Y→Z is negative, then the reaction X→Y will proceed rapidly, although by itself it would have been thermodynamically impossible.

3.4 Activated Carrier Molecules and Their Role in Biosynthesis

The energy released by the oxidation of food molecules (say glucose) must be stored temporarily before it can be used in the construction of other organic molecules needed by the cell. In most cases, the energy is stored as chemical bond energy in a small set of activated "carrier molecules" or *coenzymes.* These molecules rapidly diffuse throughout the cell and can therefore transfer energy from one site to another. The coenzymes store their energy either in (i) high energy bonds, e.g. phosphoanhydride bonds in ATP or (ii) as high energy electrons in another class of molecules that we will shortly discuss. The reaction forming an activated carrier molecule generates biological order and is, therefore, energetically unfavorable. Consequently, such molecules are formed by coupled reactions where the energetically unfavorable reaction forming an activated carrier molecule is driven by the energetically favorable reaction involved in the oxidation of glucose.

ATP is by far the most widely used activated carrier molecule in cells. The hydrolysis of ATP releases energy for doing cellular work and biosynthesis:

$$\mathrm{ATP} \longrightarrow \mathrm{ADP} + \mathrm{P_i}.$$

The reverse reaction forming ATP is driven by energy from sunlight or from food:

$$\mathrm{ADP} + \mathrm{P_i} \longrightarrow \mathrm{ATP}.$$

The two reactions above whereby ADP is phosphorylated to obtain ATP, which is then hydrolyzed to obtain energy, together constitute the mechanism by which our cells are powered all the time. In fact, ATP hydrolysis is also used to drive the reactions synthesizing several biological polymers which by themselves are energetically unfavorable.

NADH and *NADPH* are important electron carriers that supply reducing power needed for certain reactions. The structures of NADPH and its oxidized form $NADP^+$ are shown in Fig. 3.4. $NADP^+$ contains a nicotine amide ring as shown and is called nicotinamide adenine dinucleotide phosphate for reasons that are obvious from the figure. The detailed structure of these molecules is not important for our purposes. All that we need to know is that by adding a proton (H^+) and two high-energy electrons to $NADP^+$, one obtains NADPH (nicotinamide adenine dinucleotide phosphate reduced). Because of the presence of the proton and the two high energy electrons, which are easily removed, NADPH provides strong reducing power. The other important aspect to note is that NADH and NAD^+ differ from NADPH and $NADP^+$ respectively in that the bottom phosphate group is missing in the former.

NADH has a special role as an intermediate in the *catabolic* system of reactions that generate ATP through the oxidation of food molecules. NADPH,

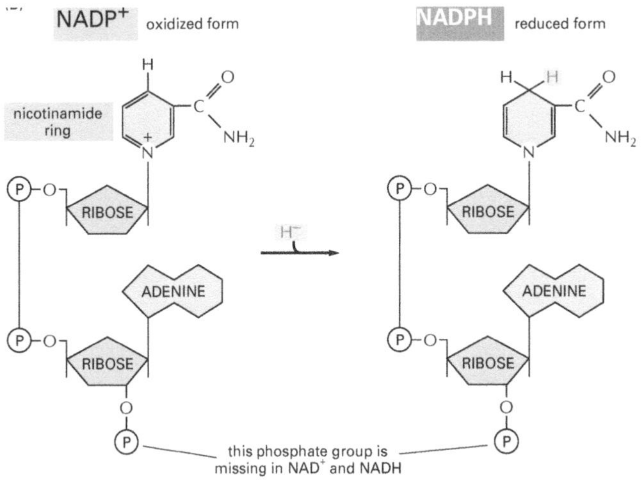

FIGURE 3.4 Structure of $NADP^+$ and NADPH [1]. Copyright 2003 from *Essential Cell Biology*, Second Edition by Alberts et al. Reproduced by permission of Garland Science/Taylor & Francis LLC.

on the other hand, operates mainly with enzymes that catalyze *anabolic* reactions, supplying the high-energy electrons needed to synthesize energy-rich biological molecules. Thus, the presence of two different sets of activated carrier molecules for two different pathways allows the cell some amount of independent control over these pathways.

Acetyl coenzyme A(acetyl CoA) is another important activated carrier molecule. The schematic structure of acetyl CoA is shown in Fig. 3.5. Note the presence of the high-energy bond between the acetyl group and the CoA molecule. Because of this high-energy bond, acetyl CoA can readily transfer the acetyl group (containing two carbon atoms) to other molecules making certain reactions possible.

It is interesting to note that all the activated carrier molecules discussed in this chapter contain a nucleotide, although only ATP is the one that is involved in actual nucleic acid synthesis. This is perhaps a relic left over from the earlier RNA world, when it is conjectured that proteins and DNA were absent, and RNA served to both store biological information (like current day DNA) and catalyze biochemical reactions (like current day proteins).

In concluding this chapter, we underscore the important role played by activated carrier molecules in biology by explicitly stating that the condensation reactions involved in the synthesis of polysaccharides, proteins and nucleic acids are all powered by the energy derived from ATP hydrolysis.

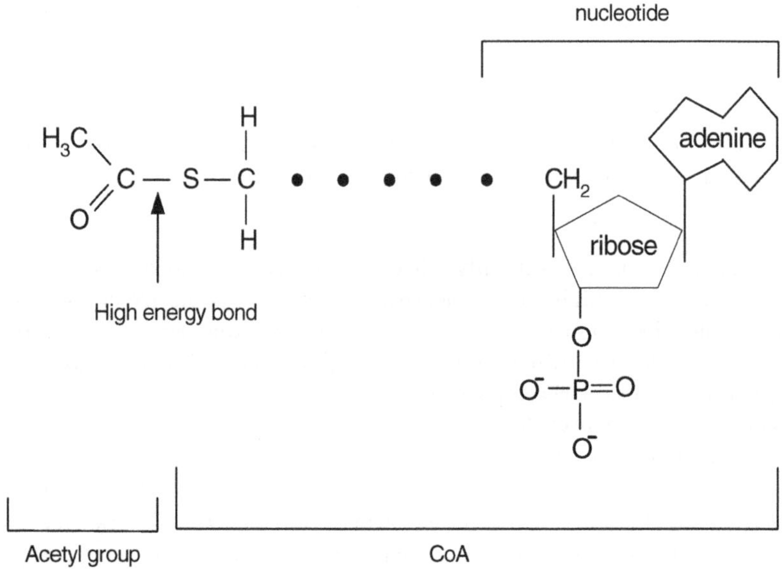

FIGURE 3.5 Schematic structure of acetyl CoA

4

Proteins

In this chapter, we provide an introduction to proteins. Proteins are by far the most versatile functional molecules found in a cell. They constitute most of the dry mass of a cell and are responsible for carrying out nearly all cellular functions. For instance, proteins called enzymes promote many chemical reactions; proteins serve as signaling molecules from one cell to another; proteins serve as moving parts of tiny molecular machines; and antibodies, toxins, hormones, etc are all proteins that have been specialized for certain functions.

4.1 Protein Structure and Function

The multiplicity of functions performed by proteins is made possible by the large number of *different three-dimensional shapes* that they adopt. The shape of a protein is specified by its amino acid sequence. Recall from Chapter 2 that there are 20 different amino acids each with a different side chain; furthermore, some of these side chains are basic, some acidic, some polar and hydrophilic and some non-polar and hydrophobic. A protein molecule is made from a long chain of amino acids, each linked to its neighbor by a *peptide bond* which is an *amide* linkage. The formation of this amide linkage has already been discussed in Chapter 2. The repeated sequence of atoms along the chain is referred to as the *polypeptide backbone*. The side chains are unique to each protein and are crucial to determining its distinctive properties. Proteins fold up into different shapes because of different sets of weak *noncovalent bonds*: (i) hydrogen bonds, (ii) ionic bonds, (iii) Van der Waals attractions (the forces that exist between two atoms when they are too far apart or too close together) and (iv) hydrophobic forces. As a simple example, consider the hypothetical protein in Fig. 4.1 which is made up of four amino acids, two with polar side chains and two with non-polar side chains. Clearly, in aqueous solution, due to the hydrophobic forces, such a protein will assume the conformation shown in Fig. 4.2. For thermodynamic reasons, a given protein will fold into a conformation of lowest energy. A protein can be unfolded or *denatured* by treatment with certain solvents. When the solvent is removed, the protein refolds spontaneously or *renatures*, into its original conformation. Thus, it follows that the final three-dimensional shape of the folded protein is specified

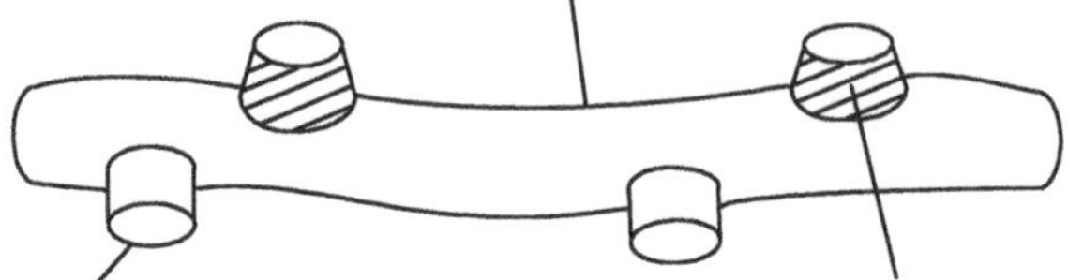

FIGURE 4.1 A simple protein with polar and non-polar side chains [1]. Copyright 2003 from *Essential Cell Biology*, Second Edition by Alberts et al. Reproduced by permission of Garland Science/Taylor & Francis LLC.

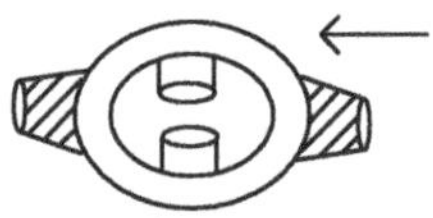

FIGURE 4.2 Three-dimensional conformation in an aqueous environment

by its amino acid sequence, which is also called its *primary structure.* An important question that arises is whether, given the amino acid sequence of a protein, one can predict its three-dimensional structure. Although computer scientists and physicists have been working hard at this problem for a number of years now, we still do not have a definitive answer to it. Protein folding inside cells is assisted by other proteins called *chaperone* proteins. Chaperone proteins do not alter the final three-dimensional shape of the protein that is being folded but simply speed up the folding process.

Proteins range in size from about 30 amino acids to more than 10,000 amino acids. However, the vast majority of proteins are between 50 and 2000 amino acids long. The conformation of a given protein is very complex, even for one that does not have too many amino acids. However, fortunately, there are a few common structural motifs that underlie the various conformations, thereby facilitating a fairly general study of protein structure.

4.1.1 The α-helix and the β-sheet

These two structures are very common because they result from hydrogen bonding between the N – H and C = O groups in the polypeptide backbone, *without involving the side chains.* Thus, they can be formed by many different amino acid sequences.

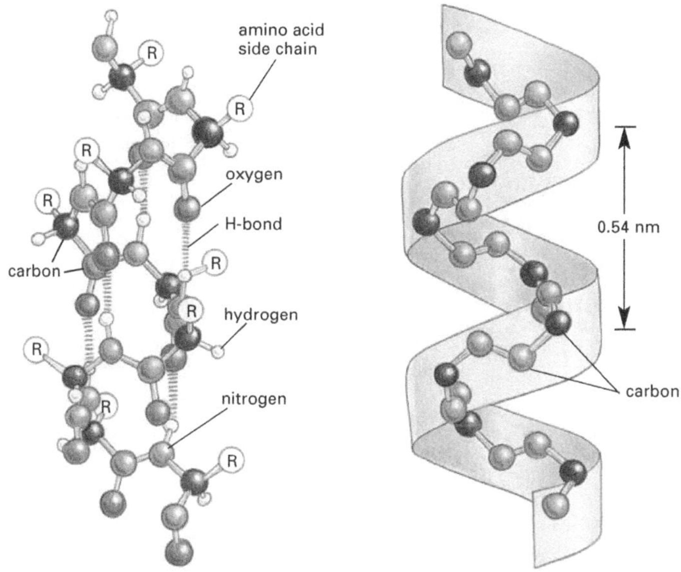

FIGURE 4.3 Protein alpha helix (See color figure following page 146) [1]. Copyright 2003 from *Essential Cell Biology*, Second Edition by Alberts et al. Reproduced by permission of Garland Science/Taylor & Francis LLC .

An α-helix is generated when a single polypeptide chain turns around itself to make a rigid cylinder. The situation is illustrated in Fig. 4.3. Here, a hydrogen bond is formed between every fourth peptide bond, linking the C=0 of one peptide bond to the $N - H$ of another. This gives rise to a regular helix with a complete turn every 3.6 amino acids. Since the interior of the lipid bilayer is hydrophobic, many of the transmembrane proteins cross the lipid bi-layer as α-helices with their hydrophobic side chains exposed to the lipid bilayer while their hydrophilic side chains face the inside of the helix. This is schematically shown in Fig. 4.4.

β-sheets can form either from neighboring polypeptide chains that run in the same orientation (i.e. parallel chains) or from a polypeptide chain that folds back and forth upon itself, with each section of the chain running in the direction opposite to that of its immediate neighbors (i.e., antiparallel chains). The two types of β sheets are schematically illustrated in Fig. 4.5. Both types of β-sheet produce a very rigid structure, held together by hydrogen bonds that connect the peptide bonds in the neighboring chains.

Some pairs of α-helices wrap around each other to form a particularly stable

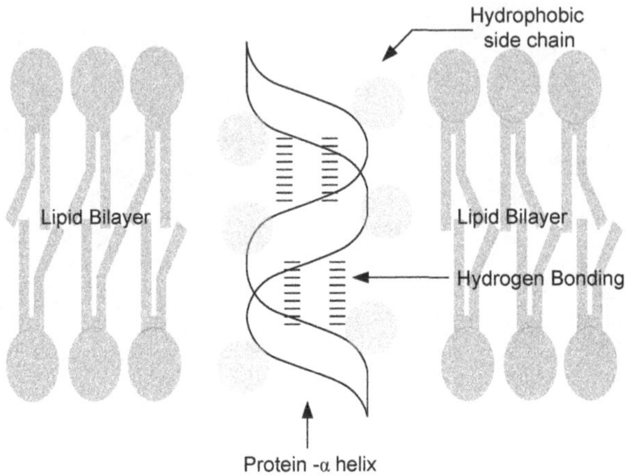

FIGURE 4.4 Schematic diagram of a transmembrane protein crossing the lipid bilayer.

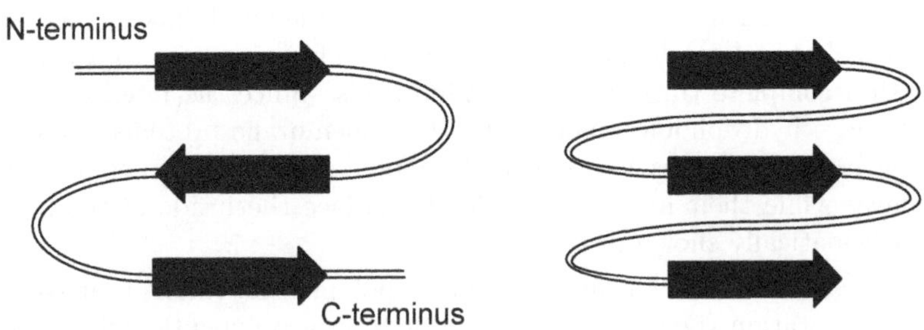

FIGURE 4.5 Schematic diagram of anti-parallel and parallel β sheets.

structure called a *coiled coil.* This structure forms when the two α-helices have most of their nonpolar (hydrophobic) side chains on one side, so that they can twist around each other with these side chains facing inward.

4.2 Levels of Organization in Proteins

The structure of a protein can be studied at four different levels. First, there is the *primary structure* which corresponds to the amino acid sequence; then, there is the *secondary structure* which corresponds to the existence of α-helices and β-sheets; then there is the *tertiary structure* which corresponds to the three-dimensional conformation; and finally, there is the *quarternary structure* which corresponds to studying the complete structure for proteins made up of *more than one* polypeptide chain.

A *protein domain* is produced by any part of a polypeptide chain that can fold independently into a compact, stable structure. A domain usually contains between 50 and 350 amino acids, and it is the modular unit from which many larger proteins are constructed.

In theory, a vast number of different polypeptide chains could be made. For instance, for a polypeptide chain 250 amino acids long, 20^{250} different polypeptide chains are possible. Only a very small fraction of this astronomically large number of polypeptide chains would adopt a stable three-dimensional conformation. Proteins that do not have a single stable conformation are not biologically useful and hence have been eliminated by natural selection during the process of evolution. Consequently, present day proteins not only have a single conformation that is highly stable, but this conformation has the exact chemical properties that enable a protein to perform its function. For instance, consider the protein hemoglobin that is found in blood. This protein has exactly the right three-dimensional shape that enables it to bind oxygen atoms and carry them from the lungs to the tissues.

Once a protein had evolved that folded up into a stable conformation with useful properties, its structure could be modified slightly during evolution to enable it to perform new functions. Such proteins can be grouped into *protein families* and the members of each such family display many similarities, including their three-dimensional conformation.

Large protein molecules often contain more than one polypeptide chain. Each polypeptide chain is called a *protein subunit.* The subunits bind to each other by weak noncovalent bonds. When the polypeptide chains are *identical,* we can have dimers (2 polypeptide chains) as shown in Fig. 4.6(a), or tetramers (4 polypeptide chains). Other proteins contain two or more different types of polypeptide chains. For instance, the protein hemoglobin contains two identical α globin subunits and two identical β globin subunits.

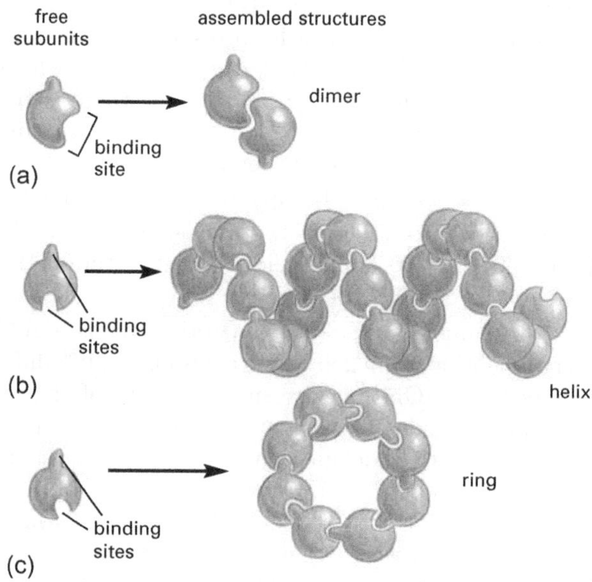

FIGURE 4.6 Identical protein units forming a dimer, a helix and a ring [1]. Copyright 2003 from *Essential Cell Biology*, Second Edition by Alberts et al. Reproduced by permission of Garland Science/Taylor & Francis LLC.

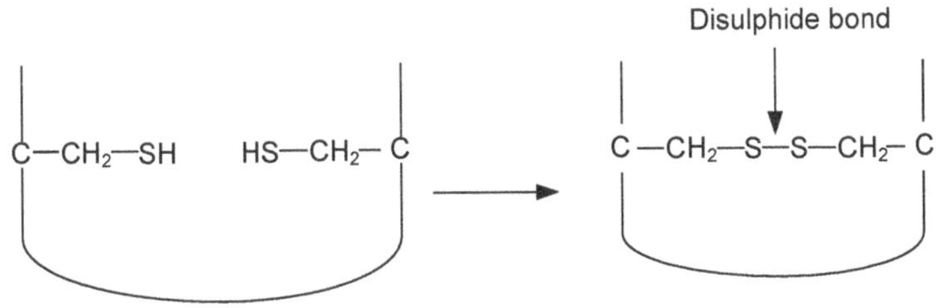

FIGURE 4.7 Disulphide bonding between cysteine side chains.

Proteins can assemble into filaments, sheets or spheres. The protein units can bind as shown in Fig. 4.6(b) to form a helix which can be very long or a circular ring as shown in Fig. 4.6(c).

Large structures such as viruses and ribosomes (protein-making machines) are built from a mixture of one or more types of proteins plus RNA or DNA molecules. All these structures can be isolated in pure form and dissociated into their constituent molecules. Often it is possible to mix the isolated components together and watch the structures reassemble spontaneously. A helix is a common motif in biological structures and results from similar units connected end to end. This is clear from Fig. 4.6(b).

Some proteins have elongated fibrous shapes, unlike the globular (ball-like) proteins discussed so far. Examples are (i) α-keratin (found in hair), (ii) collagen and (iii) elastin. In concluding this section, we note that extracellular proteins are often stabilized by covalent cross linkages such as the disulphide bonds between cysteine side chains shown in Fig. 4.7. These serve as "atomic staples" to stabilize the three-dimensional conformation of the protein outside the cell. Inside the cell, such staples are not needed and the readily available hydrogen ions make sure that no such disulphide bonds are formed.

4.3 Protein Ligand Interactions

Proteins bind to other molecules (called their *ligands*) with great specificity. The ligand must fit *precisely* into a groove on the protein surface, making it possible for the protein and its ligand to be linked together very strongly by large numbers of weak covalent bonds. It is this perfect kind of fitting that enables a protein to pick out its ligand in a crowded environment while avoiding unwanted associations with the numerous other molecules in the cell. The region of a protein that associates with a ligand is called a *binding site*. The same protein can have two or more binding sites if its function depends

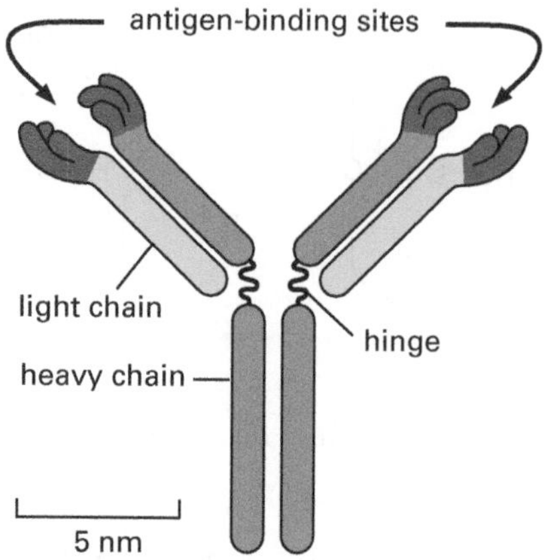

FIGURE 4.8 A typical antibody molecule [1]. Copyright 2003 from *Essential Cell Biology*, Second Edition by Alberts et al. Reproduced by permission of Garland Science/Taylor & Francis LLC.

on binding two different ligands.

We next consider *antibodies* since they are a class of proteins that have a binding capacity that seems to be most highly developed. Antibodies, or immunoglobulins, are proteins produced by the immune system in animals in response to foreign molecules, such as those on the surface of an invading microorganism. Each antibody binds to a particular target molecule extremely tightly and thereby either inactivates the target directly or marks it for destruction. A typical antibody molecule is shown in Fig. 4.8. Antibodies are Y-shaped molecules with two identical binding sites that are each complementary to a small portion of the surface of the target (antigen) molecules. An individual animal can make billions of different antibody molecules, each with a distinct antigen binding site. Each antibody can recognize its antigen with great specificity.

Antibodies defend us against infection in at least two different ways: (i) antibody and antigen aggregates are ingested by phagocytic cells (cells that clean up cellular debris); and (ii) antibody-coated bacteria or viruses are killed by special proteins in the blood.

Antibodies are made by a class of white blood cells, called *B cell lymphocytes* or *B cells*. Each resting B cell carries a different membrane-bound antibody molecule on its surface that serves as a receptor for recognizing a specific antigen. When the appropriate antigen binds to the receptor, the B cell is stimulated to divide and to secrete large amounts of the same antibody. This phenomenon can be used to raise the level of antibodies in animals and forms the basis of immunization and immunotherapy.

Antibodies can also be used to purify molecules. In *immunoaffinity column chromatography*, a mixture of A and other molecules is passed through a

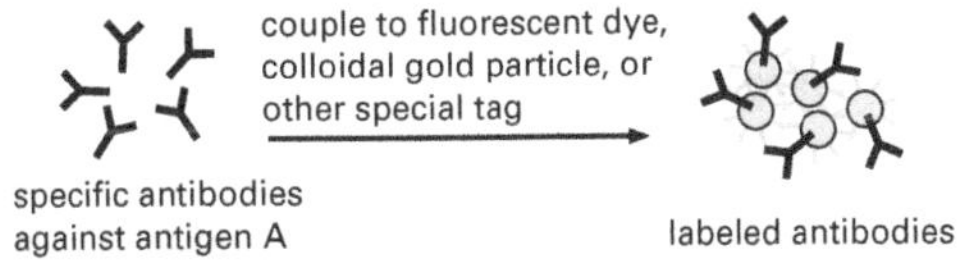

FIGURE 4.9 Antibodies as molecular markers [1]. Copyright 2003 from *Essential Cell Biology*, Second Edition by Alberts et al. Reproduced by permission of Garland Science/Taylor & Francis LLC.

column containing beads coated with anti-A antibodies. The flow through the column is discarded and the A molecules can be recovered by appropriately treating the beads. In *immunoprecipitation*, a mixture of molecules containing A is treated with anti-A antibodies, and the aggregate of A molecules plus anti-A antibodies can be collected by centrifugation.

Monoclonal antibodies are large quantities of a single type of antibody molecule. They can be obtained by fusing a B cell (taken from an animal injected with antigen A) with a cell of a B cell tumor (divides indefinitely but does not make the antibody). The resulting cell divides indefinitely and also produces the antibody.

Antibodies can also be used as molecular markers or molecular tags. To do so, one can couple a set of antibody molecules to a fluorescent dye, a colloidal gold particle or some other special tag. This results in labeled antibodies as shown in Fig. 4.9.

4.4 Isolating Proteins from Cells

Before being able to study proteins, we first need to isolate them from cells and tissues. The first step is to break open the cell by disrupting its plasma membrane. Four different approaches are commonly used: (i) break cells with high frequency sound; (ii) use a mild detergent to make holes in the plasma membrane; (iii) force cells through a small hole using high pressure; and (iv) shear cells between a close-fitting rotating plunger and the thick walls of a glass vessel. The resulting thick soup (called a *homogenate* or an *abstract*) contains large and small molecules from the cytosol as well as the membrane bounded organelles. Careful homogenization can leave the membrane bounded organelles intact. The next step is to separate out the different components. There are a number of procedures commonly used to do this.

(i) *Centrifugation*: The homogenate is placed in test tubes and rotated at high speeds in a centrifuge. As shown in Fig. 4.10, components

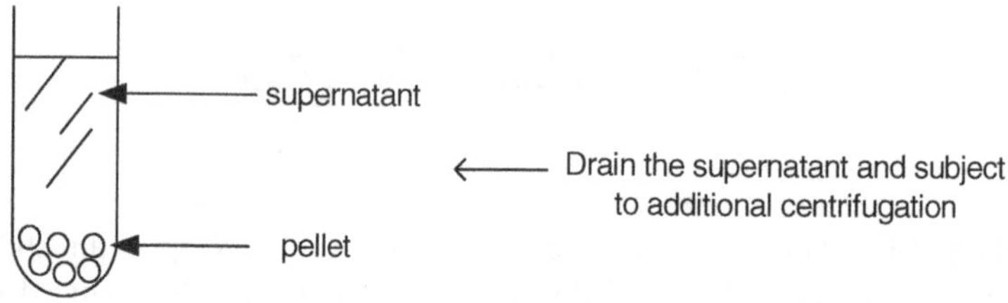

FIGURE 4.10 Separating out a mixture via centrifugation.

separate out mainly on the basis of their size and density. Repeated centrifugation at progressively higher speeds will fractionate the cell homogenate into its different components.

(ii) *Velocity sedimentation*: Subcellular components sediment at different speeds according to their size when carefully layered over a dilute salt solution. When sedimented through such a dilute solution, different cell components separate into distinct bands which can, after an appropriate time, be collected individually.

(iii) *Equilibrium sedimentation*: This procedure makes use of a centrifuge and can be used to separate cellular components on the basis of their *buoyant density*, independently of their shape or size. The sample is usually either layered on top of a steep density gradient that contains a very high concentration of sucrose or cesium chloride. Each subcellular component will move up or down when centrifuged until it reaches a position where its density matches that of its surroundings and then it will move no more. A series of distinct bands will form inside the test tubes and these bands can be separated out by collecting from the base of the test tube.

4.5 Separating a Mixture of Proteins

In this section, we discuss some experimental procedures that can be used to separate out a mixture of proteins. Proteins are very diverse molecules and differ from each other by size, shape, charge, hydrophobicity, etc. All of these properties can be used to separate proteins from each other so that they can be studied individually.

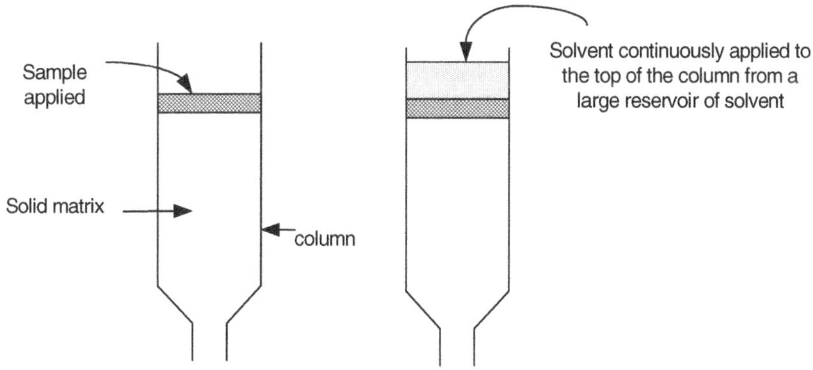

FIGURE 4.11 Column chromatography

4.5.1 Column Chromatography

In column chromatography, the mixture of proteins is forced through a vertical column containing a matrix. As shown in Fig. 4.11, the sample is positioned on top of the matrix and solvent from a reservoir is used to force the sample down the matrix. Different choices for the matrix lead to different kinds of chromatography. In *ion exchange chromatography*, the matrix consists of positively or negatively charged beads that retard proteins of the opposite charge. Thus, separation is achieved on the basis of charge. In *gel-filtration chromatography*, the matrix consists of tiny porous beads. Protein molecules that are small enough to enter the holes in the beads are delayed and travel more slowly through the column. In this case, the separation is achieved on the basis of size. In *affinity chromatography*, the matrix consists of molecules that interact specifically with the protein of interest, for instance, an antibody. Proteins that bind specifically to such a column can finally be released by a pH change or by concentrated salt solutions, and they emerge highly purified.

4.5.2 Gel Electrophoresis

This procedure is used to separate proteins on the basis of their size and net charge. In *SDS PAGE* (*sodium dodecyl sulphate polyacrylamide gel electrophoresis*), the individual polypeptide chains form a complex with negatively charged molecules of SDS and therefore migrate as a negatively charged SDS-protein complex through a slab of porous polyacrylamide gel when subjected to an electric field. The proteins migrate at a rate that is proportional to their molecular weight. Thus, in this case, protein separation is achieved on the basis of size.

Next, we examine how gel electrophoresis can be used to separate proteins on the basis of electric charge. Now, for any protein, there is a characteristic pH, called the *isoelectric point*, at which the protein has no net charge and

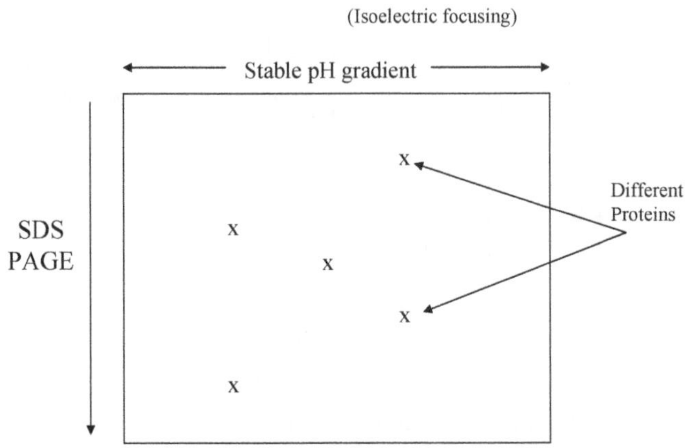

FIGURE 4.12 Two-dimensional polyacrylamide gel electrophoresis.

therefore will not move in an electric field. In *isoelectric focusing*, proteins are electrophoresed in a narrow tube of polyacrylamide gel in which a pH gradient is established by a mixture of special buffers. Each protein moves to a point in the gradient that corresponds to its isoelectric point and stays there. Thus, isoelectric focusing allows us to separate proteins on the basis of their charge.

The two procedures just discussed can be combined together to separate proteins on the basis of both size and electric charge. In *two-dimensional polyacrylamide gel electrophoresis*, first, isoelectric focusing is used to separate proteins on the basis of electric charge. Then, SDS PAGE is applied in a direction perpendicular to the earlier one to achieve further separation on the basis of size. The schematic diagram is shown in Fig. 4.12.

4.6 Protein Structure Determination

There are a number of different approaches that can be used to determine protein structure. In the first approach, a reagent is used to isolate the N-terminus amino acid from the rest of the polypeptide chain. The isolated amino acid can then be identified using a full set of known amino acid derivative standards, e.g., the position to which a particular amino acid migrates under SDS PAGE. This process can be repeatedly used to determine the entire amino acid sequence of the protein. However, it is often found that the amino-terminus amino acid is chemically blocked; in that case, this method cannot be used.

The second approach makes use of the fact that certain chemicals selectively cleave proteins at particular amino acid locations. For instance, the enzyme trypsin cleaves on the carboxyl side of *lysine* or *arginine* residues while the chemical cyanogen bromide cuts peptide bonds on the carboxyl side of *methionine*. Thus, these reagents can produce a few relatively large peptides. These can be separated (using, for instance, gel electrophoresis) to create a peptide map that is diagnostic of the protein from which the peptides are generated. The sequencing of overlapping peptides can be used to laboriously piece together the sequence of the original protein.

The third approach is based on recent advances in DNA technology that will be discussed in a later chapter. For the present, we note that knowing the sequence of as few as 20 amino acids of a protein is often enough to allow a DNA probe to be designed, so that the gene encoding the protein can be cloned. Once the gene, or the corresponding cDNA (complementary DNA) has been sequenced, the rest of the protein's amino acid sequence can be deduced by reference to the genetic code to be discussed in Chapter 6.

In addition to the above approaches, the three-dimensional structure of protein molecules has been studied using *X-ray crystallography* and *nuclear magnetic resonance spectroscopy*.

4.7 Proteins That Are Enzymes

The binding strength between an enzyme and its substrate (or an antibody and its antigen) is measured by the *equilibrium constant* K. For instance, consider the reaction where A and B associate to form the compound AB:

$$A + B \overset{\text{association}}{\longrightarrow} AB.$$

For this reaction, the reaction rate, called the *association rate* is given by

$$\text{association rate} = \text{association rate constant } (k_{\text{on}}) \times [A] \times [B]$$

where [A] and [B] denote the concentrations of A and B, respectively, in moles per liter. Next, consider the reverse reaction where AB dissociates to form A and B:

$$AB \overset{\text{dissociation}}{\longrightarrow} A + B.$$

For this reaction, the dissociation rate is given by

$$\text{dissociation rate} = \text{dissociation rate constant } (k_{\text{off}}) \times [AB].$$

At equilibrium, the association rate equals the dissociation rate and, therefore,

$$k_{\text{on}}[A_e][B_e] = k_{\text{off}}[AB_e]$$

where the subscript e denotes equilibrium concentrations. The quantity

$$K = \frac{[AB_e]}{[A_e][B_e]} = \frac{k_{\text{on}}}{k_{\text{off}}}$$

is called the *equilibrium constant.* If K is large, then $[AB_e] >> [A_e]$ and $[B_e]$, the binding between A and B is very strong, the free energy of AB minus the free energy of A+B is very negative, and the reaction is dominant in the forward direction.

Enzymes are powerful and highly specific catalysts. Each enzyme catalyzes only one very specific chemical reaction. Enzymes catalyzing similar reactions can be grouped together and have special names. A listing of such names follows: *hydrolases* are enzymes that catalyze a hydrolytic cleavage reaction; *nucleases* are enzymes that break down nucleic acids by hydrolyzing phosphodiester bonds between nucleotides; *proteases* are enzymes that break down proteins by hydrolyzing peptide bonds between amino acids; *synthases* is the general name for enzymes that synthesize molecules in the anabolic reactions by condensing two smaller molecules together; *isomerases* are enzymes that catalyze the rearrangement of bonds within a single molecule; *polymerases* are enzymes that catalyze polymerization reactions such as the synthesis of DNA and RNA; *kinases* are enzymes that catalyze the addition of phosphate groups to molecules; *phosphatases* are enzymes that catalyze the hydrolytic removal of a phosphate group from a molecule; and *oxido-reductases* is the general name for enzymes that catalyze reactions in which one molecule is oxidized while the other is reduced. Enzymes of this type are often called oxidases, reductases and dehydrogenases. *ATPases* are enzymes that hydrolyze ATP. Many proteins have an energy-harnessing ATPase activity as part of their function, e.g., motor proteins such as *myosin* (a muscle protein that is responsible for generating movement) and membrane transport proteins such as the *sodium-potassium pump* (a protein that pumps sodium and potassium ions across cell membranes in order to maintain the concentration of sodium and potassium inside cells at appropriate levels).

Let us now examine how, for a given enzyme, the rate of reaction changes with the substrate concentration. A typical situation is illustrated in Fig. 4.13. As the substrate concentration is increased, the reaction rate initially goes up as more substrate molecules are available to interact with the enzyme. However, beyond a certain substrate concentration, the reaction rate saturates since the enzyme is already functioning at its maximal speed (or velocity) and cannot process substrate molecules any faster. This limiting velocity of the enzyme is called $V_{\max}$. The substrate concentration at which a given enzyme can function at half its maximal velocity is called the K_M for that enzyme, and it measures the affinity of the particular enzyme for the substrate. If two enzymes compete for the same substrate, then the values of their $V_{\max}$'s and K_M's determine which reaction gets preference. If the substrate is limited, then the enzyme with the lower K_M gets preference while if the substrate is not a limiting factor, then the enzyme with the higher $V_{\max}$ dominates.

Despite their versatility, there are instances where a protein needs the help of a non-protein molecule for performing functions that would have otherwise been impossible. A couple of instances in this regard follow: (1) A molecule of hemoglobin needs the help of four *heme* groups (ring-shaped molecules) in

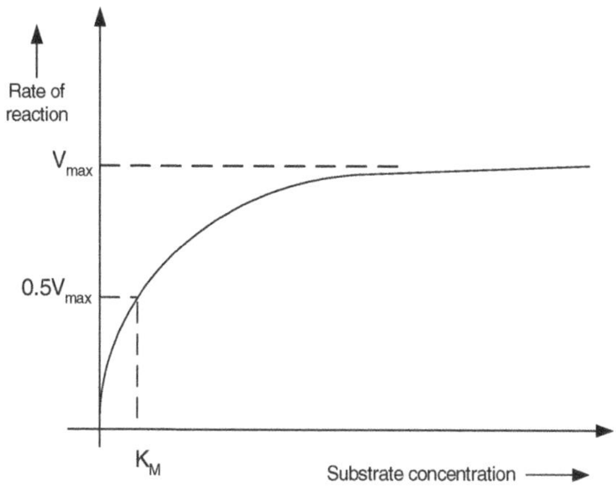

FIGURE 4.13 Reaction rate vs. substrate concentration

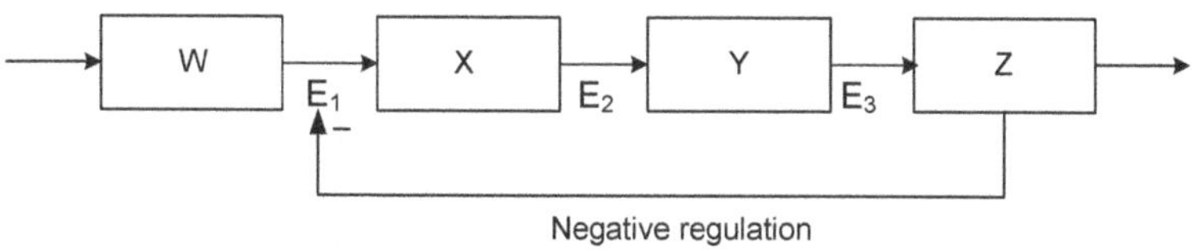

FIGURE 4.14 Feedback inhibition

order to be able to perform its function of picking up oxygen in the lungs and releasing it in the tissues; (2) the signal receptor protein *rhodopsin* in the retina detects light by means of a small molecule called *retinal* that is embedded in the protein.

The catalytic activities of enzymes are controlled at many levels. First, the cell can regulate how much of each enzyme to produce by regulating *gene expression*, a concept that will be introduced in Chapter 6. Second, the cell can control enzymatic activities by confining sets of enzymes to particular subcellular compartments. Finally, an enzyme can change its activity in response to other molecules that it encounters in the cell. An example of an enzyme changing its activity in response to the presence of other molecules is *feedback inhibition*. As shown in Fig. 4.14, in feedback inhibition, X produces Y which produces Z. However, Z inhibits the production of X. This is an example of negative feedback (single loop). However, in general, there will be multi-loop feedbacks. Another instance of negative feedback (or regulation) in a cell is when a rise in ADP in the cell (which is indicative of a low energy state) activates several enzymes involved in the oxidation of sugar molecules, thereby stimulating the cell to convert ADP to ATP. This lowers the level of ADP in the cell and restores the desired energy level.

In feedback inhibition, often times, it is found that the regulatory molecule

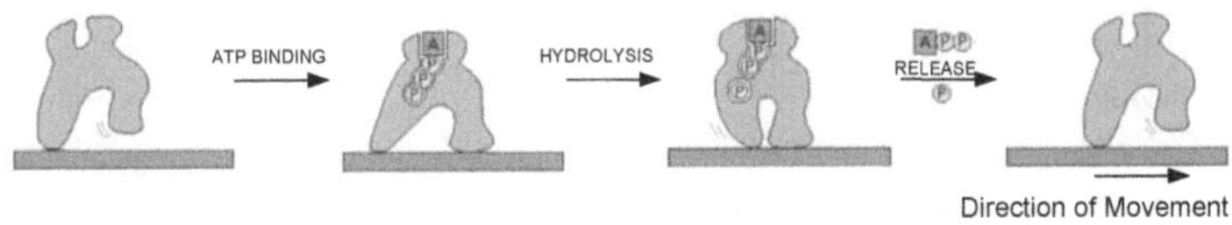

FIGURE 4.15 A motor protein that can produce directed movement [1]. Copyright 2003 from *Essential Cell Biology*, Second Edition by Alberts et al. Reproduced by permission of Garland Science/Taylor & Francis LLC.

has a shape totally different from the shape of the substrate of the enzyme. Thus, in this case, the feedback inhibition is not caused by the regulatory molecule binding preferentially to the *same binding site* on the enzyme. Instead, the enzyme has two binding sites, one for the substrate and the other for the regulatory molecule. Such proteins are called *allosteric proteins*. When the regulatory molecule binds the protein at one site, the conformation of the protein is slightly altered such that the substrate is no longer able to bind at the other binding site.

Eucaryotic cells commonly regulate a protein's function by adding a phosphate group covalently to one of its amino acid side chains. Because each phosphate group carries two negative charges, the enzyme-catalyzed addition of a phosphate group to a protein can cause a major conformational change. (Recall that the three-dimensional conformation of a protein depends on the charge distribution of its amino acids in addition to several other factors.) Removal of the phosphate group by a second enzyme returns the protein to its original conformation and restores its initial activity. So, in eucaryotic cells, protein kinases and protein phosphotases play an important role in switching proteins from inactive to active conformations and vice versa. A particular instance in point is that of GTP (guanosine triphosphate) binding proteins. These proteins undergo dramatic conformational changes when they bind GTP. The protein bound to GTP is active. The hydrolysis of the bound GTP returns the protein to an inactive configuration where it still binds GDP. After a while, the bound GDP is released, but the protein remains inactive until it binds a GTP again.

We conclude this chapter on proteins by noting that motor proteins can produce *directed movement* in cells by coupling the hydrolysis of ATP to conformational changes in the protein. This is schematically illustrated in Fig. 4.15. In the absence of ATP hydrolysis, the protein would move back and forth in different directions generating random movement. By coupling the movement to ATP hydrolysis, which is an energetically favorable reaction, the movement steps are made irreversible, thereby producing directed movement. This is the mechanism by which DNA and RNA polymerases (enzymes that synthesize DNA and RNA) move along a template DNA strand. We will encounter DNA and RNA polymerases in Chapters 5 and 6 respectively.

5

DNA

In this chapter, we discuss *deoxyribonucleic* acid (DNA), which is the molecule used in cells to store genetic information. The ability of cells to store, retrieve and translate such information is a crucial component that is needed for making and maintaining a living organism. This information must be passed on from a cell to its daughter cells at cell division, and from generation to generation of organisms through the reproductive cells. *Genes* are the information-containing elements that determine the characteristics of a species as a whole and of the individuals within it. Natural questions that come up are: (i) the chemical composition of the genes, and (ii) the mechanism by which they store information that can be replicated in an almost unlimited fashion.

In the 1940s it was discovered that the genetic information consists primarily of *instructions for making proteins.* The other crucial advance made in the 1940s was the identification of DNA as the likely carrier of genetic information. By the early twentieth century, biologists knew that genes were carried on chromosomes, which become visible during cell division. Later, biochemical analysis of chromosomes revealed that they are made up of DNA and protein. Since a DNA strand has only one of four possible nucleotides at each position while a protein has one of 20 amino acids at each position, biologists initially thought that genes had to be composed of proteins. The chemically more diverse protein molecule was much more likely to be capable of carrying so much information. This notion was put to rest by an experiment carried out in 1944.

In this experiment, two closely related strains of the bacterium *streptococcus pneumonia* were used:

1. The *R strain*, which appears rough under the microscope, is not pathogenic and does not cause pneumonia when injected into animals.

2. The *S strain*, which appears smooth under the microscope, is pathogenic and causes pneumonia when injected into animals, resulting in death.

Live R strain cells were grown in the presence of either *heat killed* S strain cells or of cell-free extract of S strain cells. Animals injected with these cultured R strain cells died leading to the conclusion that the R strain cells had been transformed to S strain cells and that molecules that carry heritable information are present in the heat killed S strain cells or in the cell-free extract from the S strain cells. To find out the precise molecules that carry

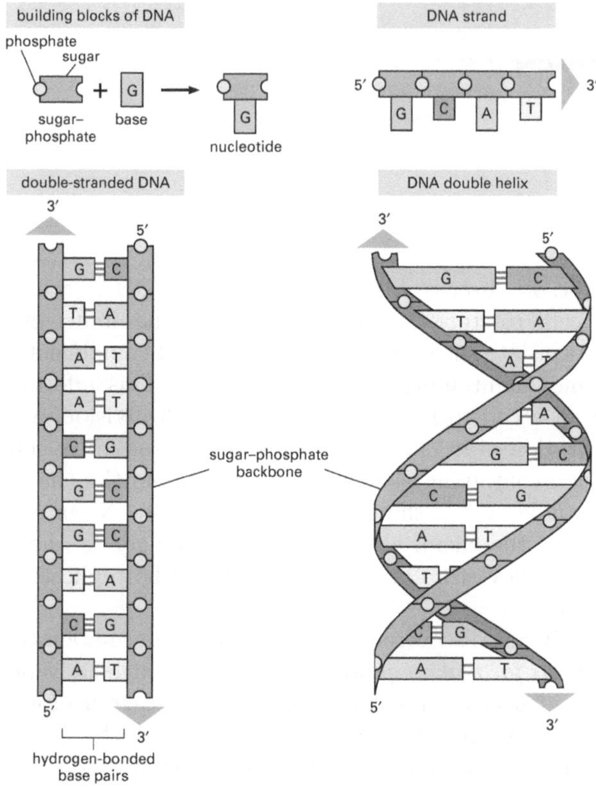

FIGURE 5.1 The structure of DNA and its building blocks [1]. Copyright 2003 from *Essential Cell Biology*, Second Edition by Alberts et al. Reproduced by permission of Garland Science/Taylor & Francis LLC.

the heritable information, the cell-free extract from these transformed R strain cells was fractionated into classes of molecules. The resulting RNA, proteins, lipids and carbohydrates were all found to be of the R strain type. Only the DNA was found to have been transformed to the S strain type. This led to the conclusion that the molecule that carries the heritable information is the DNA.

Even after the identification of DNA as the carrier of genetic information, the mechanism by which DNA is copied for transmission from cell to cell or, for that matter, the manner in which proteins are specified by instructions in the DNA remained largely mysterious until the structure of DNA was discovered by James Watson and Francis Crick in 1953. The structure of a DNA molecule is shown in Fig. 5.1.

A DNA molecule consists of two complementary chains of nucleotides. Recall from Chapter 2 that there are four different nucleotide bases namely, Adenine (A), Guanine (G), Cytosine (C) and Thymine (T) that make up DNA.

These nucleotides have a base-pairing property: A pairs with T through two hydrogen bonds while G pairs with C through three hydrogen bonds. In Chapter 2, we also learned that a nucleotide consists of a DNA base linked to the 5-carbon sugar deoxyribose and a phosphate linked to its $5'$ location. By linking nucleotides together through phosphodiester bonds, one obtains a single strand of DNA. To obtain double stranded DNA, the DNA bases on two complementary strands of DNA link to each other by hydrogen bonds. The pairing of A with T and G with C ensures that each pair is of similar width, thus holding the sugar phosphate backbones an equal distance apart along the DNA molecule. In addition, the two sugar phosphate backbones wind around each other to form a double helix. The schematic diagram in Fig. 5.1 shows how a DNA double helix is formed starting from nucleotide units. From this figure, it is clear that each DNA strand has a definite directional polarity with one end marked $5'$ and the other end marked $3'$. As discussed in Chapter 2, these markings are based on the numbering of the carbon atoms in the deoxyribose molecule.

DNA encodes information in the order, or sequence of the nucleotides along each strand. Each base A, G, C or T can be considered as a letter in a four-letter alphabet that is used to spell out biological messages in the chemical structure of the DNA. Organisms differ from one another because their respective DNA molecules have different nucleotide sequences, and consequently carry different biological messages. The linear sequence of nucleotides in a gene spells out the linear sequence of amino acids in a protein. The exact correspondence between the four-letter nucleotide alphabet of DNA and the twenty-letter amino acid alphabet of protein (the genetic code) took more than a decade to work out after the discovery of the DNA double helix and will be discussed in the next chapter. In the rest of this chapter, we will focus on the mechanisms by which DNA is replicated, possible errors that occur during DNA replication and the mechanisms used to correct them.

Most genes are short stretches of DNA encoding a single protein. However, not all of the DNA in a gene is used to encode the protein that it specifies. There are *regulatory* regions, too, that specify when and in what amounts the protein encoded by a gene is made. The complete set of information in an organism's DNA is called the *genome*. At each cell division, the cell must copy its genome in order to pass it to both the daughter cells. Because each strand of DNA contains a sequence of nucleotides that is exactly complementary to the nucleotide sequence of its partner strand, each strand can act as a *template*, or mold, for the synthesis of a new complementary strand. Referring to Fig. 5.2, the strand S can serve as a template for synthesizing a new strand S' while S' can serve as a template for synthesizing a new strand S. This suggests that the replication of a DNA double helix can proceed according to the following overall scheme: (i) the two strands separate out; (ii) each strand can be used as a template for synthesizing its complementary strand; and (iii) the newly synthesized strand and the template used to create it pair together to form a new double helix. This procedure can be repeated many times to generate

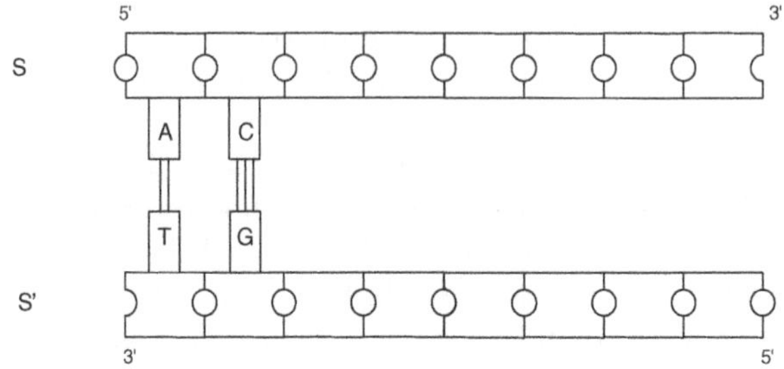

FIGURE 5.2 Complementary DNA strands

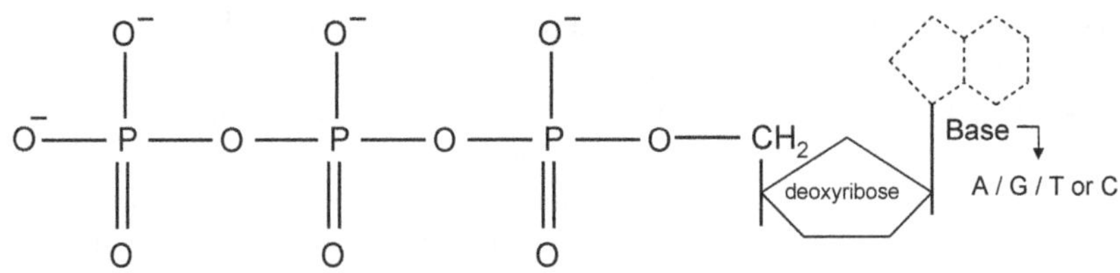

FIGURE 5.3 Energy Rich Nucleoside Triphosphate

the required number of double helices. This is, indeed, what happens and because of this, DNA replication is said to be "semi-conservative" — in each round of replication, one of the strands in each double helix is "conserved" from the previous round of replication. We next present some of the details of the DNA replication procedure.

Replication begins at *replication origins* where the two DNA strands separate. This separation of the two DNA strands is usually carried out by initiator proteins that bind to the DNA and pull the two strands apart. Since DNA that is rich in AT base pairs has fewer hydrogen bonds, it is usually the site of replication origins. A bacterial genome, which is typically small and contained in a circular DNA molecule, usually has only one replication origin. However, for larger genomes, such as human genomes, there are about $10,000$ replication origins. Simultaneous DNA replication starting at each of these origins considerably speeds up the process of replicating the entire genome. The core of the DNA replication machine is formed by an enzyme called *DNA polymerase*, which synthesizes the new DNA using one of the old strands as a template. This enzyme catalyzes the addition of nucleotides to the $3'$ end of a growing DNA strand by the formation of a phosphodiester bond between this end and the $5'$ phosphate group of the incoming nucleotide. The nucleotides enter the reaction initially as energy rich nucleoside triphosphates, shown in Fig. 5.3, which provide the energy for the polymerization reaction. The hydrolysis of one phosphoanhydride bond in the nucleoside triphosphate

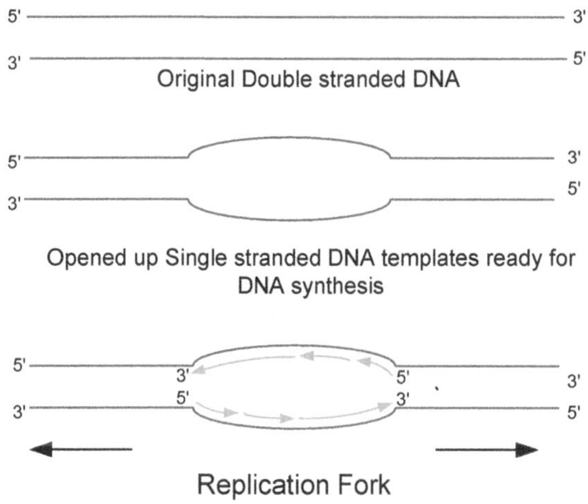

FIGURE 5.4 DNA Replication Fork (See color figure following page 146).

provides the energy for the condensation reaction that links the nucleotide monomer to the chain and releases pyrophosphate (two inorganic phosphate ions linked together by a phosphoanhydride bond). The DNA polymerase couples the release of this energy to the polymerization reaction. The pyrophosphate is further hydrolyzed to yield two inorganic phosphate ions and this makes the polymerization reaction effectively irreversible.

DNA polymerase can synthesize DNA only in the $5'$ to $3'$ direction. This causes a problem since the replication fork is asymmetrical. The situation is illustrated in Fig. 5.4. The DNA strand, whose $3'$ end has to grow as the fork moves (outward from the replication origin) is synthesized continuously — this strand is called the *leading strand*. The DNA strand whose $5'$ end must grow as the fork moves is synthesized *discontinuously*. This strand is called the *lagging strand*. The leading and lagging strands are shown by the continuous and discontinuous green lines respectively, in Fig. 5.4. The discontinuous stretches of DNA synthesized from the lagging strand template are called *Okazaki fragments*.

DNA polymerase is a self-correcting enzyme. As well as catalyzing the polymerization reaction, DNA polymerase has an error-correcting activity called *proofreading*. Before the enzyme adds a nucleotide to a growing DNA chain, it checks whether the previous nucleotide added has been correctly base-paired to the template strand. If so, the polymerase adds the next nucleotide; if not, the polymerase removes the mispaired nucleotide and tries again. Thus, DNA polymerase possesses both a $5'$-to-$3'$ polymerization activity and a $3'$-to-$5'$ nuclease activity.

Because of this proofreading activity, DNA polymerase cannot start a *com-*

pletely new DNA strand. A different enzyme is needed to begin a new DNA strand. This enzyme does not synthesize DNA but synthesizes an RNA strand (about 10 nucleotides long) using the DNA strand as a template. This RNA strand is called a *primer* and the enzyme that synthesizes it is called a *primase.* Notice from Fig. 5.4 that for the leading strand, an RNA primer is needed only to start replication at a replication origin; for the lagging strand, on the other hand, a new primer is needed every time a new DNA fragment has to be synthesized. To produce a continuous new DNA strand from the many Okazaki fragments synthesized on the lagging strand, three additional enzymes are used: (i) a *nuclease*, which disintegrates the RNA primer; (ii) a *repair polymerase*, which replaces the disintegrated RNA with DNA; and (iii) a *DNA ligase*, which joins the $5'$-phosphate end of one DNA fragment to the $3'$-hydroxyl end of the next one.

Many aspects of DNA replication are quite well understood at the present time, while some others are not. For instance, it is known that the following proteins at a replication fork cooperate to form a replication machine:

1. DNA polymerase.

2. Primase.

3. *Helicase* which is a protein that uses the energy of ATP hydrolysis to move along the DNA, opening the double helix as it moves. Obviously, this protein has to be the first one to act in the replication machine.

4. *Single strand binding protein* which clings to the single-stranded DNA exposed by the helicase and prevents it from immediately re-forming base pairs.

5. *Sliding clamp* which keeps the DNA polymerase firmly positioned on the DNA template as the polymerase slides along. On the lagging strand, for obvious reasons, the sliding clamp releases the polymerase from the DNA each time that an Okazaki fragment has been completed.

The exact structure of the DNA replication machine made up of all these proteins is not known. However, some schematic ideas as to its appearance have been proposed [1]. It is also not known if the nuclease, repair polymerase and ligase, which are all needed for synthesizing continuous DNA on the lagging strand, form a part of this replication machine. Moreover, it is not yet understood how the DNA polymerase on the leading strand is connected with that on the lagging strand in order to allow replication to proceed synchronously on both strands. Since our main focus in this book is only on getting a broad overview, the fact that some details of DNA replication are not yet figured out is not of much consequence to us.

Because all of the information specifying a living organism is contained in its DNA, cells go to great lengths to maintain the integrity of the DNA. One aspect of maintaining this integrity is that DNA replication must be carried out very accurately. The proofreading activity of DNA polymerase ensures that in DNA replication, only about one error is made for every 10^7 nucleotides copied. Elaborate DNA *repair mechanisms* ensure that this accuracy is enhanced even further so that only one error is made for every 10^9 nucleotides copied. Despite the elaborate repair mechanisms that are in place, sometimes a permanent change in the DNA, called a *mutation*, can occur. Such changes over the long run (billions of years) have led to the wide variety of living species that we see on earth today. However, in the short run, they can spell disaster for the organism. For instance, cancer is a disease that can result from the accumulation of DNA mutations.

Before presenting examples of some diseases that can result from DNA mutations, let us introduce some terminology related to genetics and heredity. A *gene* is generally a stretch of DNA that codes for a protein. An *allele* is a variant of a gene. Although different individuals may possess the same gene, the exact DNA sequences are not identical. In this case, we say that the different individuals possess different alleles of the same gene. A *wild-type* allele is the kind normally present in the population. A *mutant* allele is one that differs from the wild type. In many situations, altering parts of the genome of an organism will not result in a change at the macroscopic observational level. Such a change is called a *genotypic* change or characteristic or a change in the *genotype*. On the other hand, a characteristic that manifests itself at the observational level is called a *phenotypic* characteristic or a characteristic of the *phenotype*. In humans and many other organisms, each cell except the reproductive ones contains two copies of each gene. Consequently, in such organisms, mutating only one copy of a gene may not result in the manifestation of the associated phenotype. Such a mutation would be called a *recessive mutation*. On the other hand, a *dominant mutation* is one where mutating one copy of a gene is sufficient for the phenotype to manifest itself. When the two alleles for a particular gene in a genome are of the same type, the genome is said to be *homozygous* for that gene. On the other hand, when the two alleles are of different types, the genome is said to be *heterozygous* for that gene.

Example 5.1

Sickle cell anemia is an inherited disease that is caused by a single DNA base mutation in the β-globin gene (which is the gene that codes for one of the subunits of hemoglobin). Humans have two copies of this gene, one inherited from each parent. To have full blown sickle cell anemia, one must have two defective copies of this gene. If an individual has one defective copy, then he or she does not show symptoms of the disease although he or she can pass it on to his or her offspring (with another person having a defective copy of the same gene). Hence, a person with one defective β-globin gene is a *carrier* of sickle cell anemia. Clearly, the mutation involved in sickle cell anemia is a recessive one and the carriers of this disease have a genome which is heterozygous for the β-globin gene. □

Example 5.2

Colon cancer is caused by the failure to repair DNA damage in the colonic cells. Humans inherit two copies of a DNA repair gene, one from each parent. If one defective copy is inherited, then it is very likely that at some point in time colon cancer will result when the other copy gets accidently mutated. The incidence of colon and other types of cancers goes up dramatically with age since, unfortunately, the DNA repair mechanisms slow down as one ages. Also, enough mutations have to accumulate before cancer results and so time is required for the initiation and progression of the disease. □

Without DNA mismatch repair, mutations would develop much faster and the incidences of cancer would go up catastrophically. The DNA mismatch repair system recognizes the errors in DNA replication and corrects them and thereby prevents them from propagating. It is not yet known how the DNA repair machine preferentially acts on the newly synthesized DNA strand and leaves the original strand unchanged. In addition to replication errors, the DNA in our cells is continually suffering damage either spontaneously or due to agents such as sunlight, chemicals, etc. A few examples are given below:

1. In *depurination*, which occurs spontaneously and in large numbers, a purine base (A or G) is removed resulting in a DNA strand that has missing bases. This is shown in Fig. 5.5. When such a strand with missing bases is used as a template for DNA replication, replication errors are bound to occur.

2. In *deamination*, an amino group is lost, e.g., a cytosine is changed to a uracil as shown in Fig. 5.5. Since C pairs with G but U pairs with A, if the DNA strand after deamination is replicated then *base substitution* will occur.

3. The ultraviolet radiation in sunlight can damage DNA by causing the formation of covalent linkages between two adjacent thymine bases. This is called a *thymine dimer* and its formation is shown in Fig. 5.6.

The stability of the genome of an organism depends on DNA repair. There are a variety of DNA repair mechanisms, each catalyzed by a different set of enzymes. However, they share a number of common characteristics. Nearly all of the repair mechanisms depend on the existence of two copies of the genetic information, one in each strand of the DNA double helix. Most DNA damage creates structures that are never encountered in an undamaged DNA strand. This is thought to facilitate the identification of the damaged DNA strand.

The basic pathway for repairing damage to DNA involves the following three steps:

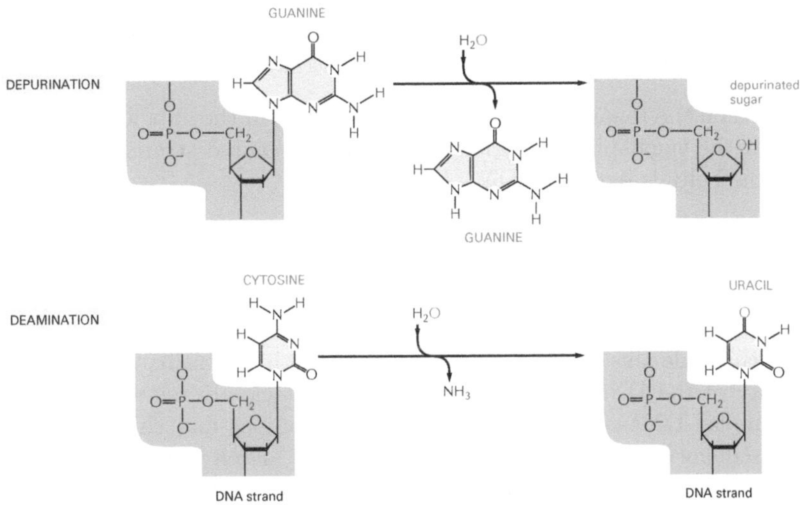

FIGURE 5.5 Depurination and deamination [1]. Copyright 2003 from *Essential Cell Biology*, Second Edition by Alberts et al. Reproduced by permission of Garland Science/Taylor & Francis LLC.

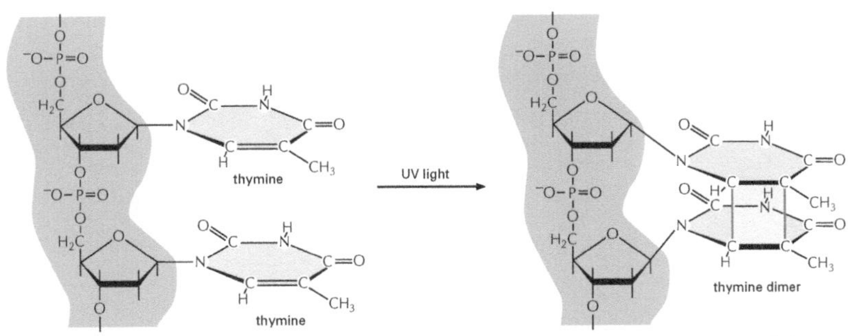

FIGURE 5.6 Formation of thymine dimers by ultraviolet light [1]. Copyright 2003 from *Essential Cell Biology*, Second Edition by Alberts et al. Reproduced by permission of Garland Science/Taylor & Francis LLC.

1. The DNA damage is recognized and removed by one of a variety of different *nucleases.*

2. A *repair DNA polymerase* binds to the $3'$ hydroxyl end of the cut DNA strand and fills in the gap by making a complementary copy of the information stored in the undamaged strand.

3. When the repair DNA polymerase has filled in the gap, a break remains in the sugar-phosphate backbone of the repaired strand. This nick in the helix is sealed by *DNA ligase.*

Failure to repair damaged DNA can result in serious diseases. For instance, failure to repair thymine dimers can result in the disease called *xeroderma pigmentosum.* People with this inherited disease are prone to severe sunburn and skin cancer.

We conclude this chapter by noting that the high fidelity with which the integrity of the DNA is maintained means that closely related species have very similar DNA sequences. For instance, humans and chimpanzees have DNA sequences that are about 98% identical. Even apparently unrelated species such as humans and whales share a lot of similarity in their DNA sequences, strengthening the hypothesis that they must have originated from a common ancestor at some point in evolutionary time.

6

Transcription and Translation

In the last two chapters, we provided an introduction to DNA and proteins. The DNA contained in the genome of an organism primarily codes instructions for making proteins. Recall from the last chapter that there are only four DNA bases A, G, C, T and so an instruction written out in the language of DNA is basically an instruction written out in a language that has a *4-letter alphabet.* Proteins, on the other hand, are made up of amino acids linked together by peptide bonds. Since there are 20 possible amino acids that occur in nature, the amino acid sequence for a protein can be thought of as a biological message written out using letters from a *20-letter alphabet.* The natural question that comes up is how does the DNA message coded in the 4-letter alphabet map into the amino acid sequence of the appropriate protein. In this chapter, we will provide a detailed answer to this question.

The DNA does not direct protein synthesis by itself but acts through certain intermediaries. When a particular protein is needed by the cell, the nucleotide sequence of the appropriate portion of the DNA molecule is first copied into another type of nucleic acid — RNA (ribonucleic acid). It is these RNA copies of short segments of the DNA that are used as templates to direct the synthesis of the protein. The process of copying the DNA into the appropriate RNA strands is referred to as *transcription* while the process of producing the protein from the information in the RNA is referred to as *translation.* When a gene is being transcribed, it is said to be *expressed* or *turned ON*. The usual flow of genetic information is from DNA to RNA to protein. All cells, from the simplest bacteria to complex organisms such as humans, express their genetic information in this way. The principle is so fundamental to all of life that it is referred to as the *central dogma of molecular biology.*

The use of an RNA intermediate makes it possible to more rapidly produce larger amounts of a particular protein than would have otherwise been possible. This is because many identical RNA copies can be made from the same gene and each RNA copy can be used to simultaneously produce many identical protein molecules. By controlling the efficiency of transcription and translation of the different genes, a cell can produce small amounts of some proteins and large amounts of others. In addition, a cell can change gene expression in response to the temporally changing needs for a particular protein.

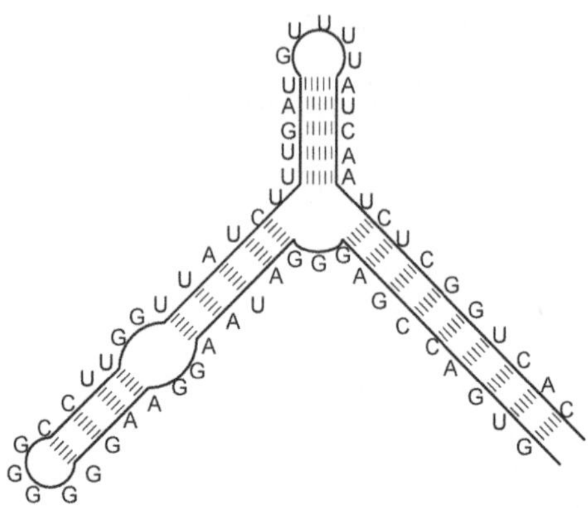

FIGURE 6.1 A single-stranded RNA molecule assuming a three-dimensional structure.

6.1 Transcription

During transcription, the cell copies the required part of the DNA sequence into a nucleotide sequence of RNA. As pointed out in Chapter 2, RNA differs chemically from DNA in two aspects:

(1) The nucleotides in RNA are *ribonucleotides*, that is, they contain the sugar ribose instead of the sugar deoxyribose that is present in DNA; and

(2) DNA contains the bases A, G, C and T while RNA contains the bases A, G, C and U.

Since U pairs with A, the complementary base-pairing properties described in the last chapter for DNA apply also to RNA. In addition to their minor chemical differences, DNA and RNA differ quite dramatically from each other in their overall structure: (i) DNA always occurs in nature as a double-stranded helix while RNA is single-stranded; and (ii) unlike DNA, a single-stranded RNA chain can fold up into a variety of shapes, just as a polypeptide chain folds up to form the final shape of a protein. Because of this latter feature, RNA can not only store information in the sequence of its nucleotides like DNA, but can also serve, in certain situations, as a catalyst or as a part of a larger molecular structure. Fig. 6.1 shows an example of a single-stranded RNA molecule assuming a particular shape resulting from the formation of hydrogen bonds between complementary base pairs.

To initiate transcription, a small portion of the DNA double helix is unwound and one of the two strands of the DNA double helix then acts as a template for the synthesis of RNA. The RNA synthesis is carried out by an enzyme called *RNA polymerase* which catalyzes the formation of an RNA strand complementary to the DNA template. The process is similar to DNA replication by DNA polymerase and uses energy rich nucleoside triphosphates which are hydrolyzed to drive the formation of the RNA strand. However, transcription and DNA replication have some major differences as noted below:

(1) Since RNA polymerase synthesizes RNA, it catalyzes the linkage of ribonucleotides and not deoxyribonucleotides.

(2) Unlike DNA polymerase, RNA polymerases do not possess nucleolytic *proof reading* activity. Consequently, RNA polymerases can start an RNA chain without the need for a primer. Also, transcription is less accurate than DNA replication. In transcription, there is 1 error for every 10^4 nucleotides copied while, as discussed in the last chapter, in DNA replication (without mismatch repair) there is 1 error in every 10^7 nucleotides copied. Since the DNA serves as a permanent storage of information while the RNA is only a temporary intermediate, it is only imperative that DNA replication be much more accurate than transcription.

(3) In transcription, the newly formed RNA strand does not remain hydrogen-bonded to the DNA template strand. Instead, just behind the region where the new ribonucleotides are being added, the DNA helix reforms and displaces the RNA chain. Thus, RNA molecules produced by transcription are single stranded. Since they are copied from only a limited region of DNA, RNA molecules tend to be much shorter than DNA molecules. Also, the immediate release of the RNA strand from the DNA as it is synthesized means that many RNA copies can be made from the same gene in a relatively short time, the synthesis of the next RNA usually being started before the first RNA is completed.

Not all RNAs produced in a cell are meant to be translated into protein. *Messenger RNAs (mRNAs)* are the RNAs that direct the synthesis of proteins. The *ribosomal RNAs (rRNAs)* form the core of the ribosomes, on which mRNAs are translated into proteins. The *transfer RNAs (tRNAs)* form the adaptors that select amino acids and hold them in place on a ribosome for their incorporation into proteins.

In eucaryotes, each gene coding for a single protein is typically transcribed into a separate mRNA molecule. In bacteria, a single mRNA is often transcribed from several adjacent genes and, therefore, contains information for several different proteins.

The way in which transcription is initiated differs somewhat between procaryotes and eucaryotes. Here, we will focus on procaryotes. Transcription initiation in eucaryotes is more complicated and will be discussed later in the next chapter.

In procaryotes, the RNA polymerase molecules slide rapidly along the DNA and tightly latch on to it when they encounter a region called a *promoter* which contains a sequence of nucleotides indicating the starting point for RNA synthesis. After the RNA polymerase makes contact with the promoter DNA and binds it tightly, it opens up the double helix immediately in front of it to expose the nucleotides on a short stretch of DNA on each strand. One of the two exposed DNA strands then acts as a template for synthesis of the complementary RNA. The RNA chain elongation continues until the RNA polymerase encounters a second signal in the DNA, the *terminator* (or stop site), where the polymerase halts and releases both the DNA template and the newly made single-stranded RNA chain.

A subunit of bacterial polymerase, called *sigma* (σ) *factor*, is primarily responsible for recognizing the promoter sequence on DNA. Once the RNA polymerase has latched onto the promoter and has synthesized approximately 10 nucleotides of RNA, the sigma factor is released. The RNA polymerase moves forward and continues the transcription process. After the polymerase reaches the terminator sequence and is released, it reassociates with a free sigma factor and is ready to start transcription again from a promoter region. Fig. 6.2 is a schematic diagram showing an RNA polymerase molecule transcribing a bacterial gene. Here, the lower DNA strand serves as the template for RNA synthesis.

Transcription can only proceed in the $5'$ to $3'$ direction for the synthesized strand. Moreover, a portion of DNA can be transcribed *only if it is preceded by a promoter sequence.* This ensures that unnecessary transcription does not take place.

Since procaryotes do not have a nucleus, procaryotic transcription takes place in the cytoplasm, which also contains the ribosomes on which protein synthesis (or translation) takes place. As mRNA molecules are transcribed, ribosomes immediately get on to the $5'$ end of the RNA transcript and protein synthesis starts.

In eucaryotes, on the other hand, the DNA is enclosed within the nucleus. Transcription takes place in the nucleus but protein synthesis takes place in the cytoplasm. So, before a eucaryotic mRNA can be translated into protein, it must be transported out of the nucleus and into the cytoplasm. Before the RNA is exported from the nucleus, it goes through several different *RNA processing* steps. Depending on which type of RNA is being produced, these transcripts are processed in different ways before leaving the nucleus. In eucaryotes, the RNA produced by transcription, but not yet processed, is often called the *primary transcript.* Primary transcripts destined to become mRNA molecules undergo the follwing two processing steps: (1) the addition of a *methyl guanine (methyl G) cap* at the $5'$ end which usually occurs just after the RNA polymerase has synthesized the $5'$ end of the primary transcript and before it has completed transcribing the entire gene; and (2) the $3'$ end of

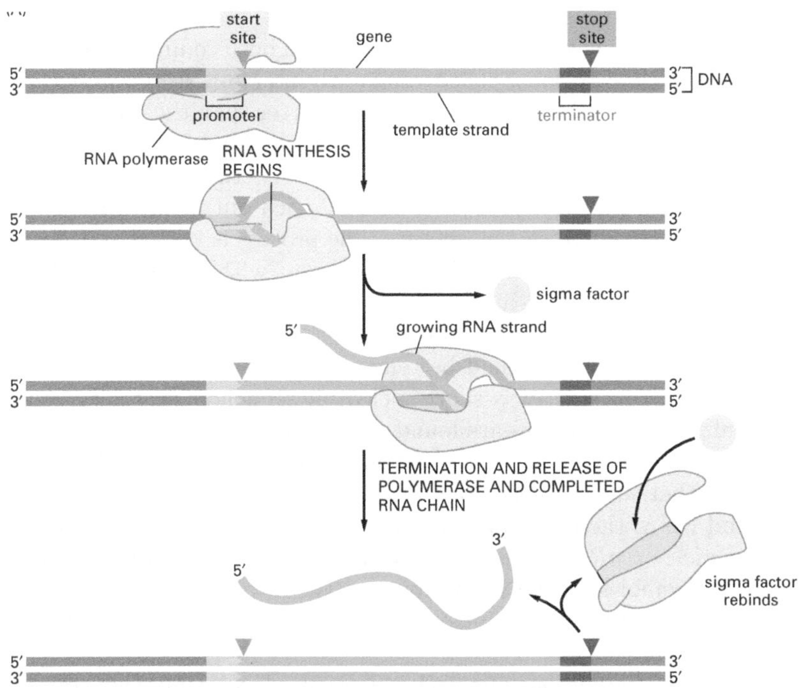

FIGURE 6.2 RNA polymerase transcribing a bacterial gene (See color figure following page 146) [1]. Copyright 2003 from *Essential Cell Biology*, Second Edition by Alberts et al. Reproduced by permission of Garland Science/Taylor & Francis LLC .

eucaryotic RNAs are trimmed at a particular sequence of nucleotides and then a series of repeated adenine (A) nucleotides are added at the cut end. This processing step is referred to as *polyadenylation.* The trimming and nucleotide addition steps are carried out by two different enzymes. The sequence of adenine nucleotides is referred to as a *poly A tail* and is generally a few hundred nucleotides long. The two modifications, capping and polyadenylation, are thought to increase the stability of the mRNA molecule and to aid its export from the nucleus to the cytoplasm. In addition, they later serve to indicate to the protein synthesis machinery that both ends of the message are present and, therefore, the message is complete.

Unlike bacterial genes, eucaryotic genes contain both coding and non-coding regions. The *coding* regions, called *exons* are interrupted by *non-coding* regions, called *introns*. The exons are usually shorter than the introns, and the coding portion of a gene is often only a small fraction of the total length of the gene. For RNAs destined to become mRNAs, the introns are removed after capping and polyadenylation. This is carried out in the nucleus before the mRNA is sent out to the cytoplasm. The process is known as *RNA splicing* and results in a functional mRNA molecule (with uninterrupted coding sequences) which can now leave the nucleus and be translated into protein.

A natural question that comes up is how the cell determines which parts of the primary transcript should be removed. This is facilitated by the fact that each intron contains a few short nucleotide sequences that act as cues for its removal. These sequences are found at or near each end of the intron and are the same or very similar in all introns. Introns are removed from RNA by enzymes that, unlike most other enzymes, are composed of a complex of protein and RNA; these splicing enzymes are called *small nuclear ribonucleoprotein particles (snRNPs)*—pronounced "snurps". At each intron, a group of snRNPs assembles on the RNA, cuts out the intron, and rejoins the RNA chain — releasing the excised intron as a *lariat*. This is schematically shown in Fig. 6.3.

The presence of numerous introns in eucaryotic DNA makes genetic recombination between exons of different genes more likely. The precise mechanism will be explained in more detail in a later chapter. Here, we simply note that eucaryotic genes for new proteins could have evolved quite rapidly by the combination of parts of pre-existing eucaryotic genes. Moreover, the primary transcripts of many eucaryotic genes, containing three or more exons, can be spliced in various ways to produce different mRNAs, depending on the cell type in which the gene is being expressed. This is referred to as *alternative splicing* and enables eucaryotes to increase the already enormous coding potential of their genomes.

The length of time that an mRNA molecule persists in the cell affects the amount of protein produced from it, since the same mRNA molecule can be translated many times. Different mRNA molecules have different lifetimes and these are in part signaled by nucleotide sequences that are present in the mRNA itself (usually between the $3'$ end of the coding sequence and the poly

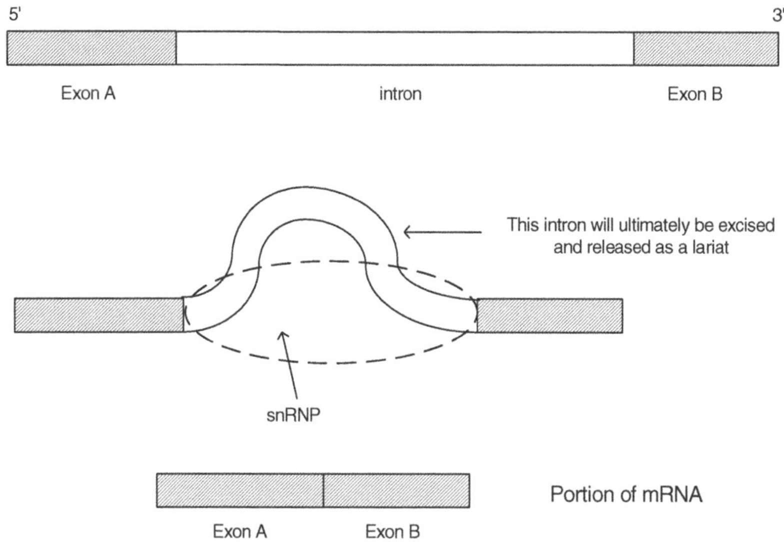

FIGURE 6.3 mRNA splicing.

A tail).

Although the genomes of current day procaryotes do not contain any introns, it is thought that their earlier ancestors may have had introns in their genes. Subsequently, procaryotic cells may have evolved and gotten rid of their introns in order to have a smaller genome which would allow faster DNA replication. This, in turn, would facilitate very rapid reproduction which is a characteristic that current day procaryotes do possess.

6.2 Translation

In translation, the sequence of nucleotides in an mRNA molecule have to be decoded to produce the appropriate proteins. The code used by nature for this purpose is called the *genetic code* and is the map from the nucleotide based language of RNA to the amino acid based language of protein. The genetic code was completely worked out in the early 1960s.

Since there are only 4 different RNA bases A, G, C and U and there are 20 amino acids that occur in nature, it is clear that, at a minimum, the code has to be a *triplet code*. Experiments carried out in the 1960s conclusively demonstrated that, in fact, the genetic code is a triplet one.

The sequence of nucleotides in the mRNA molecule is read consecutively in groups of 3, each such group specifying either the start of a protein, the end of

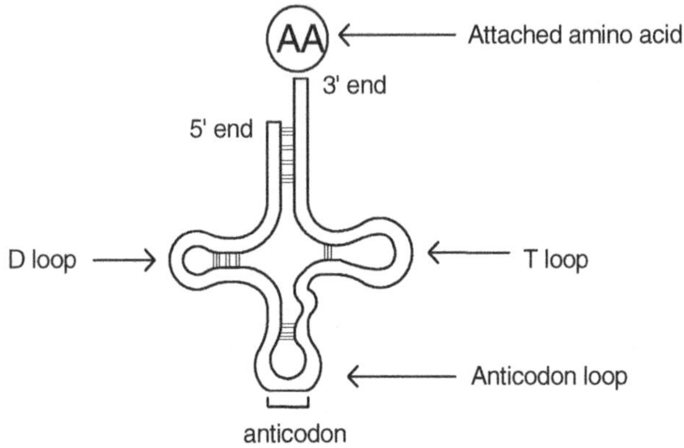

FIGURE 6.4 A typical tRNA molecule

a protein, or an amino acid. Each nucleotide triplet is called a *codon.* There are 20 possible amino acids and 64 possible codons. The codons UAA, UAG, UGA are *stop codons* and signify the end of translation. The codon AUG specifies the amino acid methionine and also indicates the start of a protein. All the other codons code for some amino acid. Since there are 20 different amino acids and 61 codons coding for amino acids, there must be multiple codons coding for the same amino acid. In other words, the genetic code is *degenerate.* The genetic code is also *universal* in the sense that all known life on this planet makes use of the same genetic code. We have not included the entire genetic code here, since it is not crucial for our purposes. The interested reader can consult any standard text on molecular biology, such as [1] for a complete description.

Since each triplet of nucleotides codes for an amino acid, an RNA sequence can be translated in any one of three *reading frames*, depending on where the decoding process begins. The choice of the reading frame is important since advancing the reading frame by even one nucleotide can alter all the amino acids being coded by the RNA sequence leading to a protein that is completely different from the intended one. We will see later how a punctuation signal at the beginning of each message sets the correct reading frame.

The triplets in an mRNA molecule specify the sequence of amino acids that have to be linked together to produce the particular protein. To actually produce the protein, some kind of a decoder or adaptor molecule is required. tRNAs serve as adaptor molecules that can recognize and bind both to the codon on the mRNA, and at another site on their surface, to the amino acid. A tRNA molecule consists of a chain of about 80 ribonucleotides. Parts of this chain assume particular shapes by complementary base pairing, as shown in Fig. 6.4. However, two regions of unpaired nucleotides are crucial to the function of tRNA in protein synthesis: (1) the *anticodon* which is a set of three consecutive nucleotides that pairs with the complementary codon in an mRNA molecule; and (2) a short single stranded region at the $3'$ end of the molecule, which is the site where the amino acid that matches the codon is attached to the tRNA.

The degeneracy of the genetic code implies that either there is more than one tRNA for many of the amino acids or that some tRNA molecules can base-pair with more than one codon. In fact, both situations occur. Some amino acids have more than one tRNA and some tRNAs are constructed so that they require accurate base-pairing only at the first two positions of the codon and can tolerate a mismatch (or *wobble*) at the third position. This wobble base pairing explains why many of the alternative codons for an amino acid differ only in their third nucleotide. For instance, the codons CCA, CCC, CCG and CCU all code for the amino acid proline.

Specific enzymes called *aminoacyl-tRNA synthetases* covalently couple each amino acid to its appropriate set of tRNA molecules. Specific nucleotides in both the anticodon and the amino acid accepting arm allow the correct tRNA to be recognized by the synthetase enzyme. The synthetase-catalyzed reaction that attaches the amino acid to the $3'$ end of the tRNA is powered by ATP hydrolysis, and it produces a high-energy bond between the tRNA and the amino acid. The energy of this bond is used at a later stage in protein synthesis to covalently link the amino acid to the growing polypeptide chain.

The RNA message is decoded on *ribosomes*. These are large molecular machines that facilitate accurate and rapid translation of mRNA into protein. The ribosome travels along the mRNA chain, capturing complementary tRNA molecules, holding them in position, and bonding together the amino acids that they carry so as to form a protein chain.

A ribosome is made from more than 50 different proteins (the *ribosomal proteins*) and several RNA molecules called *ribosomal RNAs* (rRNAs). A typical living cell contains millions of ribosomes in its cytoplasm. In a eucaryote, the ribosomal subunits are made in the nucleus, by the association of newly transcribed rRNAs with ribosomal proteins, which have been transported into the nucleus after their synthesis in the cytoplasm. The individual ribosomal units are then exported to the cytoplasm to take part in protein synthesis.

A ribosome is made up of two units — a small subunit and a large subunit that fit together to form a complete ribosome with a mass of several million daltons (for comparison, an average size protein has a mass of 40,000 daltons). The structure of a typical ribosome is shown in Fig. 6.5. The small subunit matches the tRNAs to the codons of the mRNA, while the large subunit catalyzes the formation of the peptide bonds that link the amino acids together into a polypeptide chain. The two subunits come together on an mRNA molecule usually near its beginning ($5'$ end) to begin the synthesis of a protein.

As shown in Fig. 6.5, a ribosome contains four binding sites for RNA molecules: one is for the mRNA and three (called the A-site, the P-site and the E-site) are for tRNAs. A tRNA molecule is held tightly at the A- and P-sites only if its anticodon forms base pairs (allowing for wobble) with a comple-

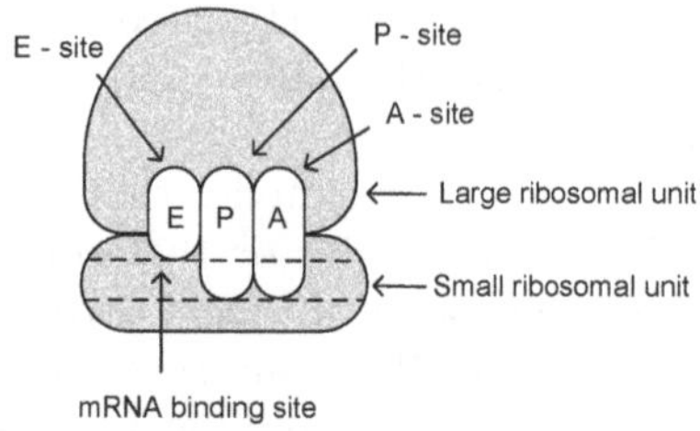

FIGURE 6.5 Structure of a typical ribosome [1]. Copyright 2003 from *Essential Cell Biology*, Second Edition by Alberts et al. Reproduced by permission of Garland Science/Taylor & Francis LLC.

mentary codon on the mRNA molecule that is bound to the ribosome. The A- and P- sites are sufficiently close together that their two tRNA molecules are forced to form base pairs with adjacent codons on the mRNA molecule.

Having introduced tRNAs and ribosomes, we are now in a position to take a detailed look at the steps involved in protein synthesis. Let us first focus on examining the mechanism by which the ribosome adds a new amino acid to a growing polypeptide chain. The three steps involved are shown in Fig. 6.6.

In the first step, the polypeptide chain containing the amino acids 1, 2 and 3 has just been synthesized, and the tRNA corresponding to amino acid 3 is still attached to amino acid 3 while being located in the P-site of the ribosome. The codon for amino acid 4 being located on the mRNA in the A-site of the ribosome, the corresponding tRNA with its bound amino acid comes and occupies the A-site of the ribosome. In the second step, the small ribosomal subunit (with the bound mRNA) moves relative to the large ribosomal subunit in such a way that the tRNAs 3 and 4 get shifted to the E and P sites respectively on the ribosome. Simultaneously, the amino acid 3 detaches from its tRNA and links up to amino acid 4 (still attached to its tRNA). In the third step, the tRNA for amino acid 3 is expelled from the E-site of the ribosome, and the small ribosomal subunit moves relative to the large ribosomal subunit so that they line up as before. This sets the stage for the tRNA corresponding to amino acid 5 to get into the A-site of the ribosome so that we are essentially back to step 1, but with a polypeptide chain one unit longer, and the entire cycle can repeat. The central reaction of protein synthesis, in which an amino acid bound to its tRNA is detached from the tRNA and linked to the growing polypeptide chain, is catalyzed by *peptidyl transferase enzyme activity*, which is part of the ribosome. The catalytic part of the ribosome in this case is thought to be not one of the proteins but rather one of the rRNAs in the large ribosomal subunit.

Let us now examine how protein synthesis starts and how it ends. Codons in the mRNA signal where to start and where to stop protein synthesis. The translation of an mRNA begins with the codon AUG, and a special tRNA is required to initiate translation. The *initiator tRNA* always carries the amino

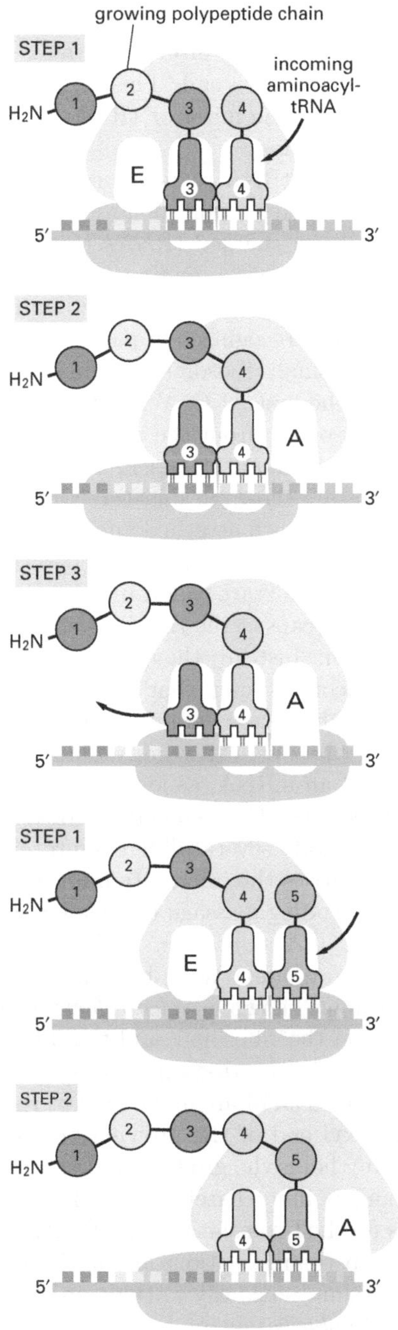

FIGURE 6.6 Steps in protein synthesis (See color figure following page 146) [1]. Copyright 2003 from *Essential Cell Biology*, Second Edition by Alberts et al. Reproduced by permission of Garland Science/Taylor & Francis LLC.

acid methionine (in bacteria, a modified form of methionine — formylmethionine — is used) so that newly made proteins all have methionine as the first amino acid at their amino-terminal end. This methionine is usually removed later by a specific protease. The initiator tRNA is distinct from the tRNA that normally carries methionine.

In eucaryotes, the initiator tRNA (which is coupled to methionine) is first loaded into the small ribosomal subunit along with additional proteins called *initiation factors.* Of all the charged tRNAs in the cell, only the charged initiator tRNA is capable of *tightly binding the small ribosomal subunit* as shown in Fig. 6.7. Next, the loaded ribosomal subunit binds to the 5′ end of an mRNA molecule, which is recognized in part by the methyl G cap present in eucaryotic mRNA. The small ribosomal subunit then moves forward along the mRNA searching for the first AUG. When this AUG is encountered, several initiation factors dissociate from the small ribosomal subunit to make way for the large ribosomal subunit to assemble and complete the ribosome. These chronological steps are shown in Fig. 6.7. Once the ribosome assembly is completed, the next tRNA with attached amino acid can bind the codon in the A-site and protein synthesis can continue as described earlier in Fig. 6.6. The mechanism for selecting a start codon in bacteria is different. Bacterial mRNA have no 5′ methyl G caps to tell the ribosome where to begin searching for the start of translation. Instead, they contain specific ribosome-binding sequences, up to six nucleotides long, that are located a few nucleotides upstream of the AUGs at which translation is to begin. Unlike a eucaryotic ribosome, a procaryotic ribosome can readily bind directly to a start codon that lies in the interior of an mRNA, as long as a ribosome-binding site precedes it by several nucleotides. Consequently, procaryotic mRNAs are often *polycistronic*, i.e., they encode several different proteins. In contrast, a eucaryotic mRNA usually carries the information for a single protein.

The end of the protein-coding message is signaled by one of several codons (UAA, UAG, or UGA) called *stop codons.* These are not recognized by a tRNA and do not specify an amino acid, but instead signal to the ribosome to stop translation. Proteins known as *release factors* bind to any stop codon that reaches the A-site on the ribosome, and this binding alters the activity of the peptidyl transferase in the ribosome, causing it to catalyze the addition of a water molecule instead of an amino acid to the peptidyl-tRNA. This reaction frees the carboxyl end of the growing polypeptide, and since only this attachment normally holds the growing polypeptide to the ribosome, the completed protein chain is immediately released into the cytoplasm. Even before a ribosome has finished translating an mRNA, other ribosomes can get on the mRNA at more upstream locations and initiate translation. Thus, several ribosomes could be simultaneously working on the same mRNA molecule forming what are called *polyribosomes.*

The amount of protein in each cell is regulated by carefully controlled protein breakdown. Proteins vary enormously in their life span. Some proteins may last for months or even years while others last for days, hours or even

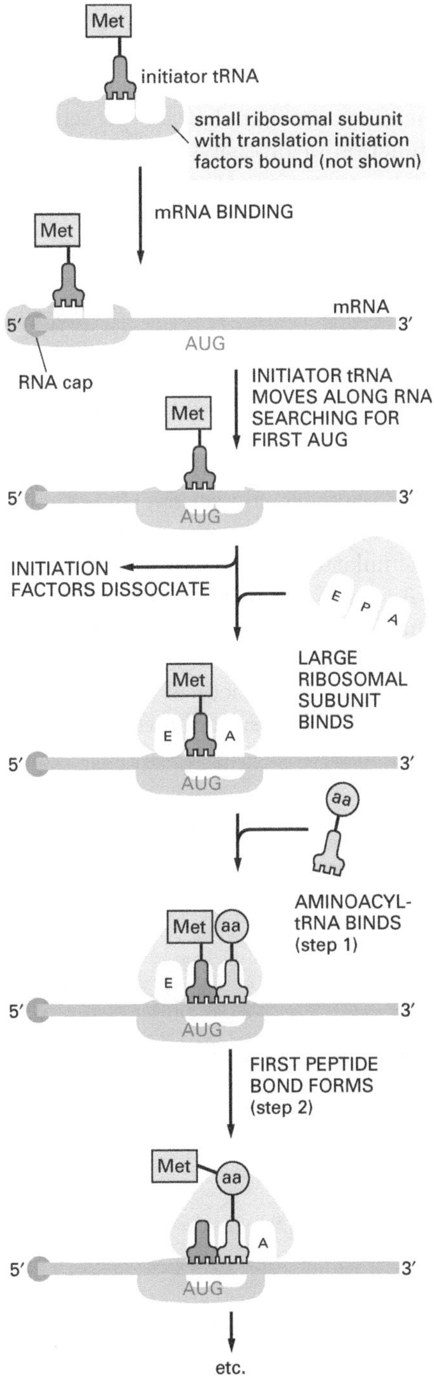

FIGURE 6.7 Initiation of protein synthesis (See color figure following page 146) [1]. Copyright 2003 from *Essential Cell Biology*, Second Edition by Alberts et al. Reproduced by permission of Garland Science/Taylor & Francis LLC.

seconds. A natural question that arises is how the cell controls these lifetimes. Cells have specialized pathways that enzymatically break proteins down into their constituent amino acids (a process termed *proteolysis*). As mentioned in Chapter 4, the enzymes that degrade proteins are known collectively as *proteases*. One function of proteolytic pathways is to rapidly degrade those proteins whose lifetimes must be short. Another is to recognize and eliminate proteins that are damaged or misfolded. Most proteins degraded in the cytosol of eucaryotic cells are broken down by large complexes of proteolytic enzymes, called *proteasomes*. The proteins imported into proteasomes for degradation have usually been marked out for destruction by the covalent attachment of a small protein called *ubiquitin*.

We conclude this chapter by noting that in all of life that we see around us today, DNA, RNA and proteins have very specialized individual roles. However, it is speculated that life initially originated in an RNA world. Since RNA is capable of both information storage as well as carrying out catalytic functions, this is a very plausible conjecture. It is thought that the initial RNA formed during the violent conditions present on the ancient earth. (Laboratory experiments simulating these conditions seem to support this hypothesis.) These initial RNA molecules, in turn, may have catalyzed the formation of additional RNA molecules. Later on, DNA and proteins took over some tasks from the RNA since DNA was better suited for information storage and proteins could perform a whole variety of other functions including serving as catalysts. This has probably resulted in the state of affairs that we see today.

7

Chromosomes and Gene Regulation

In this chapter, we study the organization of DNA inside cells and the various factors that play a role in determining whether and to what extent a particular gene is expressed in a cell. The genome of an organism encodes all of the RNA and protein molecules that are needed to make its cells. However, not every gene needs to be expressed all the time. Even the simplest single-celled bacterium can use its genes selectively, switching genes ON and OFF so that it makes different metabolic enzymes depending on the food sources that are available to it.

In multicellular organisms, *gene expression* is under even more elaborate control. These organisms have many different types of cells, which appear totally different from each other, e.g. a neuron and a lymphocyte. However, almost all of the cells of a multicellular organism (with the exception of the reproductive cells) contain the same genome and the apparent differences are caused by the fact that different cell types express different genes. For instance, the β-cells in the pancreas make the hormone *insulin*, the α-cells in the pancreas make the hormone *glucagon*, the red blood cells make the oxygen-transport protein *hemoglobin*, and so on. In the course of embryonic development, a single fertilized egg cell gives rise to many cell types that differ dramatically from each other in terms of both structure and function. This process is known as *cell differentiation* and it is achieved by the control of gene expression.

Encoding all the information needed to make just a single-celled bacterium requires large amounts of DNA. Clearly, encoding the instructions needed for the development of a multicellular organism would require far larger amounts of DNA. Packaging such a large amount of DNA within the tiny nucleus of a cell in a way so that it does not become an unmanageable tangle is, indeed, a remarkable feat of nature. In eucaryotic cells, enormously long double-stranded DNA molecules are packaged by association with specialized proteins into *chromosomes* that fit readily inside the nucleus and can be apportioned correctly between the two daughter cells at each cell division. To get a quantitative feel for the degree of compaction necessary, we note that the nucleus of a typical human cell is about 5 - 8μm in diameter and contains about 2 meters of DNA. This DNA must be folded in such a way that it not only does not become an unmanageable tangle, but must also be accessible to all of the enzymes and other proteins required for transcription, DNA replication and DNA repair.

7.1 Organization of DNA into Chromosomes

In eucaryotes, the DNA in the nucleus is distributed among a set of different chromosomes. Each chromosome consists of an enormously long DNA molecule that is folded and compacted by certain proteins. The complex of DNA and protein is called *chromatin.* In addition to the DNA packaging proteins, chromosomes are also associated with proteins involved in DNA replication, DNA repair and gene expression.

In procaryotes, on the other hand, the DNA is organized into *one circular chromosome.* This is carried out by some proteins but not too much is known about the details. Consequently, in this chapter, we will focus on eucaryotic chromosomes.

Human cells, with the exception of the germ cells (egg and sperm) each contain two copies of each chromosome, one inherited from the mother and one from the father. The two copies are called *homologous chromosomes*, the one inherited from the father is called the *paternal homolog* while the one inherited from the mother is called the *maternal homolog.* The only nonhomologous chromosomes are the sex chromosomes in males, where a Y chromosome is inherited from the father and an X chromosome is inherited from the mother. Females, on the other hand, have homologous sex chromosomes since they inherit X chromosomes from both parents.

The standard way of distinguishing one chromosome from another is to stain them with dyes that bind to certain types of DNA sequences. These dyes mainly distinguish DNA that is rich in A-T nucleotide pairs from DNA that is rich in G-C nucleotide pairs. This produces a characteristic pattern of bands along each chromosome and such a pattern is called a *karyotype.* Since the pattern for each chromosome is unique, these bands can be used to distinguish one chromosome from another. They can also be used to search for chromosomal abnormalities which characterize certain inherited birth defects at the prenatal stage and predispose individuals to certain types of cancers.

Chromosomes exist in different states during the life of a cell. The so called *cell cycle* can be broadly divided into *interphase* and *mitosis*. During interphase, transcription, translation and DNA replication take place and so the chromosome is in an extended state, and cannot be easily distinguished under a light microscope. Chromosomes in such a state are referred to as *interphase chromosomes.* During mitosis (or nuclear division), the chromosomes have already replicated and one copy needs to be delivered to each daughter cell. Consequently, during this phase, the chromosomes are highly compacted and are visible under a light microscope. The highly condensed chromosomes in a dividing cell are referred to as *mitotic chromosomes.*

Chromosomes contain specialized DNA sequences whose role is to ensure that chromosomes replicate efficiently and are correctly apportioned between the two daughter cells during cell division. The first among these are the *replication origins.* As discussed in Chapter 5, the DNA replication begins at a replication origin. Most eucaryotic chromosomes contain many replication origins to ensure that the entire chromosome can be replicated rapidly. The second specialized DNA sequence is the *centromere.* The presence of this sequence allows one copy of each duplicated chromosome to be pulled into each daughter cell when a cell divides. During mitosis, a protein complex called a *kinetochore* forms at the centromere and attaches the duplicated chromosomes to the mitotic spindle, allowing them to be pulled apart towards the opposite ends of the dividing cell. We will take a more detailed look at mitosis in a later chapter. The third specialized DNA sequence is called a *telomere.* This specialized DNA sequence is found at each of the two ends of a chromosome. Telomeres contain repeated nucleotide sequences that enable the ends of chromosomes to be replicated. Recall from Chapter 5 that DNA replication must be primed using an RNA primer. Furthermore, on the lagging strand, several primers have to be used to prime the different Okazaki fragments. Thus, at the end of a linear DNA molecule, there is no room to lay out an RNA primer. Some DNA could be easily lost from the ends of a linear DNA molecule each time it is replicated. This problem does not arise in the case of bacterial cells since their DNA is organized into a circular chromosome. In the case of eucaryotes, this problem is solved by having telomeres, which attract an enzyme called *telomerase.* This enzyme adds multiple copies of the same telomere DNA sequence to the ends of the chromosomes, thereby producing a template that allows replication of the lagging strand to be completed.

Nucleosomes are the basic units of chromatin structure. An individual nucleosome core particle consists of a complex of 8 *histone* proteins — two molecules each of histones H2A, H2B, H3 and H4 — and a double-stranded DNA around 146 nucleotide pairs. The histone octamer forms a protein core around which the double-stranded DNA helix winds. This is schematically illustrated in Fig. 7.1. The term *nucleosome* refers to a *nucleosome core particle* plus an *adjacent DNA linker* (about 50 nucleotide base pairs long). The formation of nucleosomes converts a DNA molecule into a chromatin thread approximately *one-third of its initial length.* Histones are small proteins with a high proportion of *positively charged amino acids.* These positive charges allow the histones to bind tightly to the negatively charged sugar phosphate backbone of DNA, regardless of the precise nucleotide sequence. There are additional levels of chromatin packing, one of which is facilitated by a fifth histone H1 which is thought to pull the nucleosomes closer into a regular repeating array. However, exactly how the packing takes place in these additional levels is not clear at the current time and the suggested mechanisms are purely speculative. The interested reader can consult [1] to get a flavor for some of the mechanisms that have been proposed.

The chromatin in an interphase chromosome is not in the same packing state throughout the chromosome. The most highly condensed form of interphase chromatin is called *heterochromatin.* Heterochromatin typically makes up about 10% of an interphase chromosome, and in mammalian chromosomes,

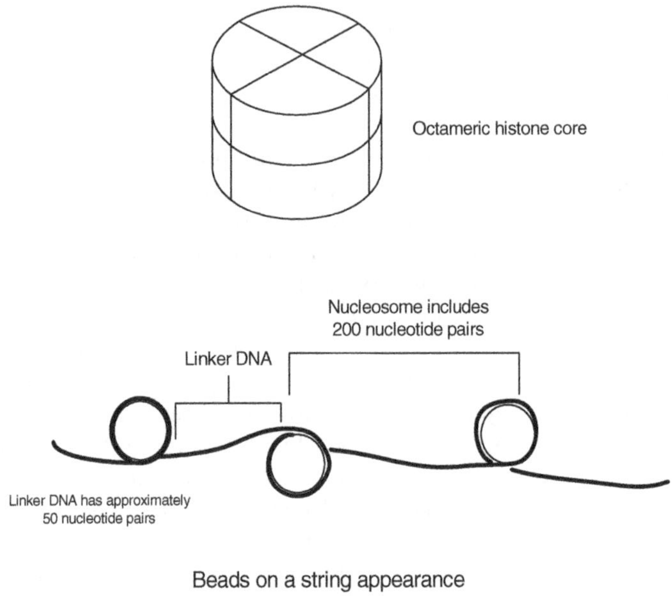

FIGURE 7.1 First level of DNA organization

it is typically concentrated around the centromere and telomere regions. Heterochromatin is *transcriptionally inactive.*

The most striking example of heterochromatin is found in the interphase X chromosomes of female mammals. Recall that male mammals have only one X chromosome while female mammals have two X chromosomes. Because a double dose of the corresponding protein in females could possibly be disastrous, female mammals permanently inactivate one of the two X chromosomes by condensing it into heterochromatin early in embryonic development. Thereafter, in all of the many progeny of the cell, the condensed and inactive state of that X chromosome is propagated. The rest of the interphase chromatin, which is in a variety of extended states, is called *euchromatin.* In a typical differentiated eucaryotic cell, some 10% of the euchromatin is in a state in which it is either actively being transcribed or is easily available for transcription; this is known as *active chromatin* and is the least condensed form of chromatin in the interphase chromosome.

From the above discussion, it appears that if a particular gene is moved from a region of heterochromatin to a region of active chromatin or vice versa, its transcriptional activity is likely to be altered. Indeed, this has been experimentally observed in a number of situations, as illustrated in Fig. 7.2. In the first instance, a gene *ADE2*, which is normally active in yeast cells and produces an enzyme needed for adenine synthesis, is moved by artificial means (to be discussed later in this book) near the transcriptionally inactive telomere region. The result is that instead of the normal white colony of yeast cells,

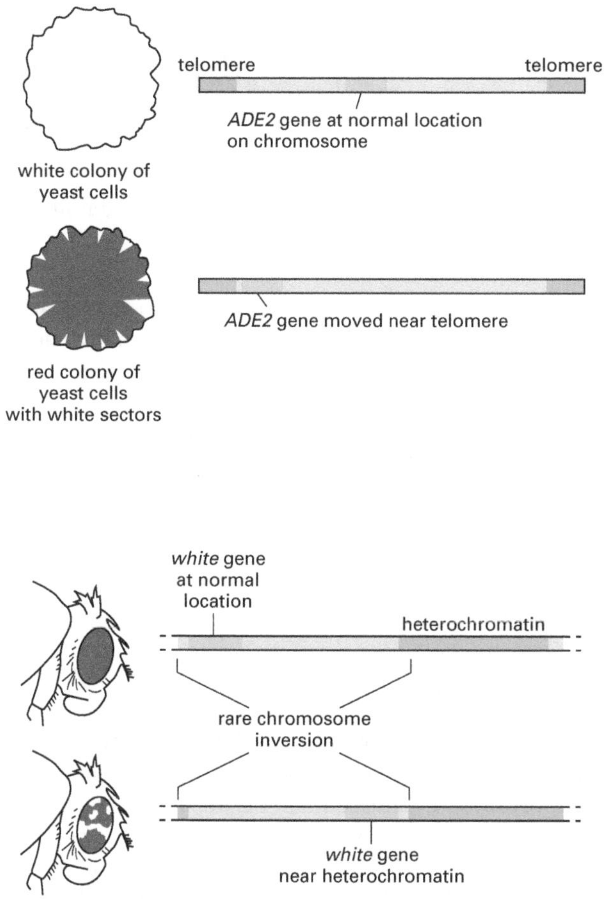

FIGURE 7.2 Position effects on gene expression (See color figure following page 146) [1]. Copyright 2003 from *Essential Cell Biology*, Second Edition by Alberts et al. Reproduced by permission of Garland Science/Taylor & Francis LLC.

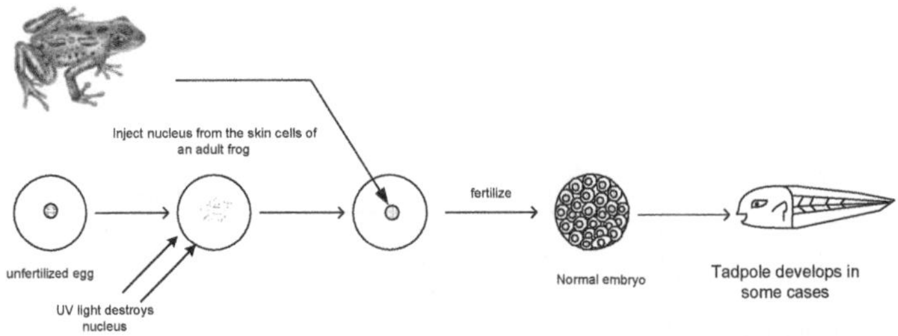

FIGURE 7.3 Different cell types contain the same genome.

we now get a colony of yeast cells that is predominantly red with some white patches here and there. The white patches arise as the movement of the ADE2 gene to the region near the telomere may not have *completely* silenced it. In the second instance, the active *white* gene in the fruit fly *Drosophila*, which normally leads to red eyes, is inactivated during development by moving it near a region of heterochromatin. This leads to flies with eyes that are patched red and white as shown in Fig. 7.2.

7.2 Gene Regulation

A central question in the study of gene regulation is how a cell specifies which of its many thousands of genes should be expressed as proteins. This question is especially important for multicellular organisms because as the organism develops, many different cell types are created starting essentially from the same precursor cells. This *differentiation* arises because cells make and accumulate different sets of RNA and protein molecules. We next describe an experiment that was carried out to conclusively demonstrate that different cell types from the same organism contain the same genome.

In this experiment, schematically illustrated in Fig. 7.3, the nuclei from the skin cells of an adult frog were injected into unfertilized frog eggs whose original nuclei had been destroyed by exposure to ultraviolet light. By fertilizing such eggs, it was observed that normal tadpoles did develop in *some cases.* This led to the conclusion that the skin cells of an adult frog did contain the genetic information necessary to produce a viable embryo.

Many of the proteins in cells are found in all cell types. For instance, the proteins needed to make DNA polymerase, RNA polymerase, ribosomes, etc. have generic functions and are found in all cell types. These proteins are called *housekeeping proteins* and the genes that encode them are called *housekeeping genes*. On the other hand, each different cell type also produces specialized proteins that are responsible for the cell's distinctive properties. For instance, hemoglobin is made in reticulocytes, the cells that develop into red blood cells, but it cannot be detected in any other cell type, and insulin is made exclusively by the β cells in the pancreas. A cell can control the proteins it makes by: (1) controlling when and how often a given gene is transcribed, (2) controlling how the primary transcript is spliced or otherwise processed (recall alternative splicing from Chapter 6), (3) selecting which mRNAs are translated by ribosomes or (4) selectively activating or inactiviting proteins after they have been made. Although examples of control in each of these four ways are documented, for most genes, the control of transcription i.e. the control of gene expression is the primary method. This makes economic sense since it ensures that no unnecessary intermediates are produced by the cell. In the rest of this chapter, we will discuss the mechanisms by which gene expression is regulated in procaryotic and eucaryotic cells.

Transcription is controlled by proteins binding to regulatory DNA sequences. The promoter region of a gene attracts the enzyme RNA polymerase and correctly orients it to begin its task of making an RNA copy of the gene. The promoters of both bacterial and eucaryotic genes include the *initiation site* and a sequence of about 50 nucleotides that extends "upstream" from the initiation site. This region contains sites that are required for the RNA polymerase to bind to the promoter. In addition to the promoter, nearly all genes, whether bacterial or eucaryotic, have *regulatory DNA sequences* that are needed in order to switch the gene ON or OFF. Some regulatory DNA sequences are as short as 10 nucleotide pairs and act as simple gene switches that respond to a single signal. This occurs primarily in bacteria. Other regulatory sequences can be very long (as many as 10,000 base pairs) and act as molecular microprocessors, taking in several inputs to determine the transcription rate. This occurs in eucaryotic cells. The regulatory DNA sequences do not act by themselves. They must be bound by *gene regulatory proteins* that recognize them. Gene regulatory proteins can either supress transcription or enhance it. *Repressors* turn genes OFF while *activators* turn them ON. We next present an example of a repressor protein and an activator protein.

In bacteria, it is common to find several genes that are transcribed as a single mRNA. Each such cluster is called an *operon*. As shown in Fig. 7.4, in E. coli, there are five genes, say A, B, C, D and E, that code for proteins needed for synthesizing the amino acid tryptophan. These five genes are transcribed from the same promoter and constitute an operon called the *tryptophan operon.* Inside the promoter region, there is also a DNA regulatory sequence. When a repressor molecule binds to this sequence, RNA polymerase cannot get on to the promoter and so the transcription of this operon stops. However, the repressor molecule can bind this sequence only if it also binds several molecules of tryptophan.

So when the level of tryptophan in the cell is high, the transcription of the operon stops. On the other hand, when the level of tryptophan in the cell falls, the repressor is no longer able to bind the regulatory sequence and so

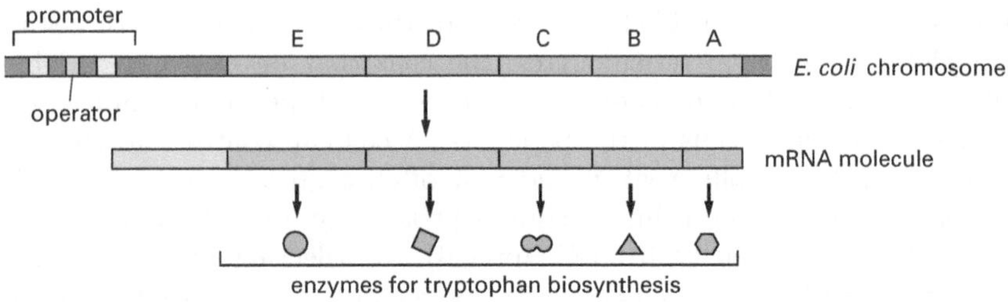

FIGURE 7.4 The schematic diagram of the Tryptophan operon [1]. Copyright 2003 from *Essential Cell Biology*, Second Edition by Alberts et al. Reproduced by permission of Garland Science/Taylor & Francis LLC.

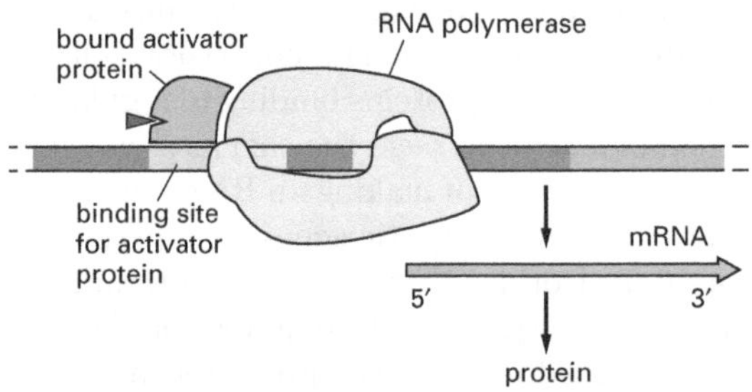

FIGURE 7.5 Role of activator proteins [1]. Copyright 2003 from *Essential Cell Biology*, Second Edition by Alberts et al. Reproduced by permission of Garland Science/Taylor & Francis LLC.

RNA polymerase can start transcription of the operon so that more tryptophan can be produced. The tryptophan repressor protein is *always present in the cell* though at a low level. Such unregulated gene expression is known as *constitutive* gene expression as opposed to *induced* gene expression, which occurs in response to some stimulus. Due to the constitutive nature of the gene expression of the repressor protein, E. coli can respond very rapidly to a rise in the tryptophan level. The tryptophan repressor is an example of an allosteric repressor protein that undergoes a conformational change when it has bound several molecules of tryptophan. Activator proteins, on the other hand, act on promoters that are only marginally functional in binding RNA polymerase on their own. These poorly functioning promoters can be made fully functional by proteins that bind to a nearby site on the DNA and contact the RNA polymerase in a way that helps it initiate transcription. This is schematically illustrated in Fig. 7.5. An example of an activator protein is the catabolic activator protein (CAP). CAP binds DNA only after binding to cyclic AMP. Cyclic AMP is a molecule that is formed when the phosphate ion attached to the 5′ location on the sugar in an adenosine monophosphate (AMP) molecule links up with 3′ hydroxyl group on the same sugar molecule. A high concentration of AMP or cyclic AMP indicates that the cell is low on energy. (Recall from Chapters 2 and 3 that ATP and ADP are the molecules that have high energy phosphoanhydride bonds.) When cyclic AMP concentration goes up, genes activated by CAP are switched on, which signals to the bacterium that glucose, its preferred energy source, is no longer available. Consequently, enzymes capable of degrading other sugars are made.

The initiation of gene transcription in eucaryotic cells is a complex process. We next discuss differences between initiation of transcription in procaryotes and eucaryotes. Procaryotic cells contain a single type of RNA polymerase while eucaryotic cells have three different RNA polymerases called RNA polymerase I, RNA polymerase II and RNA polymerase III. RNA polymerases I and III transcribe the genes encoding transfer RNA, ribosomal RNA, and small RNAs that play a structural role in the cell. RNA polymerase II, on the other hand, transcribes the vast majority of eucaryotic genes, including all those that encode proteins.

Eucaryotic RNA polymerase II cannot initiate transcription on its own. To initiate transcription, eucaryotic RNA polymerase requires the help of a large set of proteins called *transcription factors.* These are thought to position the RNA polymerase correctly at the promoter, to aid in pulling apart the two strands of DNA to allow transcription to begin, and to release RNA polymerase from the promoter once transcription begins. The assembly of transcription factors and their role in initiating transcription at the appropriate location are shown in Fig. 7.6. Once the initiation complex has assembled, the general transcription factor TFIIH (which contains a protein kinase enzyme as one of its subunits) phosphorylates the RNA polymerase. This phosphorylation is thought to help the polymerase disengage from the cluster of transcription factors, allowing transcription to begin.

In eucaryotic cells, the gene regulatory proteins can influence the initiation of transcription even when they are bound to DNA thousands of nucleotide pairs away from the promoter. A schematic diagram showing how this might be possible is shown in Fig. 7.7.

Finally, eucaryotic transcription initiation must take into account the packing of DNA into nucleosomes and more compact forms of chromatin structure. It is believed that the nucleosomes prevent the general transcription factors of RNA polymerase from assembling on the DNA and can, therefore, hinder the initiation of eucaryotic transcription.

Most bacterial genes are controlled by a single activator or repressor protein. On the other hand, most eucaryotic gene regulatory proteins work as part of a "committee" of regulatory proteins, all of which are necessary to express the gene in the right cell, in response to the right conditions, at the right time, and at the required level. This is referred to as *combinatorial control* and is similar to Boolean logic familiar to most engineers. Indeed, using Boolean functions to model relationships between expressed genes is a theme that we will encounter in a later chapter.

Just as the expression of a single eucaryotic gene can be regulated by a "committee" of proteins, the expression of several different genes can be co-

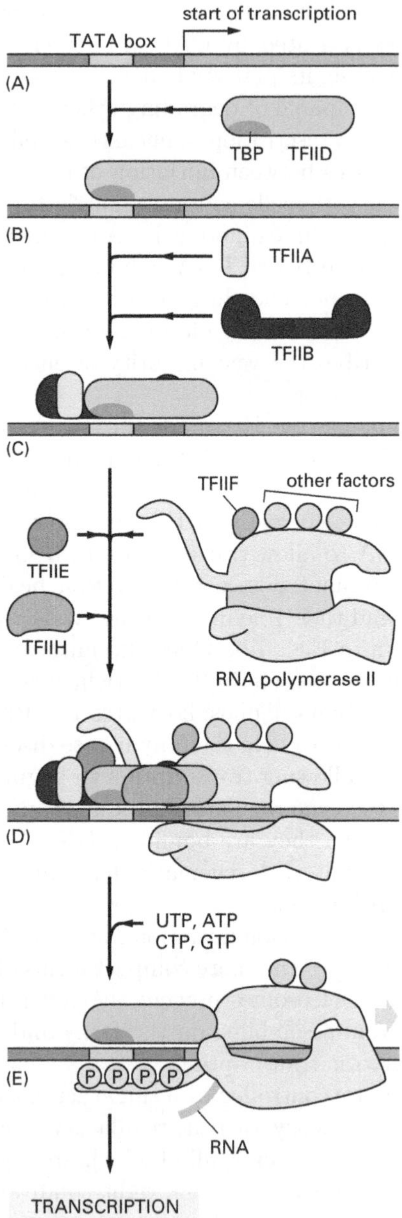

FIGURE 7.6 Role of transcription factors [1]. Copyright 2003 from *Essential Cell Biology*, Second Edition by Alberts et al. Reproduced by permission of Garland Science/Taylor & Francis LLC.

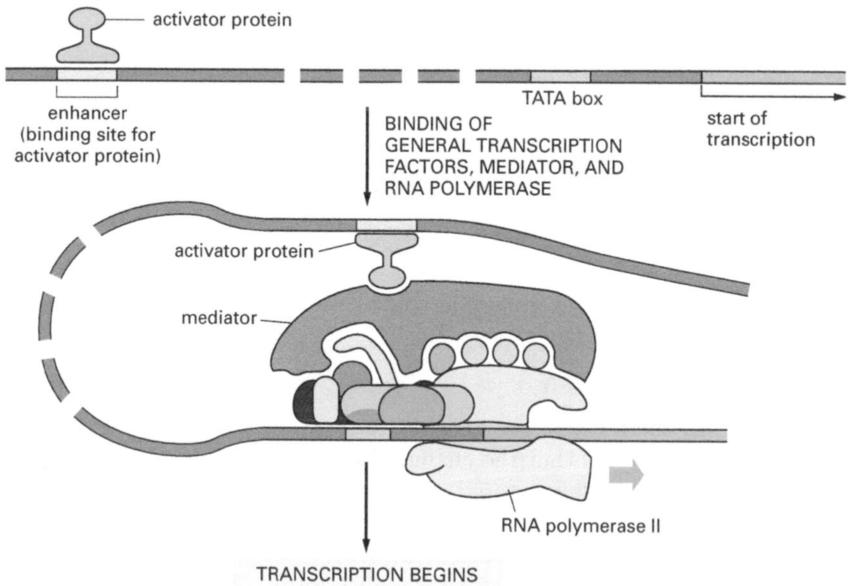

FIGURE 7.7 Transcription controlled from a distance [1]. Copyright 2003 from *Essential Cell Biology*, Second Edition by Alberts et al. Reproduced by permission of Garland Science/Taylor & Francis LLC.

ordinated by a single protein. In bacteria, this can be achieved by having all the relevant genes clustered together in a single operon, as in the case of the tryptophan operon discussed earlier. On the other hand, in eucaryotes, this can be achieved if the single protein is the *last element that is needed* in committees of regulatory proteins for turning different genes ON or OFF.

In eucaryotic cells, combinatorial control of gene expression can create different cell types. For instance, fibroblasts can be converted into myoblasts by expressing the gene Myosin D (MyoD). It appears that fibroblasts, which are derived from the same broad class of embryonic cells as muscle cells, have already accumulated all the other necessary gene regulatory proteins required for the combinatorial control of the muscle-specific genes and addition of MyoD completes the unique combination that directs the cells to become muscle. To further drive home this point, consider the hypothetical example shown in Fig. 7.8. Here there are three genes of interest, each of which could be in one of two states, ON denoted by 1 and OFF denoted by 0. Clearly, there are $8(= 2^3)$ possible gene expression patterns in this case and, by selectively expressing one or more genes in the progeny of this cell, we can end up with up to eight different possible cell types after three rounds of cell division.

Some highly specialized cells never divide again once they have differentiated, for instance, nerve cells or neurons, but there are many other differentiated cells, such as liver cells, that will divide many times in the life of an

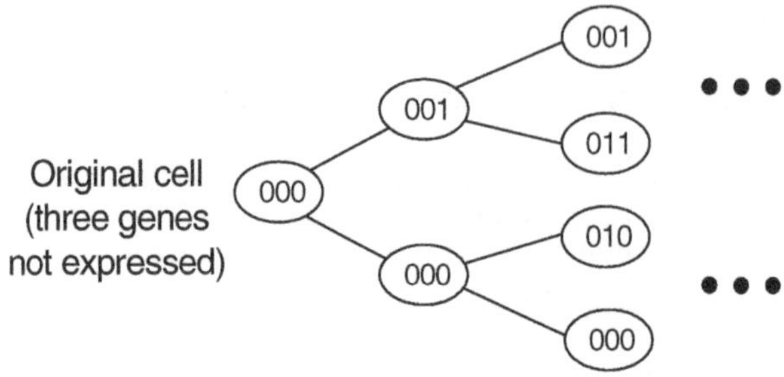

FIGURE 7.8 Combinatorial gene control can be used for generating cell types.

individual. All these cell types give rise only to cells like themselves when they divide. This means that the changes in gene expression that give rise to a differentiated cell must be stored in memory and passed on to its progeny through all the subsequent cell divisions. Cells have several ways of accomplishing this. One possible mechanism involves a *positive feedback loop* where a key gene regulatory protein activates transcription of its own gene in addition to that of other cell type specific genes. The transcription of this gene regulatory protein ensures that this particular protein is turned ON in the new daughter cells so that they inherit the same cell type as that of their parent. In another mechanism, the condensed chromatin structure is faithfully propagated from parent to daughter cell even though DNA replication intervenes. An instance of this is the inactivation of one X chromosome in female mammals, something that we have already discussed in this chapter.

We conclude this chapter by presenting a fascinating experiment where the alteration of the expression status of a single gene can trigger the development of an entire organ and, that too, at an abnormal location. Now, a gene called 'Ey' in flies and 'Pax-6' in vertebrates is crucial for eye development. In this experiment involving fruit flies the gene 'Ey' is expressed early in development using artificial means in cells that normally go on to form legs. The result is that in the corresponding fruit flies, eyes develop in the middle of the legs. Thus this experiment clearly shows that the action of just one gene regulatory protein can turn on a cascade of gene regulatory proteins whose actions can result in the formation of an organized group of many different types of cells.

8

Genetic Variation

In this chapter, we discuss mechanisms by which the DNA in organisms can undergo changes over a long period of time. Although cells go to great lengths to maintain the integrity of the DNA, permanent changes in the DNA, called *mutations*, do accumulate over time. In fact, the vast diversity of life that we see around us today has arisen through changes in the DNA that have accumulated over evolutionary time. This genetic variation or diversity refers to the genomic differences between different species as well as between members of the same species. In the latter context, we note that even members of the same species have genomes that are quite different from each other. In fact, no two human beings have identical genomes unless they happen to be identical twins.

It is generally believed that the conditions on this earth have undergone dramatic changes over billions of years. Thus, in order for life to propagate, it is essential that the "survivors" be able to adapt to changing conditions. It is believed that genetic variation, though not always beneficial, is responsible for conferring survivability in a changing environment.

There are three main mechanisms by which genetic variation can arise: (i) rare mistakes in DNA replication and repair; (ii) *DNA recombination* and the activity of viruses and *mobile genetic elements* that can move into and out of the DNA; and (iii) the reassortment of the gene pool of the species into new combinations during sexual reproduction. We next consider each of these mechanisms in some detail. For clarity of presentation, we will separately discuss genetic variation in procaryotes and eucaryotes.

8.1 Genetic Variation in Bacteria

As already mentioned in Chapter 1, among procaryotes E. coli is a model organism for genetic studies. E. coli is said to have a *haploid* genome since it has only one copy of each gene in its genome. Consequently, the effect of a mutation at the gene level will manifest itself in the phenotype. In contrast, organisms such as ourselves are called *diploid* organisms, since our genomes contain two copies of each gene. Consequently, a DNA mutation in *one chromosome* will not necessarily result in an observable phenotype. In this context, recall from Chapter 5 how an individual with only one defective β-globin gene does not exhibit the full blown symptoms of sickle-cell anemia.

E. coli, like other procaryotes, reproduces by fission-type cell division. The DNA replicates and the two identical strands of DNA move to the two ends of the growing bacterium. The bacterium then splits in two, producing two daughter cells, each containing a genome identical to that of the parent cell. In the presence of sufficient nutrients, a population of E. coli doubles in number every 20 to 25 minutes. Thus, in less than a day, a single E. coli can produce more than 5 billion descendants. Every time that a cell divides, the DNA has to be replicated which means that there is a possibility of some replication errors. The rapid rate of division of E. coli means that very large populations of E. coli cells in which DNA mutations have occurred can be produced quite rapidly. Indeed, E. coli bacteria can be quite easily used to demonstrate how DNA mutations confer survivability in a changing environment. For instance, consider the antibiotic *rifampicin*. Like many other antibiotics, this drug binds tightly to RNA polymerase inside the E. coli cell and prevents it from transcribing DNA to RNA. This inhibition eventually blocks the synthesis of new proteins, and the E. coli bacterium dies. However, in a large population of E. coli, say 10^9 cells, there will be some cells that are rifampicin resistant. If such a population is treated with rifampicin, most of the cells will die. However, the rifampicin-resistant cells will survive and ultimately take over the population. The rifampicin resistance here comes from mutations in the DNA which allow RNA polymerase to transcribe even in the presence of rifampicin.

The principle that we have encountered here, namely the ability of antibiotics to block bacterial transcription, is the basis for current day treatment of bacterial infections using antibiotics. The reason that this treatment is ineffective against viral infections is that a virus does not have its own transciption and translation machinery which the antibiotic could have targeted. Instead, as we shall see, a virus relies on hijacking the replication machinery of the host cell in order to propagate itself.

Bacterial cells can acquire genes from other bacteria. For instance, if we mix a laboratory strain of E. coli that lacks one of the enzymes for making an essential amino acid with another strain that lacks one of the enzymes for making another essential amino acid, and if the mixture is allowed to grow together for a few hours and then transferred to a medium that lacks both amino acids, many rapidly growing bacteria can be found in the new medium. This new bacterial strain occurs with a frequency higher than what could be accounted for based on mutations due to replication errors. Instead, the genome of the new bacterial strain is composed of normal genes for the synthesis of both the essential amino acids.

8.1.1 Bacterial Mating

Genes can be transferred from one bacterium to another by a process called bacterial mating. The ability to initiate mating and gene transfer, seen in

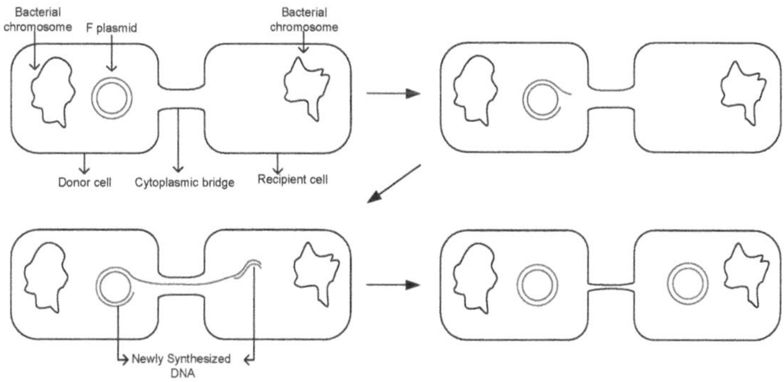

FIGURE 8.1 Gene transfer by bacterial mating (See color figure following page 146).

some bacteria, is conferred by genes contained in bacterial *plasmids*. Plasmids are small, circular double stranded DNA molecules that are separate from the larger bacterial chromosome. A plasmid that commonly initiates mating in E. coli is the F plasmid, or fertility plasmid. When a bacterium carrying the F plasmid (the *donor*) encounters a bacterium lacking the plasmid (the *recipient*), a cytoplasmic bridge is formed between the two cells. The F plasmid DNA is replicated and transferred from the donor through the bridge to the recipient. The schematic steps are shown in Fig. 8.1. Finally, the bridge breaks down and the two bacteria, both of which now contain the F plasmid, can act as donors in subsequent encounters with recipient bacteria. The F plasmid is necessary for mating because it carries genes that encode some of the proteins required to make the cytoplasmic bridge and to transfer the DNA.

Bacterial mating via the F plasmid does not generate much genetic variation since the F plasmid that is transferred contains only a small number of genes. However, occasionally the F plasmid can get integrated into the bacterial chromosome with the result that when it now initiates mating and gene transfer, it can now take parts of the bacterial chromosome with it.

Bacteria can take up DNA from their surroundings. For instance, DNA can be taken up from other dead bacteria in the surroundings. This process is called *transformation* since it can transform one strain of bacteria into another. In this context, it is appropriate to recall the experiment from Chapter 5 where one strain of bacteria got transformed into another.

The most important route by which DNA becomes incorporated into a bacterial genome is called *homologous recombination*. Homologous recombination can take place between two DNA molecules of similar nucleotide sequence. Two double-stranded DNA molecules that have regions of very similar (homologous) DNA sequence align so that their homologous sequences are in register. This is shown in Fig. 8.2. The same figure also shows how they can then cross over. We note that what is shown in Fig. 8.2 is the net result

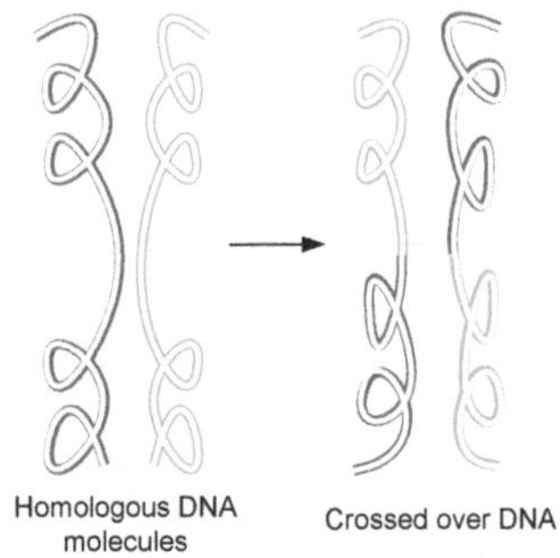

FIGURE 8.2 Homologous recombination (See color figure following page 146).

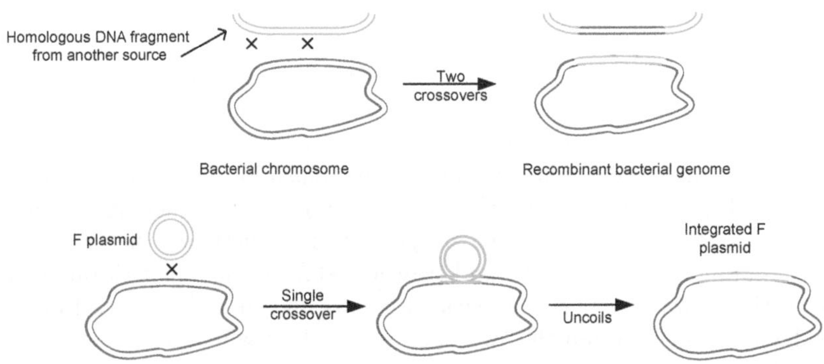

FIGURE 8.3 Two crossovers leading to variation in a bacterial genome (See color figure following page 146).

of homologous recombination. The intermediate steps are not relevant for our purposes and are, therefore, not discussed here. The interested reader is referred to [1] for a detailed description. As shown in Fig. 8.3, two such exchanges can neatly replace a long stretch of bacterial DNA with a homologous but not identical DNA fragment from another source. The same mechanism can result in the F plasmid DNA getting integrated into the bacterial chromosome, and this is also schematically shown in the same figure. Cells utilize specialized proteins to facilitate homologous recombination. Such recombination enzymes are well characterized in bacteria, but are only beginning to be understood in eucaryotic cells.

8.1.2 Gene Transfer by Bacteriophages

Bacterial viruses, or *bacteriophages*, are viruses that invade bacterial cells. A virus is usually composed of DNA enclosed inside a protein coat. A virus enters a bacterial cell and uses the cell's DNA replication, transcription and translation machinery to produce (i) more copies of its own DNA and (ii) its coat protein. The replicated DNA is packaged into additional protein coats to produce progeny viruses which can leave the current bacterial cell and invade other cells. Viral reproduction is generally lethal for the infected cell and the cell bursts open (*lyses*) as a result of the infection.

When a virus invades a bacterial cell one of two things can happen: (i) the cell lyses and a large number of copies of the virus are released; or (ii) the viral genome gets integrated into the bacterial chromosome and at each subsequent bacterial cell division, the viral genome is also replicated along with the bacterial genome and passed on. An environmental insult such as exposure to ultraviolet light can induce the viral genome to leave the host chromosome and begin a lytic phase of viral replication. Most of the time, route (i) above is followed but on rare occasions, route (ii) is also followed. The viral genome usually integrates with the bacterial genome by *site specific recombination* that is carried out by an enzyme called *integrase*. On leaving a host chromosome, the viral DNA will occasionally remove itself inaccurately and bring along a neighboring piece of host DNA in place of part of its own DNA. This host DNA will be packaged into a virus particle along with the viral DNA. So when the new virus infects a new host, it introduces, along with the viral DNA, DNA derived from the previous host. This bacterial DNA can become part of the new host chromosome in at least a couple of ways: (i) the incoming virus integrates into the new host chromosome; or (ii) if the incoming virus does not destroy the host, the passenger bacterial DNA can become a permanent part of the host's genome by homologous recombination. This process is called *transduction*.

8.1.3 Transposons

Many bacterial and eucaryotic genomes contain stretches of DNA called *transposable elements* (or *transposons*), which can move from place to place within the chromosome by a process called *transposition*. Transposons move within the DNA of their host by means of special recombination enzymes called *transposases*, encoded by the transposable element, to create great genetic diversity. As schematically illustrated in Fig. 8.4, in *non-replicative transposition*, the stretch of DNA from the donor is removed and incorporated into the recipient while in *replicative transposition*, the donor DNA remains intact while a copy is made for incorporation into the recipient DNA.

In concluding our discussion on genetic variation in bacteria, we note that most alterations in the genome are harmful to the individual bacterium and these are quickly eliminated from the population. However, alterations also confer survivability in a changing environment, as was demonstrated by the example involving the antibiotic rifampicin.

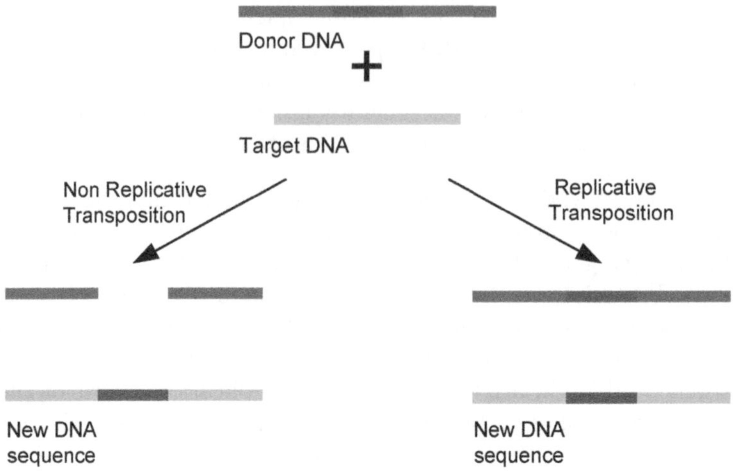

FIGURE 8.4 Nonreplicative and replicative transposition brought about by transposons (See color figure following page 146).

8.2 Sources of Genetic Change in Eucaryotic Genomes

Unlike bacterial DNA, eucaryotic DNA has both coding and non-coding regions. In procaryotes, the rate of cell division is very high and hence there is a strong selective pressure to minimize the amount of superfluous DNA in the genome. This is probably the reason that procaryotic genomes have gotten rid of most of the spacer DNA, as it were.

8.2.1 Gene Duplication

Eucaryotic genomes are also characterized by a large amount of *gene duplication* that has occurred over evolutionary time. As a result, there can be several genes belonging to the same family. The most well-documented example is the β-globin gene family and the human genome has a total of 5 β-globin genes. These genes encode the β subunits of the various hemoglobins produced at different times during embryonic, fetal and adult life. Each is especially well suited to the stage in development in which it is expressed.

Gene duplication is thought to occur from a rare recombination event between two homologous chromosomes. Let us consider the case of the β-globin gene duplication. Instead of aligning properly for a crossover, the two homologous chromosomes align in an improper fashion as shown in Fig. 8.5. After the crossover, the long chromosome has two copies of the globin gene, while the short chromosome lacks the original globin gene. Consequently, the indi-

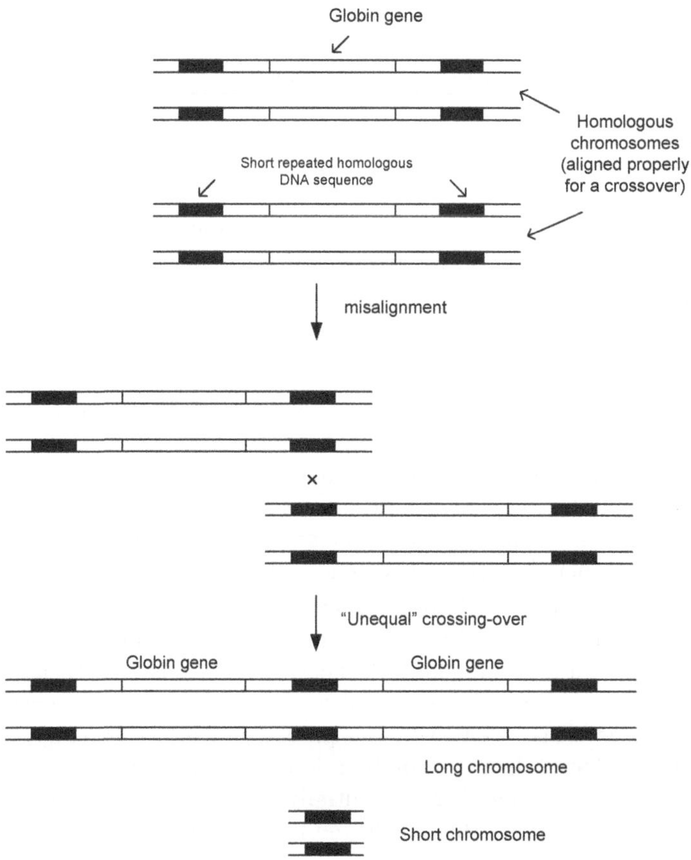

FIGURE 8.5 Unequal Crossover can cause duplication and deletion of genes.

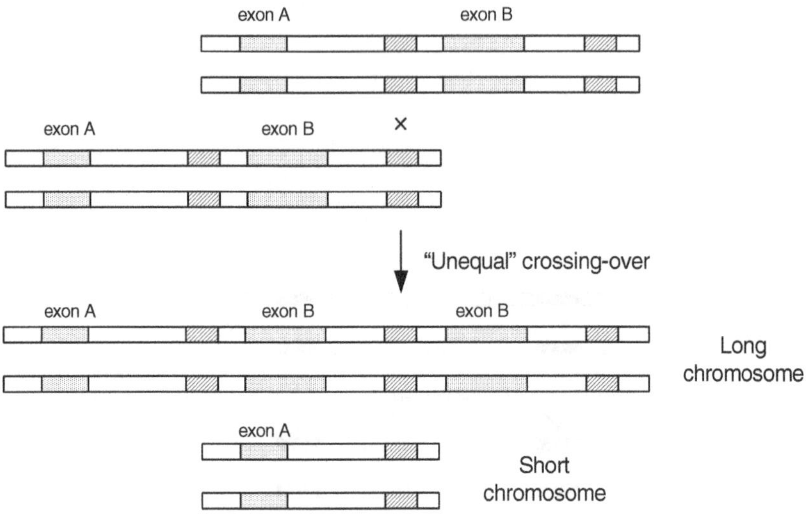

FIGURE 8.6 Duplication of exons

viduals that inherit the short chromosome would be expected to be eliminated from the population while the individuals that inherit the long chromosome would be expected to have two β-globin genes instead of one. Genes encoding new proteins can also be created by the recombination of exons. The general scheme, which is shown in Fig. 8.6, is the same as before except that *an exon within a gene*, rather than the entire gene is duplicated.

Without introns there would be very few sites on the original gene at which a recombinational exchange between homologous chromosomes could duplicate the domain without damaging it. Therefore, introns greatly increase the probability that DNA duplications will give rise to functional genes encoding functional proteins. Moreover, the presence of introns greatly increases the probability that a chance recombination event can generate a functional hybrid gene by joining together two initially separate exons coding for quite different protein domains. This is referred to as *exon shuffling.*

8.2.2 Transposable Elements and Viruses

About 10% of the human genome consists of two families of *transposable* sequences. Transposable DNA elements move from place to place by the mechanisms discussed earlier for procaryotic transposons. However for eucaryotes, there are also *retrotransposons* for which an RNA copy is first made using RNA polymerase following which DNA copies are made using *reverse transcriptase.* It is these DNA copies that are then inserted into the target. Transposition by reverse transcription is shown in Fig. 8.7. Examples of human retrotransposons are the so-called *L1 transposable element* and the *Alu*

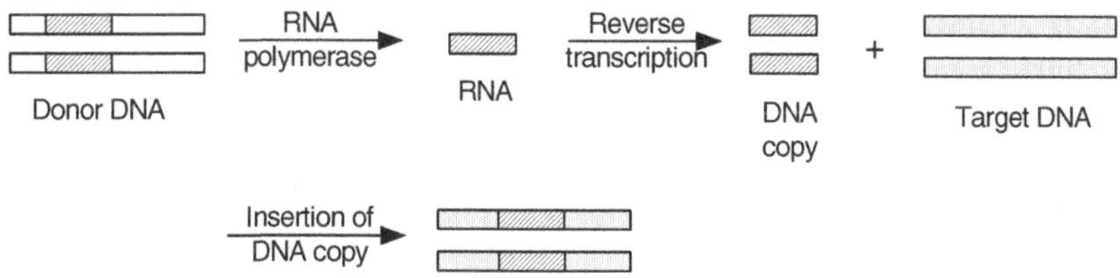

FIGURE 8.7 Transposition by reverse transcription.

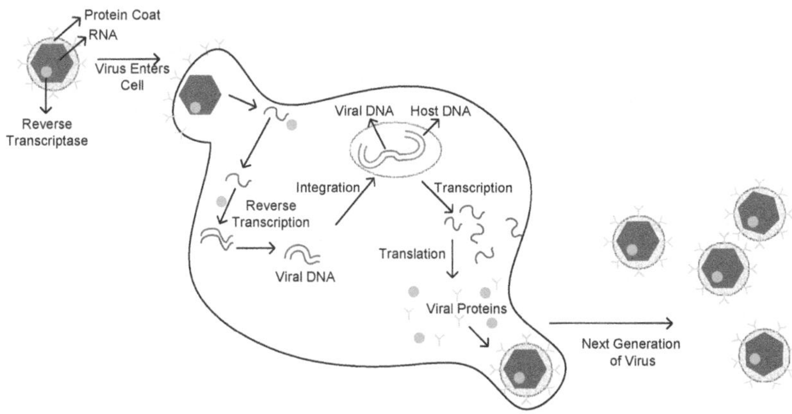

FIGURE 8.8 Retrovirus hijacking a host cell (See color figure following page 146).

sequence. The evolution of genomes has been greatly accelerated by transposable elements. The insertion of a transposable element in a regulatory region for a gene will often, by disrupting or adding short regulatory sequences, have a dramatic effect on gene expression.

Another source of genetic variation in eucaryotes is the activity of viruses. Like bacteriophages, viruses that infect eucaryotic cells are fully mobile elements that can move into or out of cells. Viral genomes can be made of DNA or RNA and can be single-stranded or double-stranded. An important class of viruses is *retroviruses*, which reverse the normal flow of genetic information. These viruses have a genome that is made of RNA, and have a protein coat that encapsulates the genome and the enzyme *reverse transcriptase*. The schematic diagram for a retrovirus hijacking a host cell is shown in Fig. 8.8.

Here, the enzyme reverse transcriptase is first used to make a (single-stranded) complementary DNA copy of the viral genome. The enzyme DNA polymerase (present in the host cell) is then used to create a double stranded DNA copy of the viral genome, which is then integrated into the host DNA. Transcription and translation from the integrated genome produces copies of the viral RNA, reverse transcriptase and the coat protein and all of these can be packaged together to produce additional retroviruses.

Retroviruses that have picked up host genes can make cells cancerous. Very few human cancers are caused by retrovirus infection, but these viruses are a prominent cause of cancers in some other animals. Perhaps, the most well known example of a cancer-causing virus is the *Rous Sarcoma* virus. This virus can pick up a gene from its chicken host. This gene called *src* is unnecessary baggage from the point of view of the virus, but it has profound consequences for cells that are infected by the virus. The normal src gene in the chicken genome encodes a protein kinase that is involved in the control of cell division and is called a *proto-oncogene*. However, the src gene carried by the virus is not quite identical to the normal cellular gene and the difference gives rise to its ability to cause a cancer. This mutated src gene is called an *oncogene* and it causes the infected cell to divide uncontrollably. The terms oncogene and proto-oncogene will be explained in more detail in Chapter 11, where we will focus on getting a basic understanding of cancer.

8.2.3 Sexual Reproduction and the Reassortment of Genes

Bacteria reproduce asexually and this gives rise to offspring that are identical to the parent. Sexual reproduction, on the other hand, involves the mixing of genomes from two individuals to produce offspring that are genetically distinct from one another and from both their parents. Sexual reproduction occurs in *diploid* organisms, in which each cell contains two sets of chromosomes, one inherited from each parent. The actual cells that carry out sexual reproduction in diploid organisms are called the germ cells or *gametes*. These gametes, which are haploid cells, are of two types — the egg and the sperm in animals. The haploid germ cells are generated from their diploid precursors by a special type of cell dvision called *meiosis*. We will be discussing meiosis in some detail in Chapter 10. Here, our main focus is on showing the staggering amount of genetic variation that results from meiosis.

A schematic diagram showing how meiosis produces haploid gametes from a diploid precursor cell is shown in Fig. 8.9. For clarity of presentation, we focus attention on only one pair of homologous chromosomes. During meiosis, the chromosomes first duplicate and then the homologous paternal and maternal chromosomes line up next to each other. Next, crossovers take place by homologous recombination and two rounds of cell splitting occur to produce four different haploid gametes.

Clearly, in the absence of a crossover, each pair of homologous chromosomes will give rise to two different types of gametes depending on whether it contains the paternal homolog or the maternal homolog. Thus, with 23 chromosome pairs in the human genome, each individual can produce 2^{23} *possible gametes*. The actual number of possible gametes is much higher because of crossovers. (On an average, there is at least one crossover on each chromosome.) To develop a new organism, two such gametes of the opposite kind (egg and sperm) have to fuse to form the single diploid cell that will undergo repeated mitosis to develop into the new organism. Thus, the number

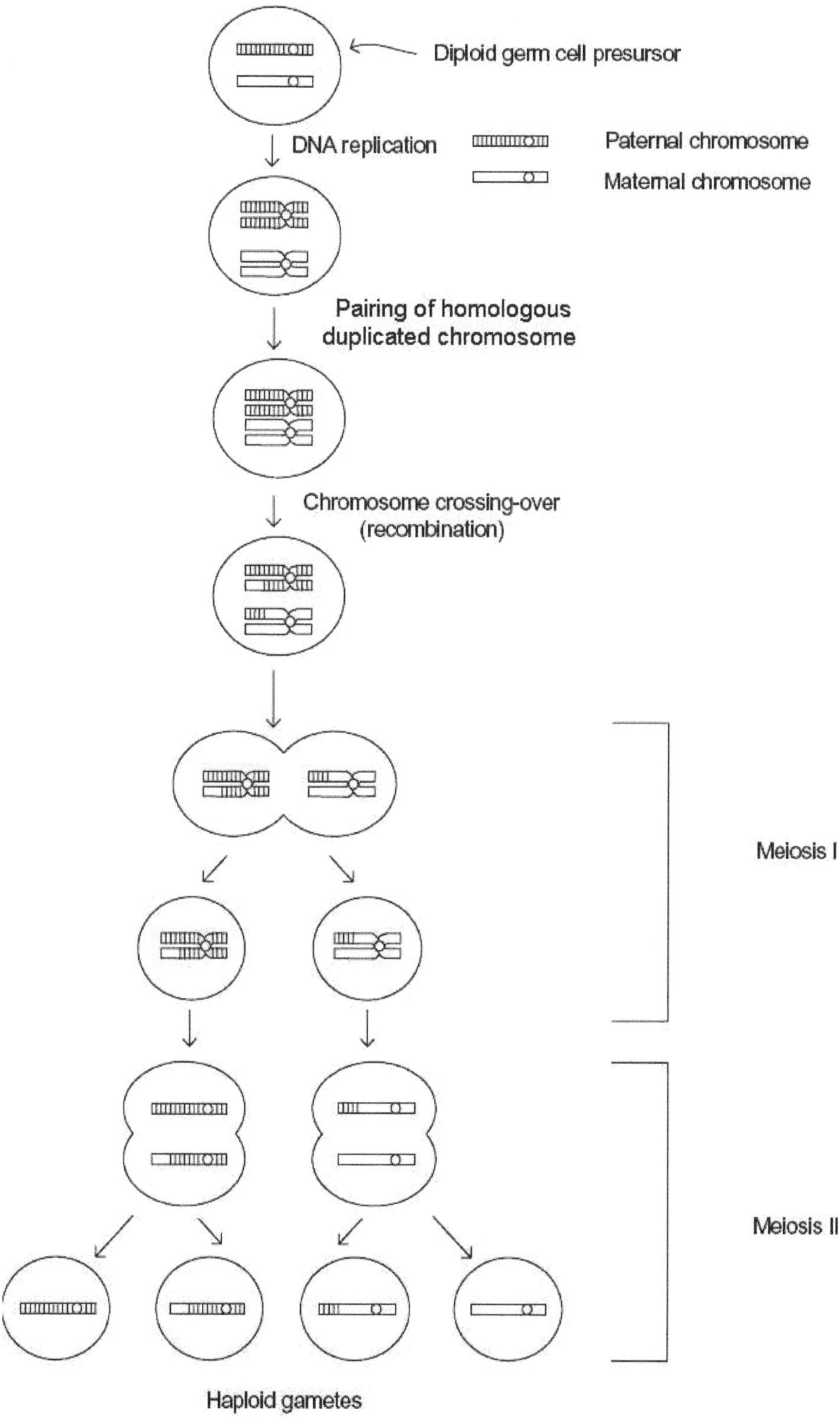

FIGURE 8.9 Meiosis.

of possible ways in which the genetic material gets reshuffled during sexual reproduction is staggering. This is the reason that it is not at all surprising that no two individuals (except identical twins) have the same genome.

9

DNA Technology

In the last chapter, we examined different mechanisms that have led to the genetic variation that we see in life around us today. In this chapter, we will look at some of the modern technologies that enable us to create genetic variation in the laboratory setting. Humans have been unknowingly experimenting with DNA for thousands of years. For instance, the cross breeding between plants and that between animals are really instances of genetic reshuffling.

In the early 1970s, it became possible for the first time to isolate a given piece of DNA out of the many millions of nucleotide pairs in a typical chromosome. This in turn made it possible to create new DNA molecules in the test tube and introduce them back into living organisms. This is referred to as "Recombinant DNA Technology" or "Genetic Engineering."

9.1 Techniques for Analyzing DNA Molecules

The development of current day DNA technology has been made possible by certain technological advances. The first among these is the discovery of *restriction nucleases*, which are enzymes that cleave DNA at certain specific nucleotide sequences. Fig. 9.1 presents a few examples of restriction nucleases cleaving DNA at specific locations. Restriction nucleases cleave DNA in a *very predictable* and *replicable manner*, unlike mechanical shear, which fragments DNA, but there is no way to predict where the fragmentation will occur. Furthermore, the fragmentation locations due to mechanical shear will change from one run of the experiment to another.

The second technique that has played a crucial role in the advancement of DNA technology is *gel electrophoresis*, which can be used to separate DNA fragments on the basis of their sizes. In gel electrophoresis, fragments of DNA are loaded on to a slab of agarose gel and subjected to an electric field as shown in Fig. 9.2. Since DNA is negatively charged, the fragments will migrate towards the bottom and the distance that each fragment travels will be inversely proportional to its size. If one incorporates the radioactive ^{32}P isotope of phosphorous in DNA before electrophoresis, then the DNA fragment positions after electrophoresis can be detected by using the technique of *autoradiography*. By comparing the sizes of the DNA fragments produced

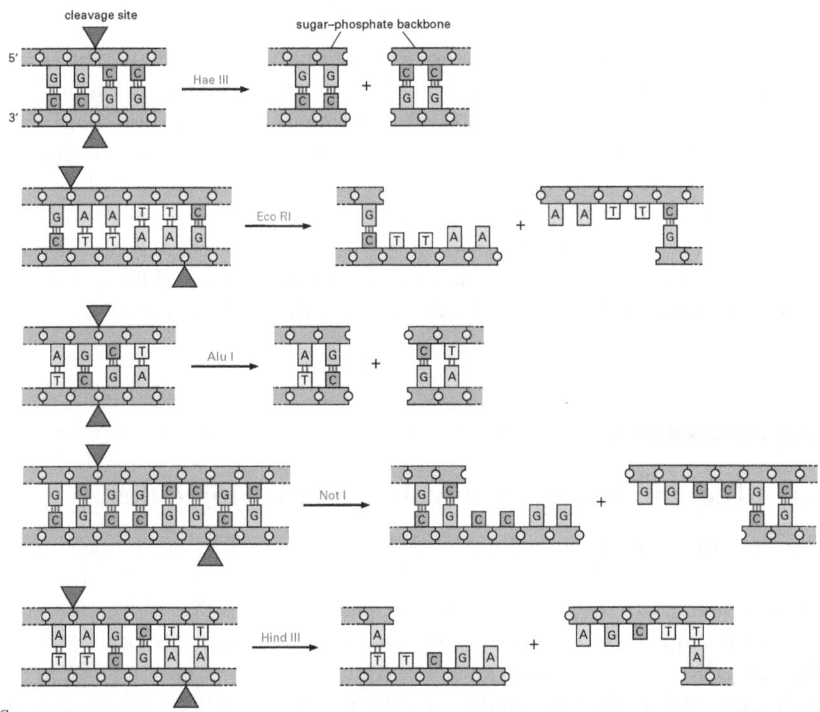

SS

FIGURE 9.1 Specific Cleaving of DNA by Restriction Nucleases [1]. Copyright 2003 from *Essential Cell Biology*, Second Edition by Alberts et al. Reproduced by permission of Garland Science/Taylor & Francis LLC.

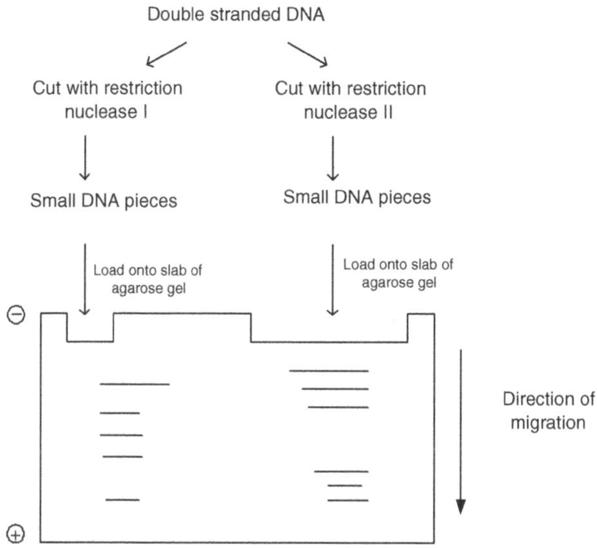

FIGURE 9.2 Gel Electrophoresis.

from a particular region of DNA after treatment with different combinations of restriction nucleases, a physical map of the region can be constructed showing the location of each cutting site. Such a map is known as a *restriction map*. A restriction map, however, does not provide the complete sequence information for the entire DNA sequence.

To determine the complete DNA sequence of a given DNA strand, one can run a four-lane gel electrophoresis following four separate DNA synthesis reactions as shown in Fig. 9.3. The key idea behind this method is that if during DNA synthesis, a dideoxyribonucleoside triphosphate (see the first part of Fig. 9.3) is incorporated into the growing DNA strand instead of a deoxyribonucleoside triphosphate, then the $3'$ end of the DNA chain is chemically blocked and, therefore, the chain cannot elongate any further. Thus, if we add dATP, dCTP, dGTP, dTTP in excess and a small amount of ddATP, then some DNA strands complementary to the given strand and terminating at the various A locations will be produced. This is illustrated in the second part of Fig. 9.3. Based on this, we outline the following procedure to determine the complete nucleotide sequence of a given DNA strand.

Take the double-stranded DNA strand. Pick one of the two strands as the DNA to be sequenced and use its complementary strand as the template. Four different chain-terminating dideoxyribonucleoside triphosphates (ddATP, ddCTP, ddGTP, ddTTP) are used in four separate DNA synthesis reactions on copies of the same single-stranded DNA template. Each reaction, which is primed using an oligonucleotide (synthetic) primer, produces a set of DNA copies that terminate at different points in the sequence. The products of

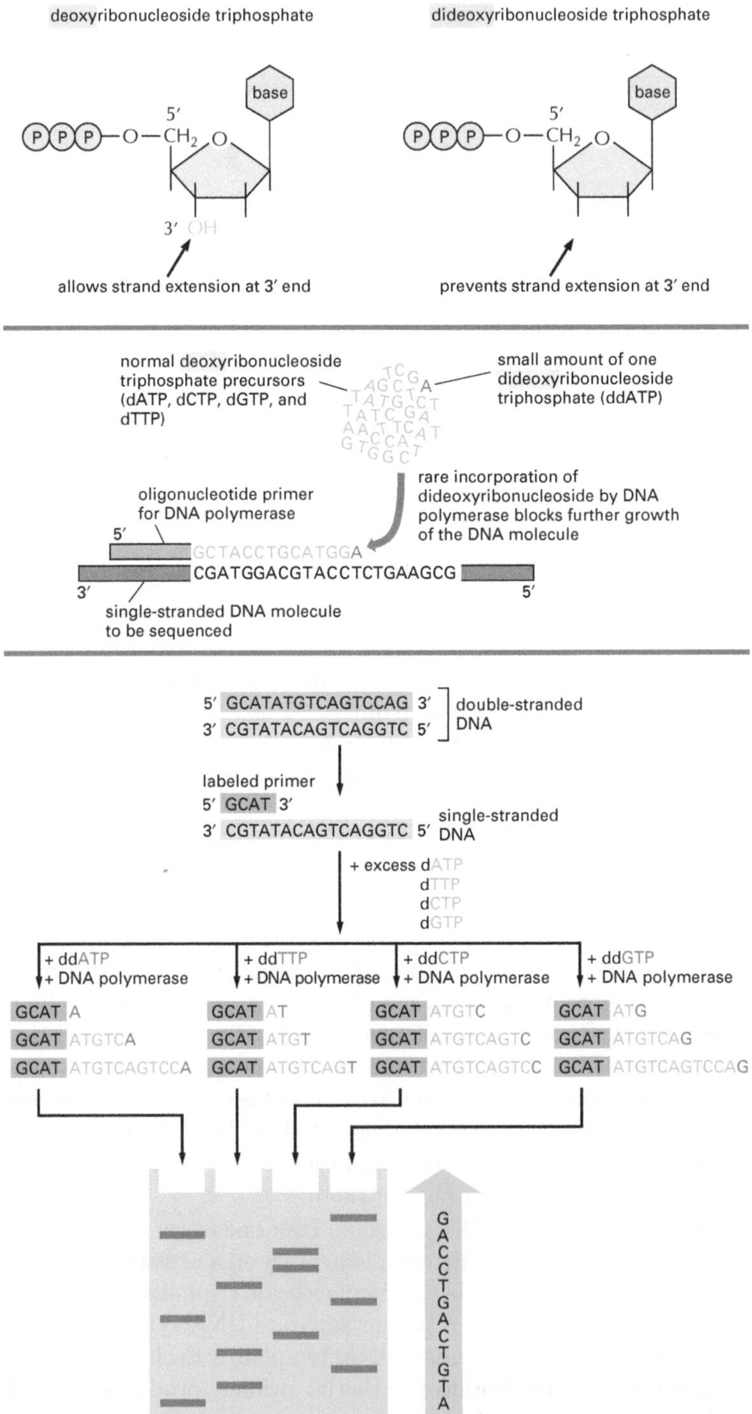

FIGURE 9.3 Sequencing DNA using the dideoxymethod (See color figure following page 146) [1]. Copyright 2003 from *Essential Cell Biology*, Second Edition by Alberts et al. Reproduced by permission of Garland Science/Taylor & Francis LLC.

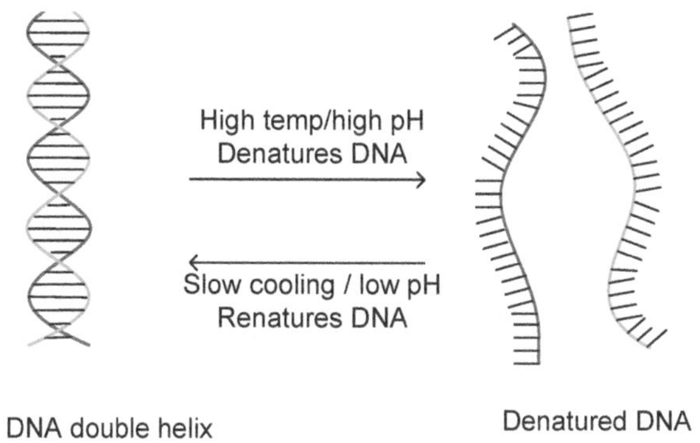

FIGURE 9.4 Denaturation and Renaturation of DNA molecules (See color figure following page 146).

these four reactions are separated by electrophoresis in four parallel lanes of an agarose gel and the positions of the DNA fragments in each of the lanes can be used to piece together the sequence of the DNA strand that we were originally interested in. The entire procedure is schematically illustrated in the third part of Fig. 9.3.

9.2 Nucleic Acid Hybridization and Associated Techniques

As already discussed in Chapter 5, DNA always occurs in nature as a double helix. By subjecting double-stranded DNA to high temperatures or high pH, the two strands of DNA can be made to separate as shown in Fig. 9.4. This is referred to as *denaturation* of the DNA. By slowly cooling the DNA or lowering the pH, the DNA can be made to *renature*, i.e. reform the double helix. The nucleic acid hybridization based techniques make use of the complementary base pairing properties of nucleotides: A pairs with T and G pairs with C.

One application of DNA hybridization is in the pre-natal diagnosis of genetic diseases, for instance, sickle cell anemia. Examining a single gene in the human genome requires searching through a total genome of over three billion nucleotides, which is an incredibly daunting task. However, the tremendous specificity of DNA hybridization makes it possible to do so in a fairly tractable fashion. For instance, for sickle cell anemia, the exact nucleotide change in the mutant gene is known. For prenatal diagnosis of sickle cell anemia, DNA is extracted from fetal cells. Two *DNA probes* are used to test fetal DNA — one corresponding to the normal gene sequence in the region of the mutation and the other corresponding to the mutant gene sequence. (A DNA probe is a short single-stranded DNA, an *oligonucleotide* (or *synthetic* DNA strand) typically 10–1000 nucleotides long, that is used in hybridization reactions to detect nucleic acid molecules containing a complementary sequence.) DNA samples from the fetus are first treated with restriction nucleases and all the resulting DNA fragments are electrophoresed through a gel. The gel is then treated with a DNA probe that detects only the restriction fragment that carries the β-globin gene. Using the two DNA probes, it is possible to distinguish whether the fetus contains one, two or no defective β-globin genes.

The laboratory procedure used to visualize the hybridization is known as *Southern blotting* and it involves the following six steps:

(i) Cleave double-stranded DNA using restriction nucleases.

(ii) Electrophorese the fragments to separate the fragments by length.

(iii) A sheet of nitrocellulose paper is laid over the gel and separated DNA fragments are transferred to the sheet by blotting. As this occurs, the DNA is denatured and the single-stranded DNA fragments adhere firmly to the surface of the nitrocellulose sheet.

(iv) The nitrocellulose sheet is carefully peeled off from the gel.

(v) The sheet containing the bound single-stranded DNA fragments is placed in a sealed plastic bag together with buffer containing a radioactively labeled DNA probe specific for the required DNA sequence. This gives the probe a chance to hybridize with its complement, if the latter is present on the sheet.

(vi) The sheet is removed from the bag and washed thoroughly so that only probe molecules that have hybridized to the DNA on the paper remain attached. After autoradiography, the DNA that has hybridized to the labeled probe will show up as bands on the autoradiograph.

In the context of DNA technology, one often hears the term *cloning*. DNA cloning has two possible meanings in biology:

(i) The act of making many identical copies of a DNA molecule.

(ii) The isolation of a particular stretch of DNA from the rest of the cell's DNA. This isolation is facilitated by making many identical copies of the DNA of interest, thereby amplifying it relative to the rest of the DNA.

Just as restriction nucleases can be used to break DNA into smaller fragments, an enzyme called DNA ligase, which we encountered earlier in Chapter 5 in the context of DNA replication and DNA repair, can be used to join DNA fragments together to produce a recombinant DNA molecule. Since DNA has the same chemical structure in all organisms, the use of this enzyme allows DNAs from any source to be joined together.

Restriction nucleases and DNA ligase play an important role in cloning DNA. This can be carried out using a bacterial plasmid. The purified plasmid DNA is exposed to a restriction nuclease that cleaves it in just one place, and the DNA fragment to be cloned is covalently inserted into it using DNA ligase. This recombinant DNA molecule is then introduced into a bacterium (usually E. coli) by transformation and the bacterium is allowed to grow in the presence of nutrients, where it doubles in number about every 20 minutes. After just a day, more than a billion copies of the plasmid would have been produced. The bacteria are then lysed, and the much smaller plasmid DNA is purified away from the rest of the cell contents, including the large bacterial chromosome. The DNA fragment can be recovered by cutting it cleanly out of the plasmid DNA using the appropriate restriction nuclease and separating it from the plasmid DNA by gel electrophoresis.

9.3 Construction of Human Genomic and cDNA Libraries

Human genes can be isolated by DNA cloning. Dealing with the unfragmented 3 billion nucleotides of the complete human genome is a daunting task. This can be avoided by breaking up the total genomic DNA into smaller, more manageable pieces to make it easier to work with. To do so, the total DNA extracted from a tissue sample or a culture of human cells is cut up into a set of DNA fragments by restriction nuclease treatment, and each fragment is cloned using bacterial plasmids as described earlier. The collection of cloned DNA fragments thus obtained is known as a *DNA library*. In this case, it is called a *genomic library*, as the DNA fragments are derived directly from the chromosomal DNA.

While producing a genomic library, one must make sure that each colony of E. coli produces clones of only one DNA fragment. This can be done by carrying out the entire procedure under the following favourable conditions:

- DNA fragments are inserted into plasmid vectors under conditions that favor *the insertion of one DNA fragment for each plasmid molecule.*

- These recombinant plasmids are mixed with a culture of E. coli at a concentration that ensures that *no more than one plasmid molecule is taken up by each bacterium.*

A natural question that arises is how one can find a particular gene in this huge genomic library. If the sequence of the complementary DNA is known, one could make a probe and use it to identify the particular gene by exploiting the properties of nucleic acid hybridization. If, on the other hand, the sequence of the gene is not known, one can use *protein sequencing* to identify a few of the amino acids coded by the gene. By using the genetic code in reverse, the

DNA sequences that code for these amino acid sequences can be deduced and a suitable DNA probe can be synthetically prepared. Using this probe, those rare bacterial clones containing human DNA complementary to the probe can be identified by DNA hybridization. It is possible that for a given gene several clones may be identified, as no single clone might contain the entire gene. This is especially true for long DNA sequences having lots of introns.

For many applications of DNA technology, it is advantageous to obtain a clone that contains only the coding sequence of a gene, that is, a clone that lacks the intron DNA. It is a relatively simple matter to isolate a gene free of all its introns. For this purpose, a different type of library, called a complementary DNA *(cDNA) library* is used. The creation of a cDNA library involves the following steps. First, starting from the mRNA that is expressed in a particular tissue or cell culture, construct the complementary DNA (cDNA) using *reverse transcriptase.* Then, degrade the mRNA copy using an alkali (or a base) and use the single-stranded cDNA left over to produce a double-stranded cDNA copy of the original mRNA. The cDNA molecules can then be cloned, just like the genomic DNA fragments described earlier, to produce the cDNA library.

Based on our discussion so far, we can highlight the following differences between DNA clones and cDNA clones:

(i) Genomic clones represent a random sample of all of the DNA sequences found in an organism's genome while cDNA clones contain fragments of *only those genes that have been transcribed into mRNA* in the tissue from which the RNA came. Since cells of different tissues produce distinct sets of RNA molecules, a different cDNA library will be obtained for each type of tissue.

(ii) Genomic clones from eucaryotes contain large amounts of repetitive DNA sequences, introns, gene regulatory regions, and spacer DNA, in addition to protein coding sequences, while cDNA clones contain only coding sequences. Thus cDNA clones are particularly well suited for (a) deducing the amino acid sequence of a protein from the DNA; and (b) producing the protein in bulk by expressing the cloned gene in a bacterial or yeast cell (neither of which can remove introns from mammalian RNA transcripts). For instance, insulin needed for treating Type I diabetics can be produced in bulk in this way.

In concluding this section, we note that hybridization allows even distantly related genes to be identified. Hybridization can be carried out under conditions that allow even an *imperfect match* between a DNA probe and its corresponding DNA to form a stable double helix. This can be used to identify closely related genes using a single DNA probe or even use a DNA probe for one species to identify the corresponding gene for another species (since the analogous genes in two different species usually share quite a lot of sequence similarity.)

9.4 Polymerase Chain Reaction (PCR)

Polymerase chain reaction, or PCR, is a synthetic procedure that can be used to selectively replicate a given nucleotide sequence quite rapidly and in large amounts from any DNA that contains it. Each cycle of the PCR reaction consists of three steps as shown in Fig. 9.5. In the first step, heat is applied to separate out the two DNA strands of a double-stranded DNA molecule. In the second step, primers are hybridized on the two strands to mark the beginning of the regions of DNA to be amplified. In the third step, DNA polymerase and deoxyribonucleoside triphosphates are added so that DNA complementary to each of the two strands and originating at the two primers can be synthesized. Thus, in one cycle of PCR, the quantity of DNA is amplified by a factor of 2. If additional amplification is necessary, then the procedure can be repeated over and over again. In n cycles of PCR, the DNA would have been amplified by a factor of 2^n. In each cycle of the PCR reaction, the DNA strands are separated by heating. Consequently, the DNA polymerase used for PCR is a special kind of heat-resistant polymerase which is usually isolated from thermophilic (heat loving) bacteria. Otherwise, the DNA polymerase would have had to be replaced in each cycle of the PCR reaction.

We next describe some common applications of PCR.

(1) PCR can be used to clone directly a particular DNA fragment. The main advantage of this procedure is that it does not require any cell culturing.

(2) PCR can be used to detect viral infection at very early stages. Here short sequences complementary to the viral genome are used as primers, and following many cycles of amplification, the presence or absence of even a single copy of a viral genome in a sample of blood can be ascertained. Once the amplification has been carried out using PCR, the virus detection can be done using gel electrophoresis. The schematic diagram for the entire procedure in the case of the HIV virus is shown in Fig. 9.6.

(3) PCR has great potential in forensic science used to track down the perpetrator of a crime. The DNA sequences that create the variability used in this type of analysis contain runs of short repeated sequences, such as GTGTGT$\cdots$, which are found in various positions (loci) in the human genome. The number of repeats in each run is highly variable in the population, ranging from 4 to 40 in different individuals. A run of repeated nucleotides of this type is commonly referred to as a VNTR (*variable number of tandem repeats*) sequence. Because of the variability in these sequences, each individual will usually inherit a different variant of each VNTR locus from their mother and from their father; two

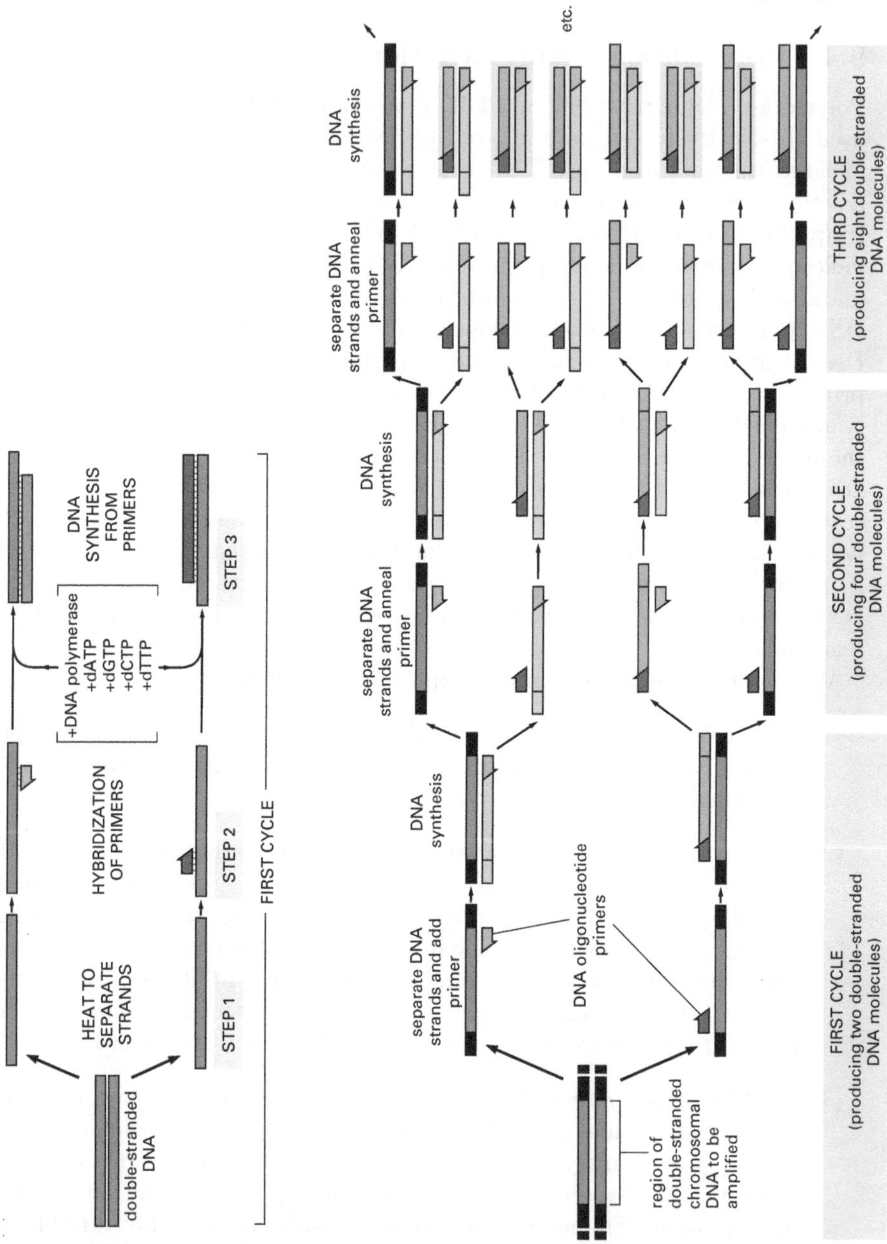

FIGURE 9.5 Steps in PCR (See color figure following page 146) [1]. Copyright 2003 from *Essential Cell Biology*, Second Edition by Alberts et al. Reproduced by permission of Garland Science/Taylor & Francis LLC.

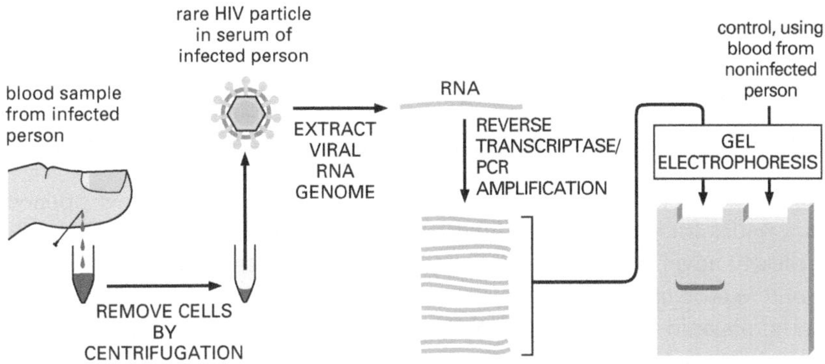

FIGURE 9.6 Use of PCR in detecting viral (HIV) infection (See color figure following page 146) [1]. Copyright 2003 from *Essential Cell Biology*, Second Edition by Alberts et al. Reproduced by permission of Garland Science/Taylor & Francis LLC.

unrelated individuals will therefore not usually contain the same pair of sequences. A PCR reaction using primers that bracket the VNTR locus produces a pair of bands of amplified DNA from each individual, one band representing the maternal variant and the other representing the paternal variant. The length of the amplified DNA will depend on the exact number of repeats at the locus. Recall from Section 9.1 that the position of a DNA fragment after electrophoresis depends on its length. Thus the position of the amplified DNA after gel electrophoresis will depend on the number of repeats that are present at a particular locus. This fact can be used to narrow down the list of suspects in a criminal investigation. The steps are as follows: (1) Select a number of VNTR loci; (2) use primers to bracket each locus; (3) amplify the DNA using PCR and then run gel electrophoresis. This process should be carried out on DNA obtained from the forensic sample and also DNA samples from each of the suspected individuals. If the bands obtained during gel electrophoresis on the forensic sample do not match the bands obtained from a suspect, then this particular suspect can reasonably be eliminated from the list of suspects.

9.5 Genetic Engineering

9.5.1 Engineering DNA Molecules, Proteins and RNAs

Advances made in DNA technology over the last three decades have opened up vast avenues for genetic manipulation. For instance, using DNA technology, completely novel DNA molecules can be constructed. To do so, one can cut a plasmid vector using a restriction nuclease and then insert a DNA fragment into the plasmid vector using DNA ligase. The procedure can be repeated to insert additional DNA fragments.

Another application of DNA technology is the production of large quantities of rare cellular proteins using cloned DNA. To do so, one can make use of a plasmid that already has a promoter sequence inserted into it. Such a plasmid is called a *double stranded expression vector*. The protein coding sequence is inserted into the region immediately following the promoter by using a restriction nuclease and DNA ligase. By introducing the recombinant plasmid into bacterial cells, these cells will be overexpressing the mRNA and the corresponding protein.

Yet another application of DNA technology is the artificial production of RNAs by *in vitro* transcription. Most RNAs found in a cell are present only in small quantities and are hard to isolate from the other cellular components. Once the gene coding for an RNA has been isolated, that RNA can be produced artificially in a test tube by letting RNA polymerase operate on the gene.

9.5.2 Engineering Mutant Haploid Organisms

The function of a gene is best revealed by an organism that has had that gene mutated in some way. Neither the complete nucleotide sequence of a gene nor the three-dimensional structure of a protein is sufficient to deduce a protein's function. Many proteins, such as those that have a structural role in the cell or normally form part of a large multienzyme complex, will have no obvious activity when viewed in isolation. Mutants that lack a particular protein may clearly reveal what the normal function or functions of that protein are. Prior to the advent of recombinant DNA technology, such mutations would have had to arise naturally and so would necessarily *arise by chance* and be *unpredictable*. Recombinant DNA technology makes it possible to introduce precise mutations by the technique of *site-directed mutagenesis*.

In recombinant DNA technology, one can start with a cloned gene and proceed to make mutations in it *in vitro*. Then, by reintroducing the altered gene back into the organism from which it originally came, one can produce a mutant organism in which the gene's function may be revealed. The steps involved in carrying out site-directed mutagenesis in a haploid organism are

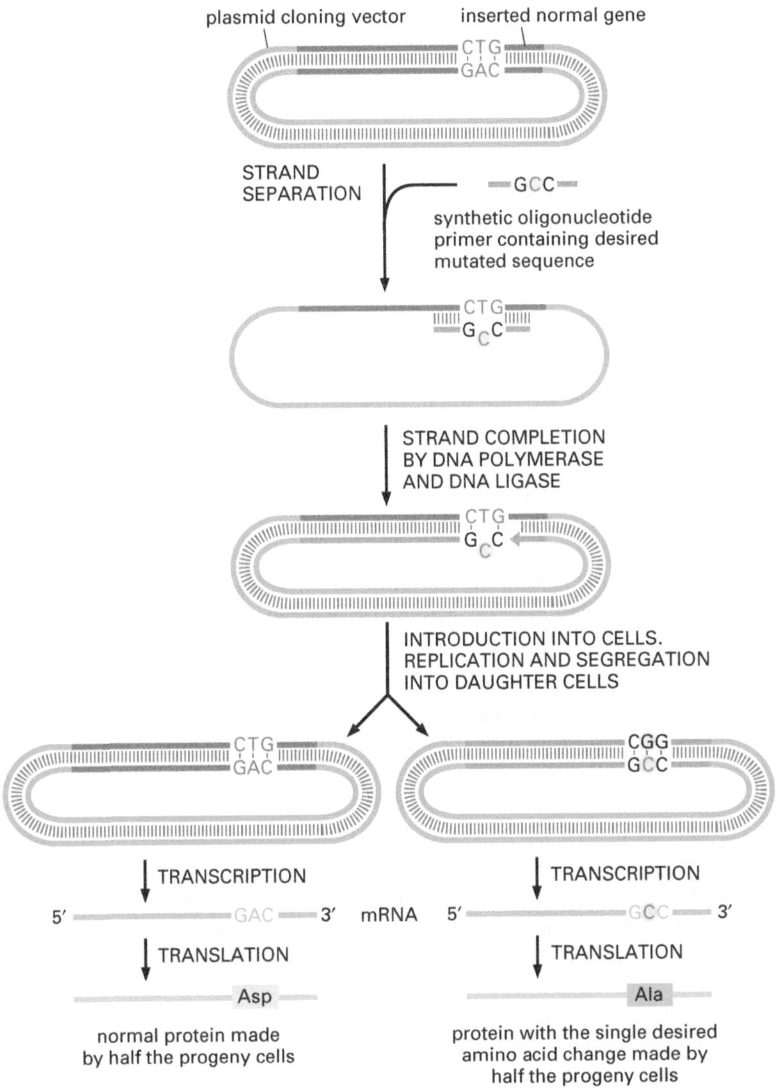

FIGURE 9.7 Site directed mutagenesis (See color figure following page 146) [1]. Copyright 2003 from *Essential Cell Biology*, Second Edition by Alberts et al. Reproduced by permission of Garland Science/Taylor & Francis LLC.

shown in Fig. 9.7. The gene to be mutated is first inserted into a plasmid and a sequence GAC (say) in one of the DNA strands in the gene is replaced by a mutated sequence GCC. The plasmid is then introduced into a cell. Due to the semi-conservative nature of DNA replication, every time the cell replicates, one half of the progeny end up inheriting the normal gene while the other half end up inheriting the mutated gene. The result is that half of the progeny produce the corresponding normal protein while the other half produce the mutated protein. Recombinant DNA technology can be used to carry out (i) gene replacement, (ii) gene knockout or (iii) gene addition. Organisms into which a new gene has been introduced, or those whose genomes have been altered in other ways using recombinant DNA techniques, are known as *transgenic organisms.*

9.5.3 Engineering Transgenic Animals

From the preceding discussion, it is clear that for haploid organisms such as bacteria or yeasts, transgenic progeny can be produced quite easily. Animals such as mice are diploid organisms. For such organisms it is more difficult, but still possible, to achieve gene replacements. This involves the following four steps:

Step 1 An altered version of the gene is introduced into cultured embryonic stem cells. A few rare embryonic stem cells will have their corresponding normal genes replaced by the altered gene through homologous recombination. Although often laborious, these rare cells can be identified and cultured to produce many descendants, each of which carries *an altered gene* in place of one of its two normal corresponding genes.

Step 2 These altered embryonic stem cells are injected into a very early mouse embryo. As a result, the cells are incorporated into the growing embryo, and a mouse produced by such an embryo will contain some somatic cells (cells other than the reproductive ones) that carry the altered gene. Some of these mice will also contain germ-line (precursors of reproductive) cells that contain the altered gene.

Step 3 When bred with a normal mouse, some of the progeny of these mice will contain the altered gene in all of their cells. This is because two haploid reproductive cells of different kinds (egg and sperm) have to fuse together to produce the diploid cell that will undergo repeated cell division to produce the entire organism.

Step 4 If two such mice are in turn bred, some of the progeny will contain two altered genes (one on each chromosome) in all of their cells. This again is a consequence of the meiotic process by which the haploid gametes are produced from their diploid precursors, and the fact that when the egg with an altered gene fuses with a sperm with the same gene altered, then the resulting diploid cell will have both copies of the gene altered. If the original gene alteration completely inactivates the function of the gene, then we will have what are called "knockout" mice.

10

Cell Division

In this chapter, we discuss cell division, which plays a crucial role in the propagation of all life. New cells can be generated only from existing ones and, therfore, the only way to make more cells is by division of those that already exist. A cell reproduces by carrying out a tightly controlled sequence of events in which it duplicates its contents and divides in two. This cycle of duplication and division, known as the *cell cycle*, is the essential mechanism by which all living things reproduce.

In unicellular organisms such as bacteria or yeast, each cell division leads to a complete new organism, while in multicellular organisms, many rounds of cell division are required to make a new individual from the single-celled egg. In multicellular organisms, quite often cell division has to be carried out to sustain the organism in the steady-state. The details of the cell cycle vary from organism to organism and at different stages in an organism's life. Certain characteristics, however, are universal. For instance, the cell that is undergoing division must replicate its DNA and pass on identical copies of the DNA to its two daughter cells.

Procaryotes (or bacteria) do not have a nucleus and reproduce by a fission type of cell division which is quite simple to describe. Here, the circular bacterial chromosome is replicated and one copy moves towards each end of the dividing bacterium. The cell wall and plasma membrane in the center of the dividing bacterium pinch inwards resulting in two daughter cells. Cell division in a eucaryotic cell is much more complicated. This is because most of the genetic information of the cell, namely its nuclear genome, is distributed between multiple chromosomes contained in the nucleus. In addition, the cytoplasm contains many organelles which must be duplicated and apportioned out equally between the two daughter cells.

There are three major questions that arise in the context of cell division: the first question concerns the mechanisms by which cells duplicate their contents; the second question concerns the mechanisms by which cells partition their duplicated contents and split in two; and the third question concerns the mechanisms employed by cells to ensure that the different steps involved in cell division take place in the proper sequence.

Partial answers to the first question have already been provided in earlier chapters. For instance, in Chapter 5, we discussed DNA replication and in Chapter 6, we discussed protein synthesis. The manufacture of other components such as membranes and organelles can be found in any book on

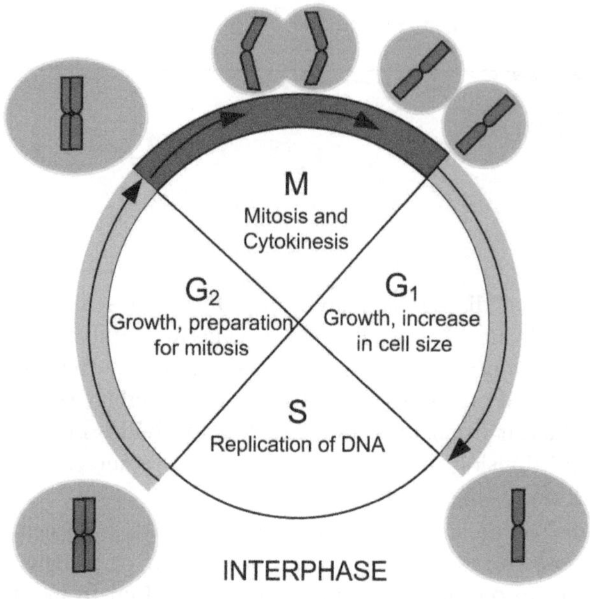

FIGURE 10.1 Eucaryotic cell cycle (See color figure following page 146).

molecular cell biology, for instance [1], but the details are not relevant for our purposes. Here, we will focus on providing an answer to the second question. The answer to the third question will be taken up in the next chapter.

The duration of the cell cycle varies greatly from one cell type to another. For instance, a single-celled yeast can divide every 90-120 minutes in ideal conditions, while a mammalian liver cell divides, on average, less than once a year. In this chapter, we will focus on the sequence of events in a fairly rapidly dividing mammalian cell, with a cell cycle time of about 24 hours.

The eucaryotic cell cycle is broadly divided into two phases: (i) *M phase*, which is composed of mitosis (nuclear division) and cytokinesis (splitting of the cell in two); and (ii) *interphase*, which is the period between one M phase and the next. Fig. 10.1 is a schematic diagram of the eucaryotic cell cycle, where the M phase and the interphase are marked in purple and green, respectively. The interphase is further divided into the remaining three phases of the cell cycle as shown in Fig. 10.1. During *S phase* (S=synthesis), DNA replication takes place. The S phase is flanked by two phases where the cell continues to grow. The G_1 *phase* (G=gap) is the interval between the completion of the M phase and the beginning of the S phase. The G_2 *phase* is the interval between the end of the S phase and the beginning of the M phase. We will see in the next chapter that there are particular times in the G_1 and G_2 phases when the cell makes a decision whether to proceed to the next phase or pause to allow more time to prepare. These times are called the G_1 and G_2 checkpoints, respectively.

Although when viewed under the microscope, interphase appears to be an uneventful period, in reality it is a very busy time for the cell. During all of interphase, a cell continues to transcribe genes, synthesize proteins, and grow in mass. Together G_1 and G_2 phases provide additional time for the cell to grow and duplicate its cytoplasmic organelles. Without the G_1 and G_2 phases, the cell would get progressively smaller at each cell division. This is, in fact, what happens to a fertilized egg during the first few cell divisions of its life, referred to as *cleavage* cell divisions.

The first readily visible sign under a microscope that a cell is about to enter M phase is the progressive *condensation* of its chromosomes. This condensation makes the chromosomes less likely to get entangled and therefore physically easier to separate during mitosis. The separation of the duplicated condensed chromosomes is carried out by a transient structure called the *mitotic spindle*, while, for animal cells, the actual division of the cell is carried out by a *contractile ring* made up of two kinds of protein filaments (actin and myosin). Like the mitotic spindle, the contractile ring is also a transient structure that assembles at the appropriate time. In plant cells, the cytoplasmic division has to be carried out by a different mechanism since each cell is enclosed within a hard cell wall.

Small organelles such as mitochondria and chloroplasts are usually present in large numbers in each cell and will be safely inherited by the daughter cells if, on the average, their numbers simply double once every cell cycle. Other larger organelles, such as the Golgi apparatus and the endoplasmic reticulum disintegrate into small fragments during mitosis, which increases the chance that the latter will be more or less evenly distributed among the daughter cells when the cell divides. Subsequently, the organelles are reconstructed from the inherited fragments present in each daughter cell. We next take a detailed look at *mitosis*, which is the type of nuclear division that most eucaryotic cells undergo.

10.1 Mitosis and Cytokinesis

Before mitosis begins, each chromosome has been replicated and consists of two identical *chromatids* (called *sister chromatids*) which are joined together along their length by interactions between proteins on the surface of the two chromatids. A typical pair of sister chromatids is shown in Fig. 10.2. Although mitosis proceeds as a continuous sequence of events, it is traditionally divided into five stages, namely *prophase*, *prometaphase*, *metaphase*, *anaphase* and *telophase*. We first provide a summary description of what happens at each of these stages before providing additional details. During *prophase*, the replicated chromosomes condense and the mitotic spindle begins

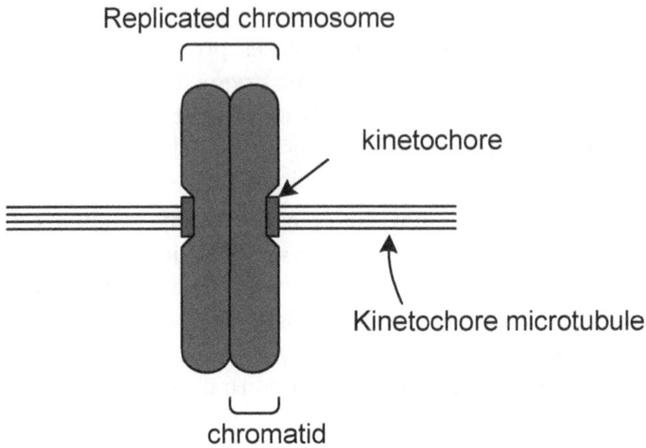

FIGURE 10.2 A typical pair of sister chromatids.

to assemble outside the nucleus. During *prometaphase*, the nuclear envelope breaks down, allowing the spindle microtubules to contact the chromosomes and bind to them. During *metaphase* the mitotic spindle gathers all of the chromosomes to the center (equator) of the spindle. During *anaphase* the two sister chromatids in each replicated chromosome synchronously split apart, and the spindle draws them to opposite poles of the cells. During *telophase* a nuclear envelope reassembles around each of the two sets of separated chromosomes to form two nuclei. In three of the stages desribed above, we have referred to the mitotic spindle. A natural question that comes up is how is the mitotic spindle formed and what exactly is its role.

The mitotic spindle starts to assemble in prophase. Towards the end of the S phase, the cell duplicates a structure called the *centrosome* to produce two daughter centrosomes, which initially remain together at one side of the nucleus. As *prophase* begins, the two daughter centrosomes separate and move to the opposite poles of the cell, driven by centrosome-associated motor proteins that use the energy of ATP hydrolysis to move along microtubules. Each centrosome serves to organize its own array of microtubules and the two sets of microtubules then interact to form the *mitotic spindle*. Fig. 10.3 shows the two centrosomes with their own array of microtubules. The microtubules that radiate from the centrosome in an interphase cell continuously polymerize and depolymerize by the addition and loss of units of a protein called *tubulin*. Individual microtubules, therefore, alternate between growing and shrinking and this process is called *dynamic instability.* The rapidly growing and shrinking microtubules extend in all directions from the centrosomes, exploring the interior of the cell. During prophase, while the nuclear envelope is still intact, some of these microtubules become stabilized against disassem-

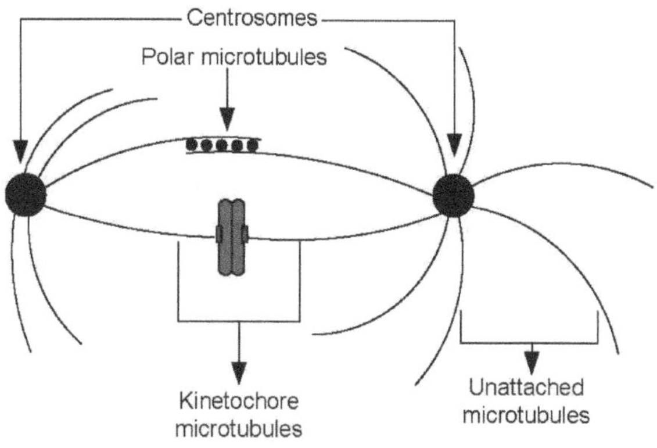

FIGURE 10.3 Three classes of microtubules in a mitotic cell

bly to form the highly organized mitotic spindle. This happens when some of the microtubules growing from opposite centrosomes interact, binding the two sets of microtubules together to form the basic framework of the mitotic spindle. This is shown in Fig. 10.3. The interacting microtubules are called *polar microtubules* (since they originate from the two poles of the spindle), and the two centrosomes that give rise to them are called the *spindle poles.*

Chromosomes get attached to the mitotic spindle during prometaphase. Prometaphase starts abruptly with the disintegration of the nuclear envelope. Following this, the spindle microtubules, which have been lying in wait outside the nucleus, suddenly gain access to the replicated chromosomes and bind to them. The spindle microtubules bind the chromosomes through specialized protein complexes called *kinetochores*, which are formed on the chromosomes during late prophase. As we already mentioned, each replicated chromosome consists of two sister chromatids joined together along their length, and each chromatid is constricted at a region of specialized DNA sequence called the *centromere.* Just before prometaphase, kinetochore proteins assemble into a large complex on each centromere. Once the nuclear envelope has disintegrated, a randomly probing microtubule encountering a kinetochore will bind to it, thereby capturing the chromosome. Such a microtubule is called a *kinetochore microtubule*, and it links the chromosome to a spindle pole. Not every microtubule emanating from the two centrosomes ends up forming a kinetochore microtubule or a polar microtubule. In fact, several of the microtubules remain unattached and are called *unattached microtubules.* All three classes of microtubules are shown in Fig. 10.3.

During prometaphase, the chromosomes, which are now attached to the mitotic spindle, appear to move as if jerked around randomly in different directions. Finally they align at the equator of the spindle, halfway between the two spindle poles, thereby forming the *metaphase plate.* This defines the beginning of *metaphase*. The precise forces that act to bring the chromosomes to the equator are not well understood. However, the continual growth and shrinkage of the microtubules and the action of microtubule motor proteins are thought to play a role. In this context, we note that a continuous balanced addition and loss of tubulin subunits is required to maintain the mitotic spindle. Thus when tubulin addition is blocked in a mitotic cell using say the drug *colchicine*, tubulin loss continues until the spindle disappears. This prevents the cell from dividing and so colchine can be used in cancer therapy to prevent a cancer cell from proliferating.

At the start of *anaphase*, the connections between the two sister chromatids are cut by proteolytic enzymes, allowing each chromatid (now called a *daughter chromosome*) to be gradually pulled to the spindle pole to which it is attached. This movement is the result of two independent processes brought about by different parts of the mitotic spindle. These processes are called *anaphase A* and *anaphase B*, and their occurrence is more or less simultaneous. In anaphase A, the kinetochore microtubules shorten by depolymerization, and the two daughter chromosomes move towards their respective spindle poles. In anaphase B, the spindle poles themselves move away from each other, further contributing to the segregation of the two groups of daughter chromosomes. The driving force for the movements of anaphase A is thought to be provided partly by the action of microtubule motor (movement generating) proteins operating at the kinetochore and partly by the loss of tubulin subunits that occurs mainly at the kinetochore end of the kinetochore microtubules. The driving force for the moving apart of the spindle poles in anaphase B is thought to be provided by the polymerization of the polar microtubules by the addition of tubulin units at their free ends.

The last step in mitosis is *telophase*. During telophase a nuclear envelope reforms around each group of chromosomes to form the two daughter nuclei. With the creation of the two daughter nuclei, the process of nuclear division (or mitosis) is complete.

M phase involves more than just the segregation of the daughter chromosomes and the formation of new nuclei. It is also the time during which other components of the cell — membranes, organelles, and proteins — are distributed more or less evenly between the two daughter cells. This is achieved by *cytokinesis* and usually begins in anaphase but is not completed until after the two daughter nuclei have formed.

Regarding cytokinesis, we note the following: (i) the mitotic spindle determines the plane of cytoplasmic cleavage, which is usually orthogonal to the orientation of the mitotic spindle; (ii) the *contractile ring* of animal cells, which is responsible for carrying out cytokinesis in such cells, is made up of the proteins *actin* and *myosin*; and (iii) cytokinesis in plant cells involves the formation of a new cell wall (presumably because the plant cells in general are surrounded by a tough cell wall).

10.2 Meiosis

In this section, we take a look at *meiosis*, which is another kind of eucaryotic cell division employed by sexually reproducing diploid organisms to produce the reproductive cells. In a diploid organism, all the cells of the body including the germ-line cells that give rise to the gametes are *diploid*; however, the gametes themselves are *haploid*. Thus the haploid gametes must be produced from their diploid precursors by a special type of cell division. This special type of cell division is called *meiosis*.

With the exception of the chromosomes that determine sex (the sex chromosomes), a diploid nucleus contains two very similar versions of each chromosome, one from the father called the paternal chromosome and one from the mother called the maternal chromosome. The two versions of each chromosome, however, are not genetically identical, as they carry different variants of many of the genes. They are, therefore, called *homologous chromosomes*, or *homologues*, meaning that they are similar but not identical. A diploid cell in a sexually reproducing organism consequently carries two similar sets of genetic information. In most cells, the paternal and maternal homologues maintain a completely separate existence as independent chromosomes.

Mitosis and meiosis are similar in certain respects but there are important differences that we will be discussing. For clarity of presentation, we will make use of Fig. 10.4, which illustrates the key differences between mitosis and meiosis using a fictitious diploid cell with only one pair of homologous chromosomes. In mitosis, each chromosome replicates and the replicated chromosomes *line up in random order at the metaphase plate*; the two sister chromatids then separate from each other to become individual chromosomes, and the two daughter cells produced by cytokinesis inherit a copy of each paternal chromosome and a copy of each maternal chromosome. Thus both sets of genetic information are transmitted intact to the two daughter cells, which are, therefore, each diploid and genetically identical.

In contrast, when diploid cells divide by meiosis they form haploid gametes with only one half the original number of chromosomes i.e. only one chromosome of each type instead of a pair of homologues of each type. Thus, each gamete acquires either the maternal copy or the paternal copy of a chromosome but not both. This reduction is needed so that when two gametes of opposite types (an egg and a sperm in animals) fuse at fertilization, the chromosome number is restored in the embryo to the original diploid number for that species. Since the assignment of maternal and paternal chromosomes to the gametes during meiosis occurs at random, the original maternal and paternal chromosomes are reshuffled into an incredibly large number of different combinations. We have already discussed the effects of this in Chapter 8. As noted in that chapter, and as shown in Fig. 10.4, in a meiotic cell division, the replicated homologous paternal and maternal chromosomes (including the two replicated sex chromosomes) pair up alongside each other before they line up on the spindle. This makes it possible for crossovers by homologous recombination to occur at this stage.

In principle, since meiosis involves a halving of the number of chromosomes, it could have occurred by a simple modification of a normal mitotic cell division, such that the DNA replication (S phase) is omitted. For unknown reasons, the actual meiotic process is more complicated and involves DNA replication followed by two cell divisions instead of one. This is shown in Fig. 10.4.

Occasionally, the meiotic process occurs abnormally and homologues fail to separate — a phenomenon known as *nondisjunction*. In this case some of the gametes that are produced lack a particular chromosome, while others have more than one copy of it. Such gametes, when combined with a normal gamete of the opposite type, form abnormal embryos, most of which do not survive. Some, however, do and *Down syndrome* in humans is an example of a disease that is caused by an extra copy of Chromosome 21. This condition is, therefore, referred to in the scientific literature as *Trisomy 21*, meaning that such a person's genome has three copies of Chromosome 21 instead of the two copies that normal humans have.

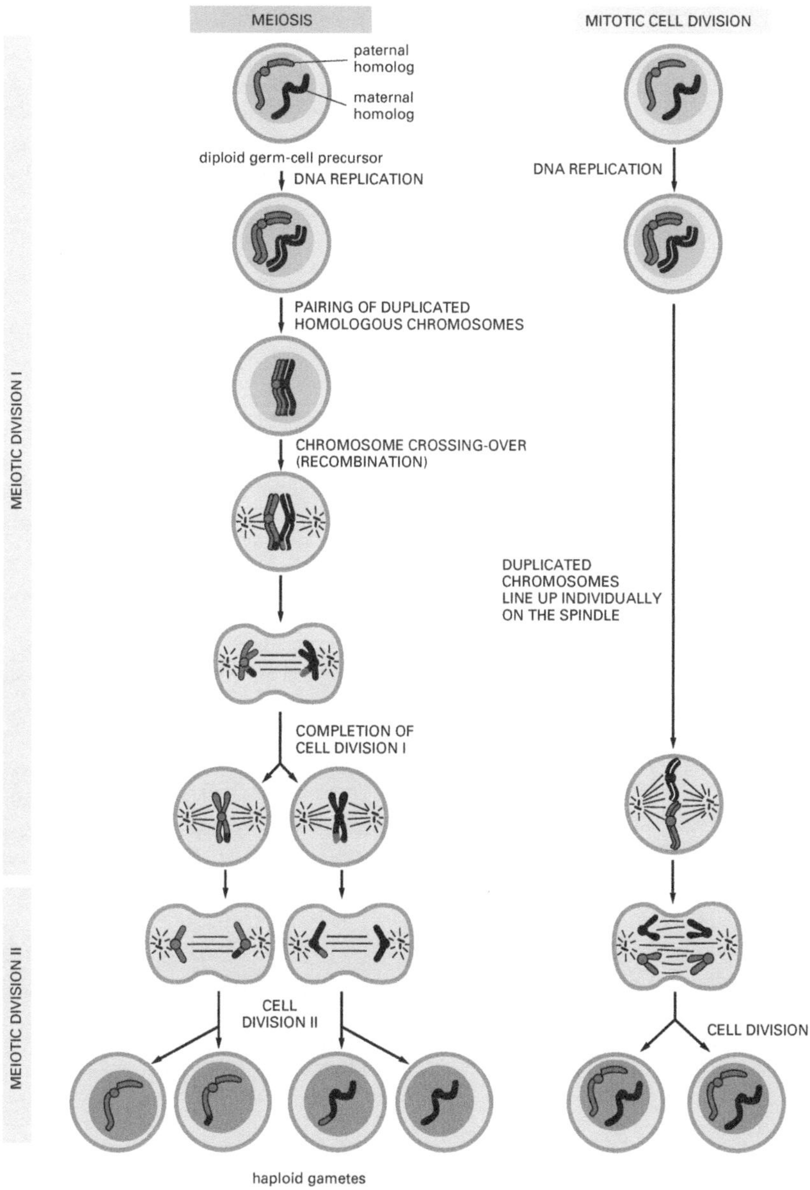

FIGURE 10.4 Comparing mitosis and meiosis (See color figure following page 146) [1]. Copyright 2003 from *Essential Cell Biology*, Second Edition by Alberts et al. Reproduced by permission of Garland Science/Taylor & Francis LLC.

11

Cell Cycle Control, Cell Death and Cancer

In this chapter, we introduce the phenomena of *cell cycle control* and *programmed cell death* in multicellular organisms and discuss how disruption in either of these can lead to the disease called *cancer*. Our discussion so far in this book has focussed on two aspects of cell division: (i) replication of the contents of the cell; and (ii) the actual partitioning of these replicated contents between the two daughter cells when the cell divides. However, there is yet another aspect of cell division that is crucially important, namely, the mechanism by which the cell controls the different chronological steps that are involved in cell division.

As we have seen in the last chapter, the events of the cell cycle occur in a fixed sequence, namely M phase followed by G_1 phase, G_1 phase followed by S phase, S phase followed by G_2 phase and finally G_2 phase followed by M phase again. This is ensured by a *cell cycle control system* which has to perform a number of functions. First, it has to activate the enzymes and other proteins responsible for carrying out each process in the cell cycle at the appropriate time, and then it has to deactivate them once the process is completed. Second, it must also ensure that each stage of the cycle is completed before the next one is begun. For instance, it has to make sure that DNA replication has been completed before mitosis begins, that mitosis has been completed before cytokinesis begins, and that another round of DNA replication does not begin until the cell has passed through mitosis and grown to an appropriate size. Third, the control system must also take into account if the conditions outside the cell are conducive for division. For instance, in a multicellular organism the control system must be responsive to signals from other cells, such as those that stimulate cell division when more cells are needed. From the preceding discussion, it is clear that the cell cycle control system plays a major role in the regulation of cell numbers in the tissues of the body. When this system malfunctions, it can result in cancer.

The operation of the cell cycle control system is very similar to that of the control system for any cyclic process. We will illustrate this by drawing the analogy with an automatic washing machine. The duty cycle of an automatic washing machine consists of the following five steps: (i) take in water, (ii) mix with detergent, (iii) wash the clothes, (iv) rinse them, and (v) spin them dry. In the eucaryotic cell cycle, these steps are analogous to (a) DNA replication, (b) mitosis, etc. Furthermore, the washing machine controller is itself regulated at certain critical points of the cycle by *feedback* from the processes that are being performed. For instance, sensors monitor the water levels and send signals back to the controller to prevent the start of the next process before the current one has been completed. Similarly, the events of the cell cycle have to occur in a particular sequence, and this sequence must be preserved even if one of the steps takes longer than usual. For instance, all of the nuclear DNA must replicated before the nucleus begins to divide and it is crucial for most cells to double in size before dividing in two as otherwise the cells would get progressively smaller at each cell division. The cell-cycle control system achieves all this by means of molecular brakes that can stop the cycle at various *checkpoints*.

The control system in most cells has checkpoints for cell size, where the cell-cycle is halted until the cell has grown to an appropriate size. In G_1, a size checkpoint allows the system to halt and the cell to grow further, if necessary, before a new round of DNA replication is triggered. Cell growth depends on an adequate supply of nutrients and other factors in the extracellular environment, and the G_1 checkpoint also allows the cell to check that the environment is favorable for cell proliferation before committing itself to the DNA replication (S) phase. A second size checkpoint occurs in G_2, allowing the system to halt before it triggers mitosis. The G_2 checkpoint also allows the cell to check that DNA replication is complete before proceeding to mitosis. Checkpoints are important in another way since they are the points in the cell cycle where the control system can be regulated by signals from other cells, such as growth factors (molecules that promote growth) and other extracellular signaling molecules.

11.1 Cyclin-Dependent Kinases and Their Role

The cell cycle control system governs the cell-cycle machinery through the *phosphorylation of key proteins* that initiate or regulate DNA replication, mitosis, and cytokinesis. Recall from Chapter 4 that the phosphorylation reactions are carried out by enzymes called *kinases*. The protein kinases of the cell-cycle control system are present in dividing cells throughout the cell cycle. They are activated, however, only at appropriate times in the cycle, after which they quickly become deactivated again. This is made possible by a second set of protein components of the control system, which are called the *cyclins*. Cyclins have no enzymatic activity by themselves, but they have to bind to the cell-cycle kinases before the kinases can become enzymatically active. As a result, the kinases of the cell cycle control system are referred to as *cyclin-dependent protein kinases*, or *Cdks*. Cyclins derive their name from the fact that their concentrations vary in a cyclical fashion during the cell cycle. For illustrative purposes, we will focus on the cyclin-Cdk complex

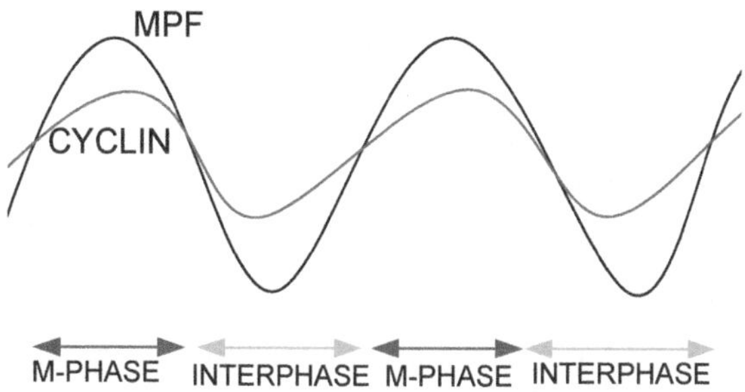

FIGURE 11.1 Variation of MPF activity and cyclin concentration during different stages of the cell cycle (See color figure following page 146).

that is responsible for driving cells into mitosis. For reasons to be explained shortly, this cyclin-Cdk complex is known as the *M-phase promoting factor* (MPF).

The cyclin-Cdk complex that drives cells into M phase was first discovered through studies of cell division in frog eggs. The fertilized eggs of many animals are especially well-suited for biochemical studies of the cell cycle because (i) they are very large cells and (ii) they divide very rapidly. This is due to the fact that they undergo cleavage divisions, i.e. M phase followed by S phase and then M phase again with little or no G_1 or G_2 phases in between.

By taking frog eggs at a particular stage of the cell cycle, an extract can be prepared that is representative of that cell-cycle stage. The biological activity of such an extract can then be tested by injecting it into a *Xenopus* oocyte (the immature diploid precursor of the unfertilized frog egg) and observing its effects on cell cycle behavior. The *Xenopus* oocyte is a convenient test system for detecting an activity that drives cells into M phase, as it has completed DNA replication and is arrested just before M phase of the first meiotic division. (Recall our detailed discussion of meiosis in Chapter 10.) The oocyte is therefore at a stage in the cell cycle that is equivalent to the G_2 phase of a mitotic cell cycle.

In such experiments it was found that an abstract from an M-phase fertilized egg instantly drives the oocyte into M phase, whereas cytoplasm from a cleaving egg at other phases of the cycle does not. When initially discovered, the chemical composition and the mechanism of action of the factor responsible for this activity were unknown, and consequently the factor was simply called the *M-phase promoting factor* or MPF. MPF activity was found to oscillate dramatically during the course of each cell cycle as shown in Fig. 11.1; it increased rapidly just before the start of mitosis and fell rapidly to zero towards the end of mitosis.

Subsequent studies revealed that MPF contains a single protein kinase, which is required for its activity. By phosphorylating key proteins, the kinase causes several phenomena associated with mitosis to occur: the condensation of the chromosomes, the disintegration of the nuclear envelope and the formation of the mitotic spindle. However, the MPF kinase is not capable of acting by itself and has to have a specific cyclin bound to it in order to function.

Biochemical experiments using cleaving clam eggs led to the discovery of *cyclin*. *Cyclin* was initially identified as a protein whose concentration rose gradually during interphase and then fell rapidly to zero as the cells went through M phase, repeating this performance in each cell cycle. This is shown in Fig. 11.1. The MPF is a protein complex containing two subunits — a regulatory subunit that is a cyclin and a catalytic subunit that is the mitotic Cdk. Many of the cell-cycle control genes have been remarkably conserved during biological evolution. Indeed, the human version of these genes will function perfectly well when introduced into a yeast cell.

The manufacture of the cyclin component of MPF starts immediately after cell division and continues steadily through interphase. The cyclin accumulates, so that its concentration rises gradually and helps time the onset of mitosis; its subsequent rapid decrease helps initiate the exit from mitosis. The sudden fall in the cyclin concentration during mitosis is the result of the rapid degradation of the cyclin by the ubiquitin-dependent proteolytic system. The MPF activation initiates a process that, after some time delay, leads to the ubiquination and degradation of the cyclin, thereby turning the kinase off.

From Fig. 11.1 we observe that the cyclin concentration increases gradually throughout interphase, whereas the MPF kinase activity switches on *abruptly* at the end of interphase. Thus, the cyclic variations in the cyclin concentration alone cannot completely explain MPF kinase activity. This is due to the fact that the kinase itself also has to be phosphorylated at one or more sites and dephosphorylated at others before it can become enzymatically active. The removal of the inhibitory phosphate groups by a specific protein phosphatase is the step that activates the kinase at the end of interphase. Once activated, a cyclin-Cdk complex can activate more cyclin-Cdk complexes by a *positive feedback* type of mechanism. In fact, this positive feedback type of mechanism is what causes the sudden explosive increase in MPF kinase activity that drives the cell abruptly into M-phase.

Here, we have focused attention on the cyclin-Cdk complex that constitutes the MPF. However, there are many varieties of cyclin and, in most eucaryotes, many varieties of Cdk that are involved in cell-cycle control. For instance, S phase cyclins trigger entry into S phase while G_1 cyclins act earlier in G_1. The latter bind to Cdk molecules to help initiate the formation and activation of the S-phase cyclin-Cdk complexes and thereby drive the cell toward S-phase. A detailed study of each cyclin-Cdk complex is beyond the scope of this text. Instead, for our purposes, it is sufficient to note that the concentration of each type of cyclin rises and then falls sharply at a specific time in the cell cycle to control the timing of a particular stage. As before, the sudden fall in the cyclin concentration is the result of cyclin degradation by the ubiquitin pathway.

As we have already seen, the cell cycle control system triggers the events of the cell cycle *in a specific order.* If one of the steps is delayed, the control system of necessity must delay the activation of the following steps so that the sequence is maintained. This remarkable feat is accomplished by the action of molecular brakes that can stop the cell cycle at specific *checkpoints.* For most cases, the detailed molecular mechanisms involved are not well understood. However, in some cases, it is known that specific *Cdk inhibitor proteins* come into play and we next discuss one such case.

One of the best understood checkpoints stops the cell cycle in the G_1 phase if the DNA is damaged, helping to ensure that a cell does not replicate damaged DNA. By an unknown mechanism, DNA damage causes an increase in both the concentration and activity of a gene regulatory protein called *p53* (protein with a molecular weight of 53,000 units). When activated, *p53* stimulates the transcription of a gene encoding a Cdk inhibitor protein called *p21* (protein with a molecular weight of 21,000 units). This increases the concentration of the p21 protein, which binds to the S phase cyclin-Cdk complexes responsible for driving the cell into S phase and blocks their action. The arrest of the cell cycle in G_1 allows the cell time to repair the damaged DNA before replicating it. If p53 is missing or defective, the unrestrained replication of damaged DNA leads to a high rate of mutation and the production of cells that tend to become cancerous. In fact, mutations in the p53 gene that permit cells with damaged DNA to proliferate play an important part in the development of many human cancers.

Although we have been discussing the different steps in the cell cycle, for a multicellular organism, not every cell in the body will keep traversing all the steps of the cell cycle. Indeed, cells can dismantle their control system and withdraw from the cell cycle altogether. In the human body, for instance, nerve cells have to persist for a lifetime without dividing. Consequently, they enter a modified G_1 state called G_0, in which the cell-cycle control system is partly dismantled in that many of the Cdks and cyclins disappear. As a result, the nerve cell remains permanently in the G_0 state. As a general rule, mammalian cells proliferate only if they are stimulated to do so by signals from other cells. If deprived of such signals, the cell cycle arrests at a G_1 checkpoint and enters the G_0 state.

11.2 Control of Cell Numbers in Multicellular Organisms

Unicellular organisms such as bacteria and yeasts tend to grow and divide as fast as they can, and their rate of proliferation depends largely on the availability of nutrients in the environment. For an animal cell to proliferate, nutrients are not enough. It must also receive stimulating signals from other cells, usually its neighbors. In other words, in multicellular organisms, cell division is under very tight control.

An important example of a brake that normally holds cell proliferation in check is the retinoblastoma (Rb) protein. The Rb protein, which is abundant in the nucleus of all vertebrate cells, binds to particular gene regulatory proteins, preventing them from stimulating the transcription of genes required for cell proliferation. Extracellular signals such as *growth factors* that stimulate cell proliferation lead to the activation of the G_1 cyclin-Cdk complexes mentioned earlier. These phosphorylate the Rb protein, altering its conformation so that it releases its bound gene regulatory proteins, which are then free to activate the genes required for cell proliferation to proceed.

The stimulating signals that act to override the brakes on cell proliferation are mostly protein *growth factors*. One example of a protein growth factor is the so called *platelet-derived growth factor* (PDGF). When blood clots (in a wound, for example), blood platelets incorporated in the clot are triggered to release PDGF, which binds to receptor tyrosine kinases in surviving cells at the wound site, thereby stimulating them to proliferate and heal the wound. Another example of a protein growth factor is the *hepatocyte growth factor*. In this case, if part of the liver is lost through surgery or acute injury, cells in the liver and elsewhere produce this protein called hepatocyte growth factor which helps to stimulate the surviving liver cells to proliferate.

Even in the presence of growth factors, normal animal cells do not keep on dividing indefinitely in culture. Even cell types that maintain the ability to divide throughout the lifetime of the animal stop dividing after a limited number of divisions. For instance, fibroblasts (a class of cells from which connective tissue is derived) taken from a human fetus stop dividing after 80 rounds of cell division, while fibroblasts taken from a 40-year-old adult stop after 40 rounds of cell division. This phenomenon is known as *cell senescence*. Fibroblasts from a mouse embryo, on the other hand, halt their proliferation after only about 30 divisions in culture. This may partly explain the difference in size between mice and humans. The mechanisms that halt the cell cycle in either developing or aging cells are not clearly understood at the present time, although the accumulation of Cdk inhibitor proteins and the loss of Cdks are likely to be involved.

11.3 Programmed Cell Death

Animal cells require signals from other cells to avoid programmed cell death. These signals are called *survival factors*. If deprived of such survival factors, the cells activate an intracellular suicide program and die by a process called *programmed cell death* or *apoptosis*. This helps ensure that cells survive only when and where they are needed. The amount of programmed cell death that occurs in both developing and adult tissues is astonishing. In the developing vertebrate nervous system, for example, more than half of the nerve cells normally die soon after they are formed. In a healthy adult human, billions of cells die in the intestine every hour. A natural question that arises is what purpose is served by this huge amount of cell death. In some cases, the answers are clear. For instance, the sculpting of hands and feet in mammals or the developmental loss of the tail of a tadpole are brought about by apoptosis. In the case of the developing verterbrate nervous system, apoptosis is used to match the number of nerve cells to the number of target cells that require innervation. In yet another case an enlarged liver can be returned to normal size via apoptosis.

We next examine how apoptosis uniquely differs from other kinds of cell death. Cells that die as a result of acute injury typically swell and burst and spill their contents all over their neighbors (a process called *cell necrosis*), causing a potentially damaging inflammatory response. By contrast, a cell that undergoes apoptosis dies very neatly, without damaging its neighbors. The cell shrinks and condenses, the cytoskeleton collapses, the nuclear envelope disassembles, and the nuclear DNA breaks up into fragments. Most important, the cell surface is altered, displaying properties that cause the dying cell to be cleaned up immediately, either by its neighbors or by a macrophage before there is any leakage of its contents.

The machinery that is responsible for this kind of controlled cell suicide seems to be similar in all animal cells. It involves a family of proteases which are themselves activated by proteolytic cleavage in response to signals that induce programmed cell death. The activated suicide proteases cleave, and thereby activate, other members of the family, resulting in an amplifying proteolytic cascade which is often referred to as an *apoptotic cascade*. The activated proteases then cleave other key proteins in the cell, killing it quickly and neatly.

11.4 Cancer as the Breakdown of Cell Cycle Control

Cancers are the products of mutations that set cells free from the usual controls on cell proliferation and survival. A cell in the body mutates through a series of chance events and acquires the ability to proliferate without the normal restraints. Its progeny inherit the mutations and give rise to a tumor that can grow without limit. *Oncogenes* (that turn on cell division) and *tumor-suppressor genes* (that function as brakes on cell division) play an important role in causing cancer. An example of an oncogene is a gene produced from the gene for the platelet derived growth factor (PDGF). Here the normal PDGF gene is called a *proto-oncogene* and the corresponding cancer causing gene is called an oncogene. While the proto-oncogene stimulates cell proliferation when more cells are needed, the oncogene causes cell proliferation even when no new cells are needed. An example of a tumor suppressor gene is the retinoblastoma gene whose product normally acts as a brake on cell division. When this gene is mutated, the braking action may no longer be present leading to excessive cell proliferation.

In a normal diploid cell, there are two copies of each gene in the genome. Since a proto-oncogene accelerates cell division when called upon to do so, only one copy of a proto-oncogene needs to be mutated to an oncogene before excessive cell proliferation, and possibly cancer, can result. On the other hand, a tumor suppressor gene is associated with braking action on cell division and so, both copies have to be mutated before a loss of function, possibly leading to cancer, can occur.

In almost all adult tissues, cells are continually dying and being replaced; through all the hurly-burly of cell replacement and tissue renewal, the architecture of the tissue must be maintained. In other words, we have some kind of a dynamic equilibrium between cell proliferation and apoptosis. Three main factors contribute to make this possible:

(i) Cell communication: this ensures that new cells are produced only when and where they are required;

(ii) Selective cell-cell adhesion: The selectivity of adhesion prevents the different cell types in a tissue from becoming chaotically mixed; and

(iii) Cell memory: Cells autonomously preserve their distinctive character and pass it on to their progeny. This preserves the diversity of cell types in the tissue.

Different tissues in the body are renewed at different rates. For instance, nerve cells in the body never divide while liver cells divide about once a year. The cells lining the inside of the intestine, on the other hand, are turned over about once every three days. Many of the differentiated cells that need continual replacement are themselves unable to divide. More of such cells are generated from a stock of precursor cells, called *stem cells*, that are retained in the corresponding tissues along with the differentiated cells. Stem cells can divide without limit: some of the progeny of the stem cells will remain stem cells, while others will embark on a course leading irreversibly to terminal differentiation.

Cancer cells are defined by two heritable properties: they and their progeny (1) reproduce in defiance of the normal social constraints and (2) invade and colonize territories normally reserved for other cells. Cells that have only property (1) will result in a *benign tumor* while cells that have properties (1) and (2) will result in a *malignant tumor* (cancer). Malignant tumor cells can break loose from the primary tumor, enter the blood stream or lymphatic vessels, and form secondary tumors, or *metastases*, at other sites in the body. In everday language, this is referred to as the spreading of cancer.

It is important to note that cancer is a consequence of mutation and natural selection within the population of cells that form the body. Indeed, the mutations in cancer cells give them a competitive advantage over their normal neighbors. Unfortunately, it is this very fact that leads to disaster for the multicellular organism, since the controls on cell proliferation and survival, which are at the heart of maintaining tissue architecture, are disrupted.

About 10^{16} cell divisions take place in the human body over the lifetime of an individual. Based on this, it can be shown that every single gene is likely to have undergone mutations on about 10^{10} separate occasions in any individual human being. Typically, at least 5 or 6 independent mutations must occur in one cell to make it cancerous. This is because the body of a multicellular organism has different levels of protection against cancer. For instance, to become cancerous, an epithelial stem cell in the skin or the lining of the gut must undergo changes that not only enable it to divide more frequently than it should, but also let its progeny escape being sloughed off in the normal way from the exposed surface of the epithelium, enable them to displace their normal neighbors, and let them attract a blood supply sufficient to nourish tumor growth. Thus the mutations that give rise to cancer accumulate over a long time and that is why cancer is typically a disease of old age. Occasionally, however, individuals are encountered who have inherited a germ-line mutation in a tumor suppressor gene or an oncogene. For these people, unfortunately the number of mutations required is less and the disease occurs more frequently and at an earlier age. The families that carry such mutations are, therefore, prone to cancer.

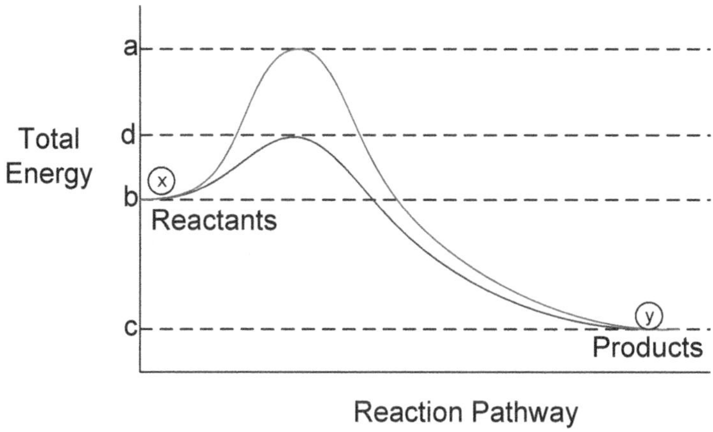

FIGURE 3.2 Activation energy requirement in the absence (red) and presence (blue) of enzymes.

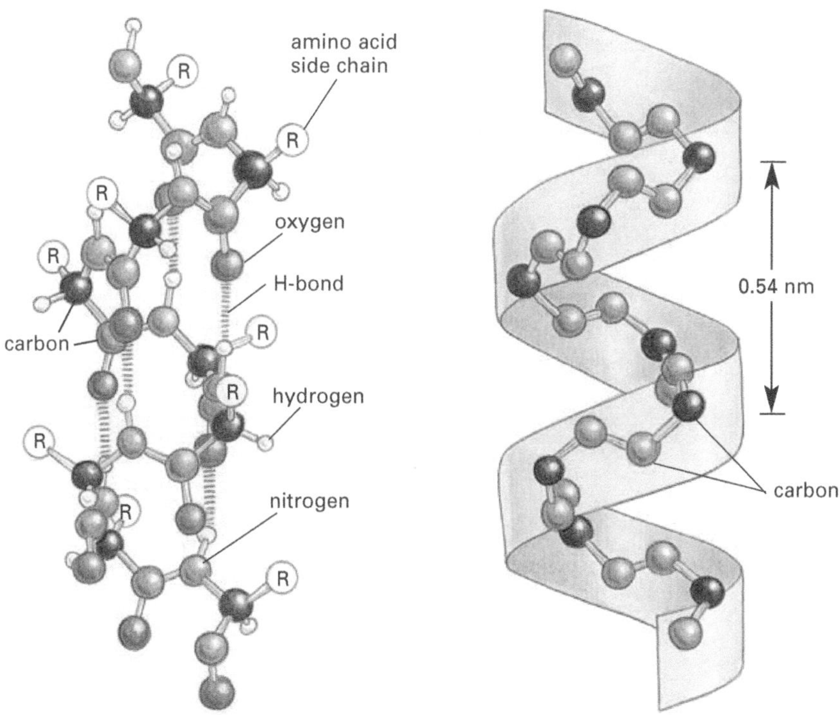

FIGURE 4.3 Protein alpha helix [1]. Copyright 2003 from *Essential Cell Biology*, Second Edition by Alberts et al. Reproduced by permission of Garland Science/Taylor & Francis LLC.

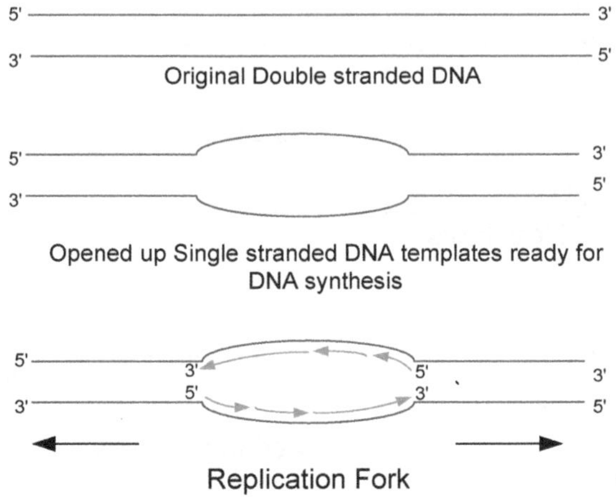

FIGURE 5.4 DNA Replication Fork

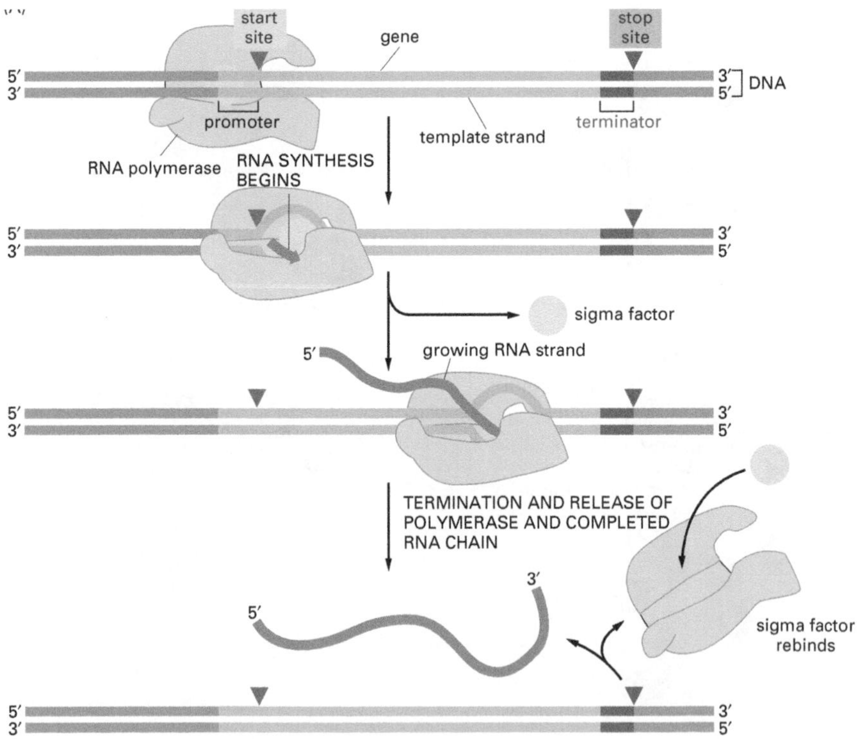

FIGURE 6.2 RNA polymerase transcribing a bacterial gene [1]. Copyright 2003 from *Essential Cell Biology*, Second Edition by Alberts et al. Reproduced by permission of Garland Science/ Taylor & Francis LLC.

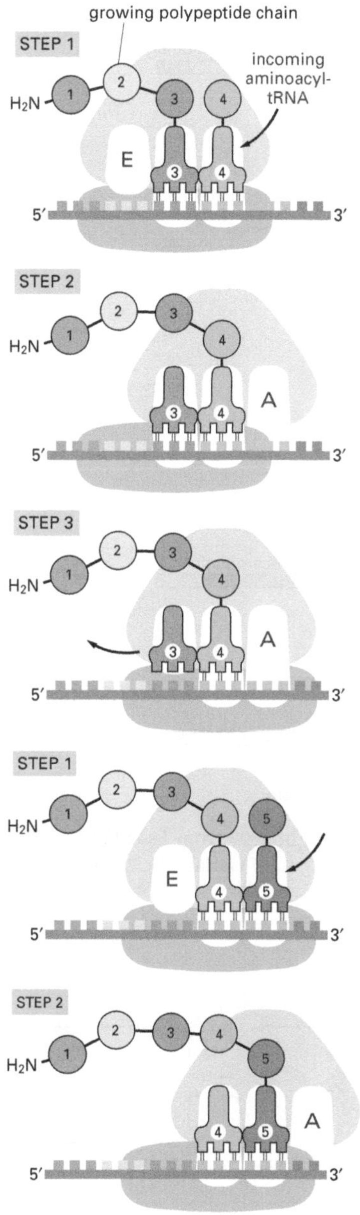

FIGURE 6.6 Steps in protein synthesis [1]. Copyright 2003 from *Essential Cell Biology*, Second Edition by Alberts et al. Reproduced by permission of Garland Science/Taylor & Francis LLC.

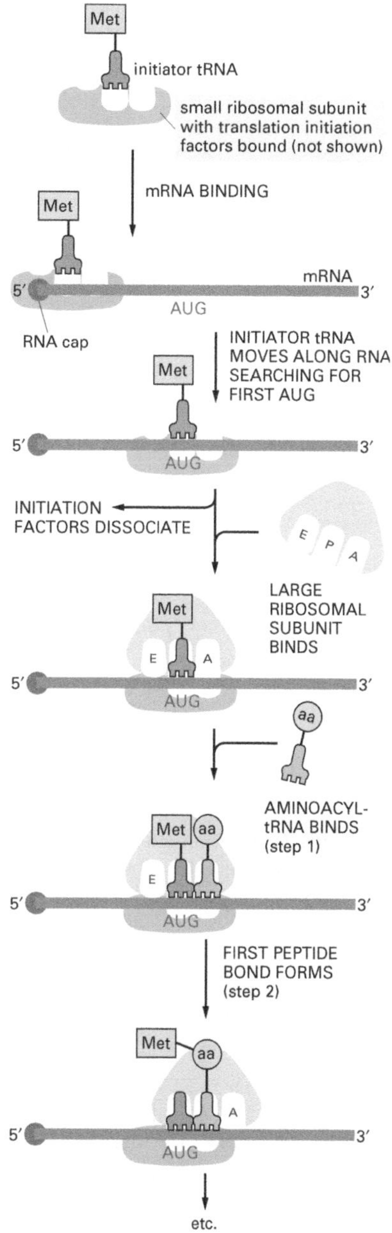

FIGURE 6.7 Initiation of protein synthesis [1]. Copyright 2003 from *Essential Cell Biology*, Second Edition by Alberts et al. Reproduced by permission of Garland Science/Taylor & Francis LLC.

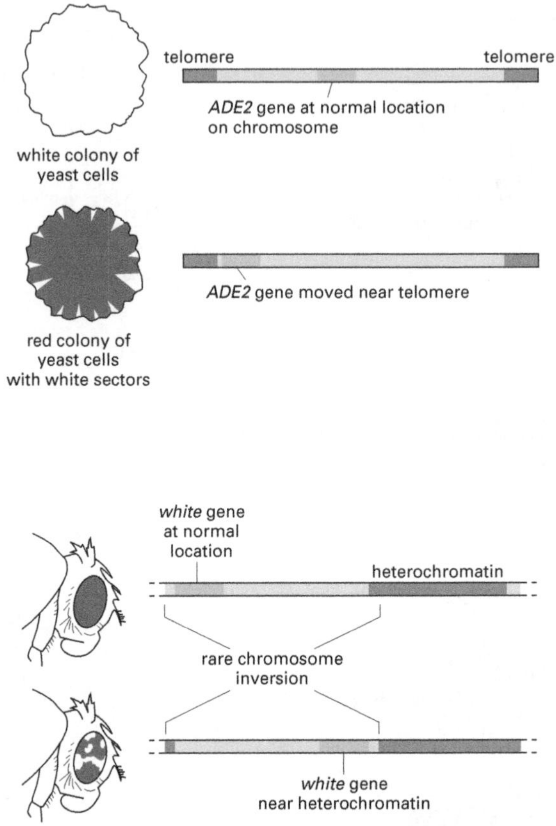

FIGURE 7.2 Position effects on gene expression [1]. Copyright 2003 from *Essential Cell Biology,* Second Edition by Alberts et al. Reproduced by permission of Garland Science/Taylor & Francis LLC.

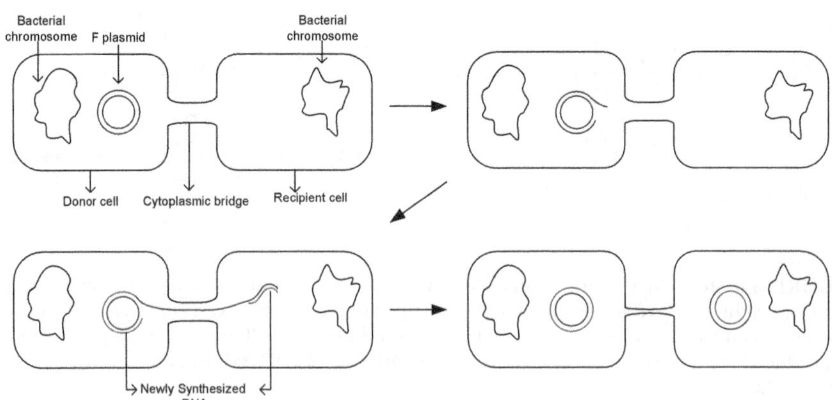

FIGURE 8.1 Gene transfer by bacterial mating.

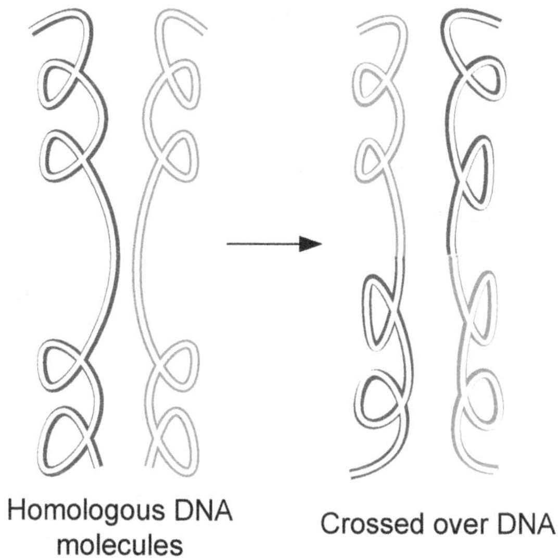

FIGURE 8.2 Homologous recombination.

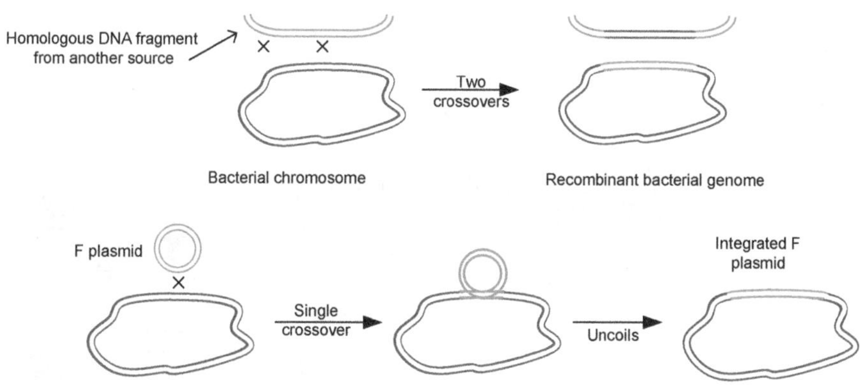

FIGURE 8.3 Two crossovers leading to variation in a bacterial genome.

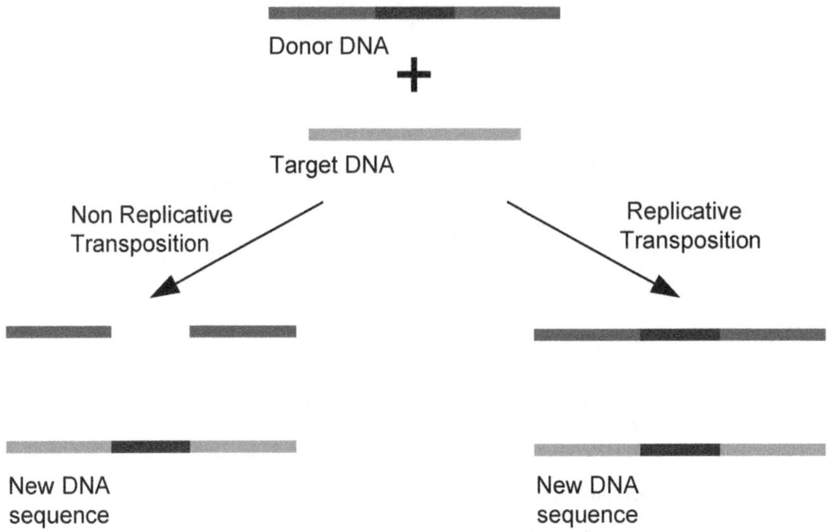

FIGURE 8.4 Nonreplicative and replicative transposition brought about by transposons.

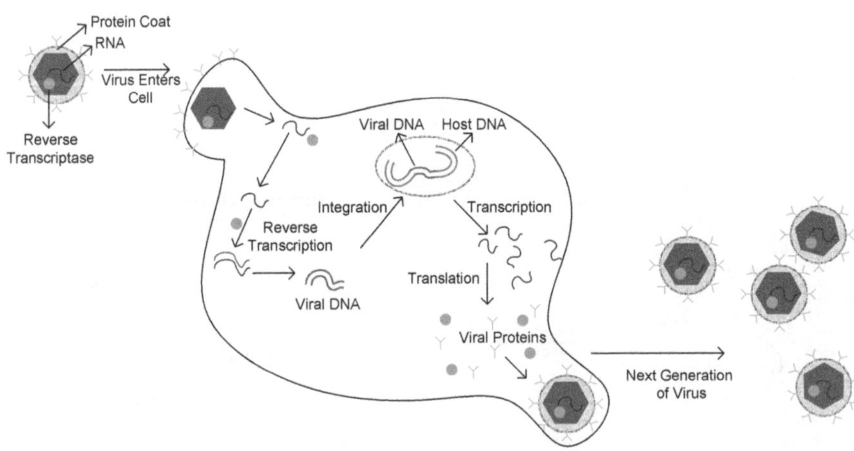

FIGURE 8.8 Retrovirus hijacking a host cell.

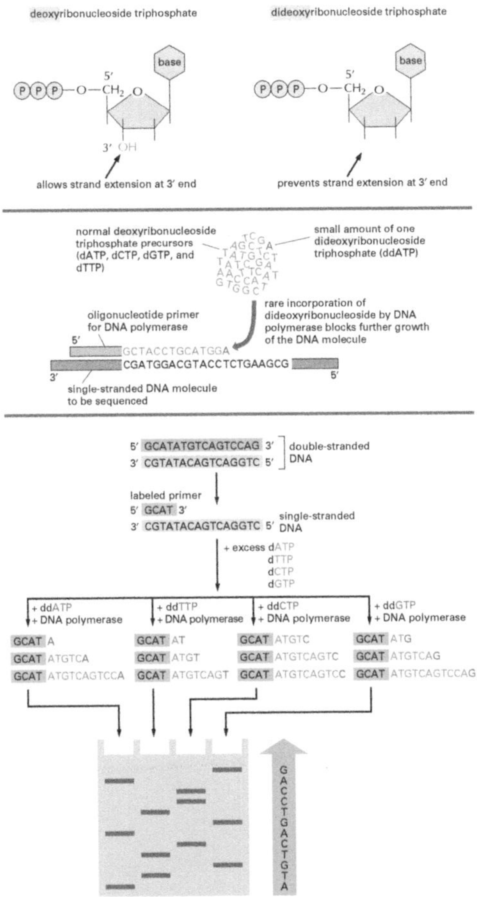

FIGURE 9.3 Sequencing DNA using the dideoxymethod [1]. Copyright 2003 from *Essential Cell Biology,* Second Edition by Alberts et al. Reproduced by permission of Garland Science/ Taylor & Francis LLC.

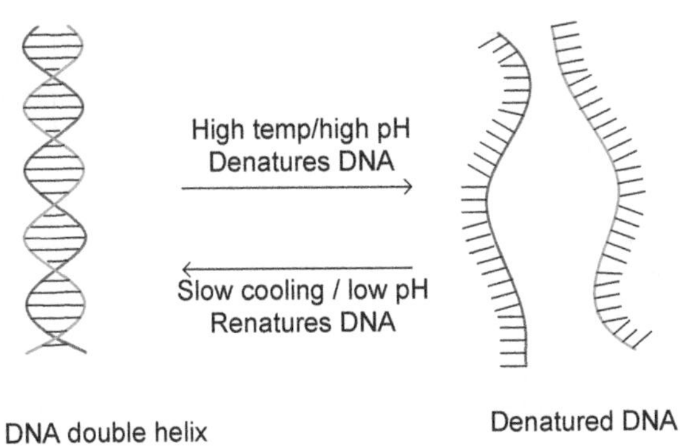

FIGURE 9.4 Denaturation and Renaturation of DNA molecules.

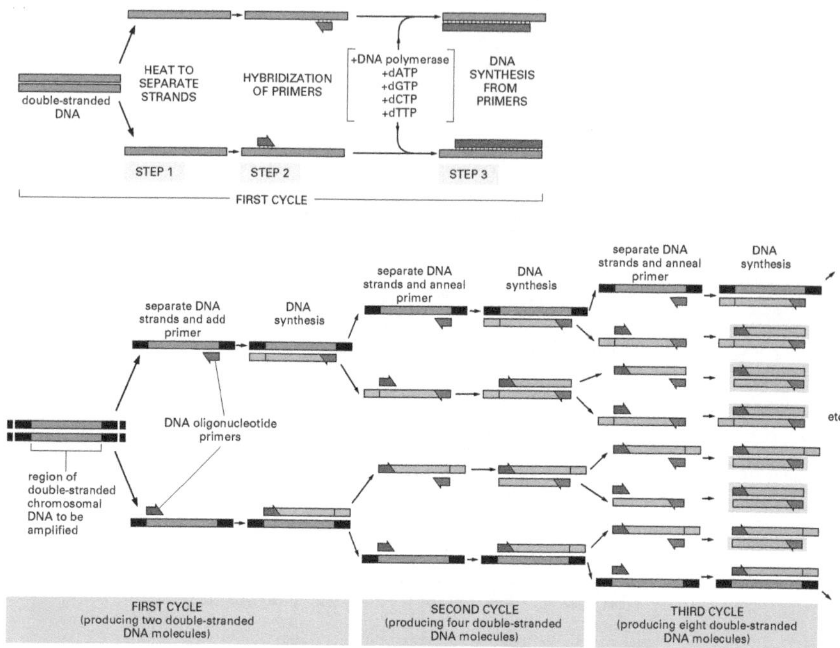

FIGURE 9.5 Steps in PCR [1]. Copyright 2003 from *Essential Cell Biology,* Second Edition by Alberts et al. Reproduced by permission of Garland Science/Taylor & Francis LLC.

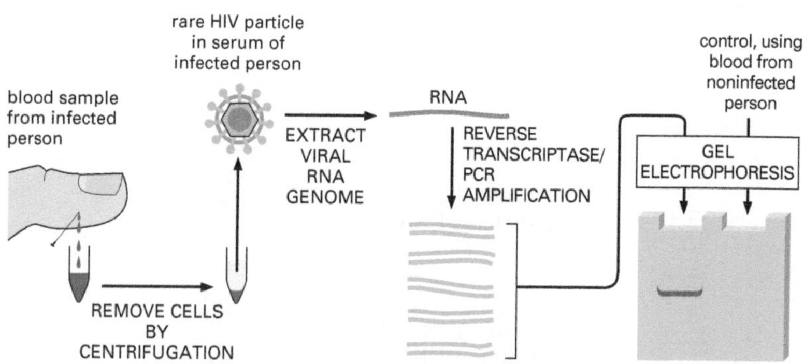

FIGURE 9.6 Use of PCR in detecting viral (HIV) infection [1]. Copyright 2003 from *Essential Cell Biology,* Second Edition by Alberts et al. Reproduced by permission of Garland Science/Taylor & Francis LLC.

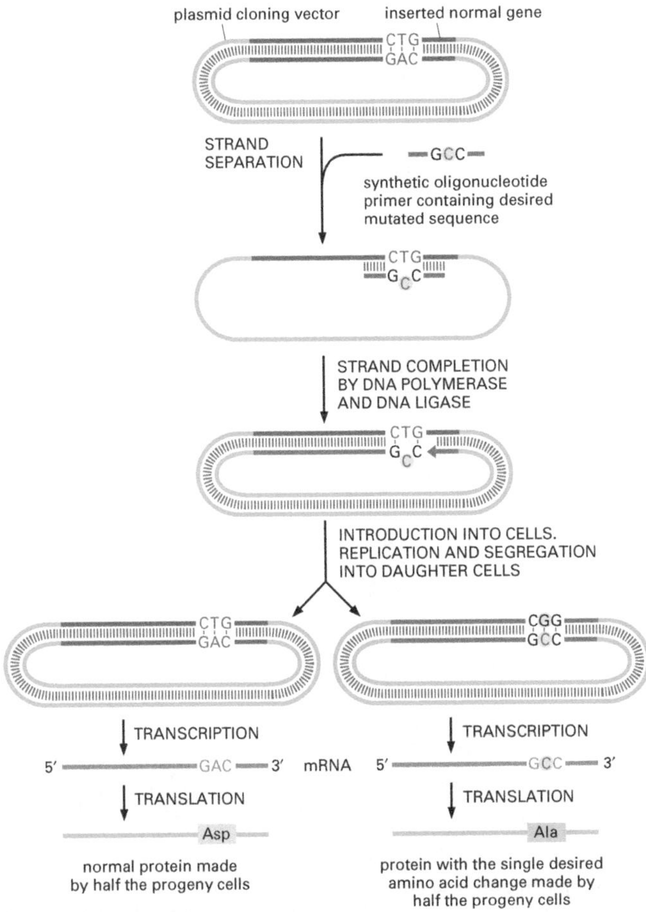

FIGURE 9.7 Site directed mutagenesis [1]. Copyright 2003 from *Essential Cell Biology,* Second Edition by Alberts et al. Reproduced by permission of Garland Science/Taylor & Francis LLC.

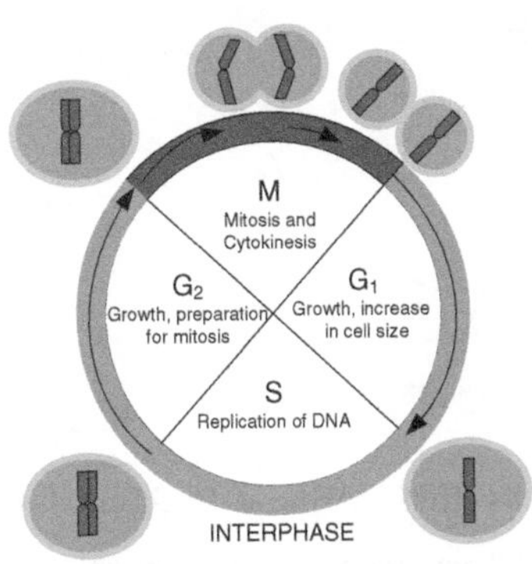

FIGURE 10.1 Eucaryotic cell cycle.

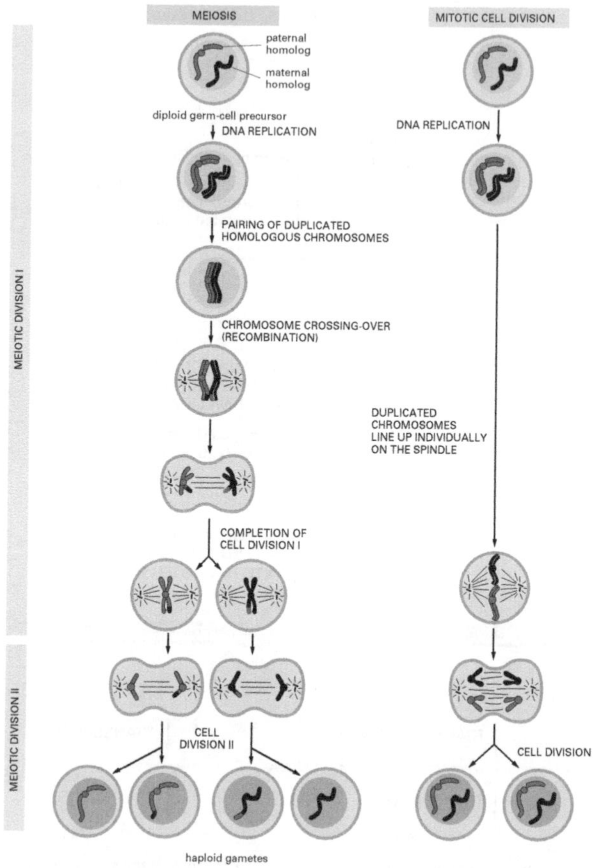

FIGURE 10.4 Comparing mitosis and meosis [1]. Copyright 2003 from *Essential Cell Biology*, Second Edition by Alberts et al. Reproduced by permission of Garland Science/Taylor & Francis LLC.

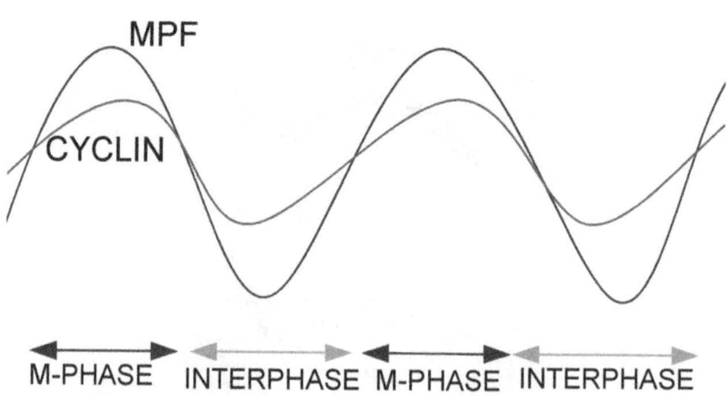

FIGURE 11.1 Variation of MPF activity and cyclin concentration during different stages of the cell cycle.

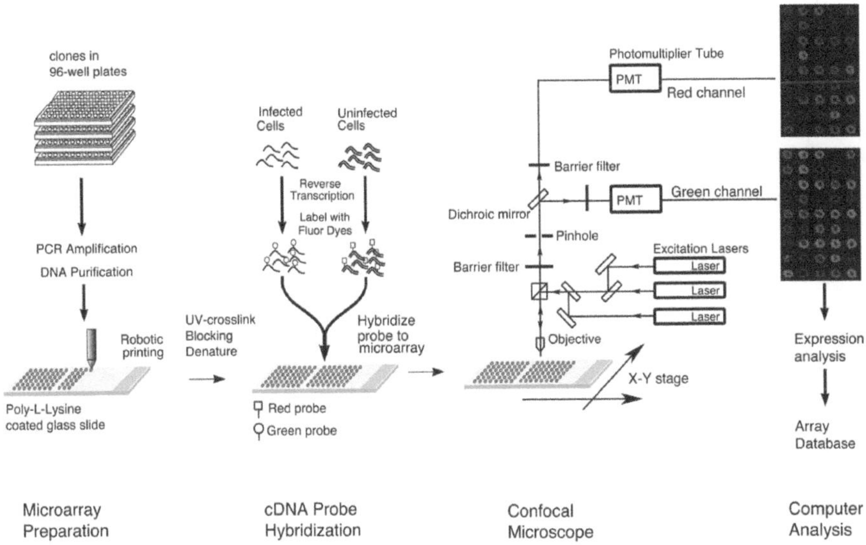

FIGURE 12.1 Microarray flow chart.

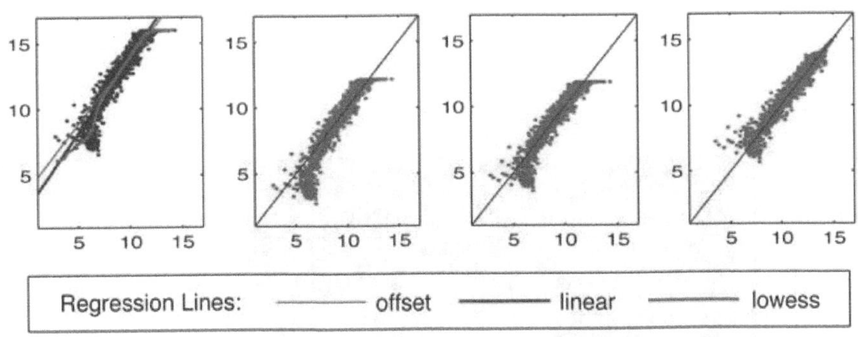

FIGURE 12.2 Effects of normalization: the left-hand scatter plot shows the regression lines and the red scatter plots show the normalized scatter plots, offset, linear, and lowess, from left to right.

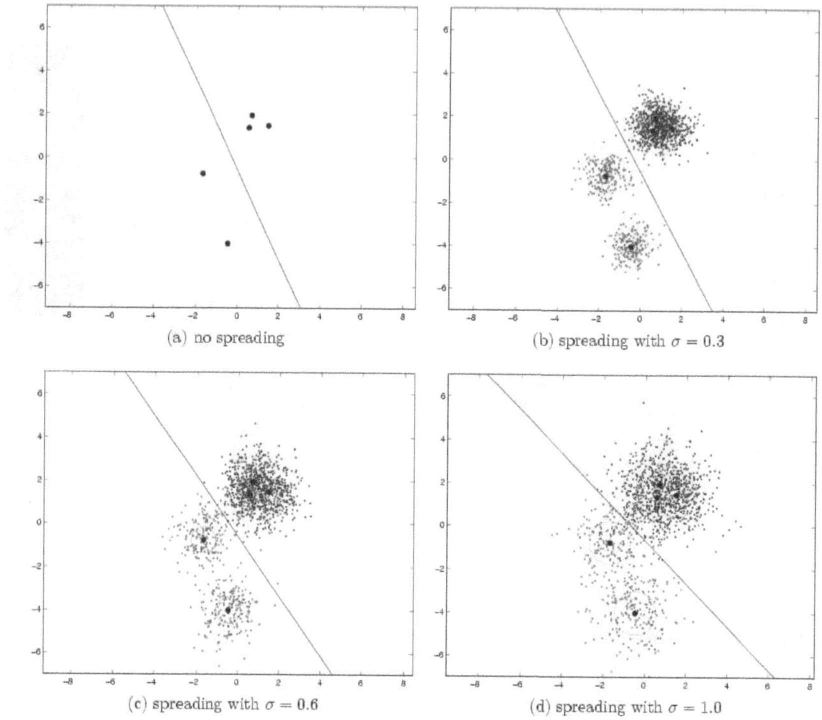

FIGURE 13.5 Noise injection for LDA with different spreads.

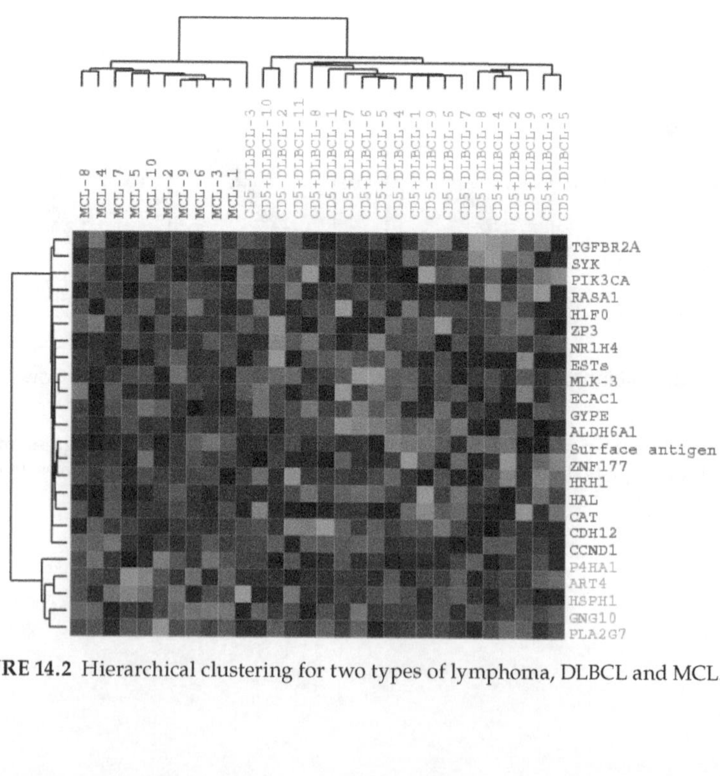

FIGURE 14.2 Hierarchical clustering for two types of lymphoma, DLBCL and MCL.

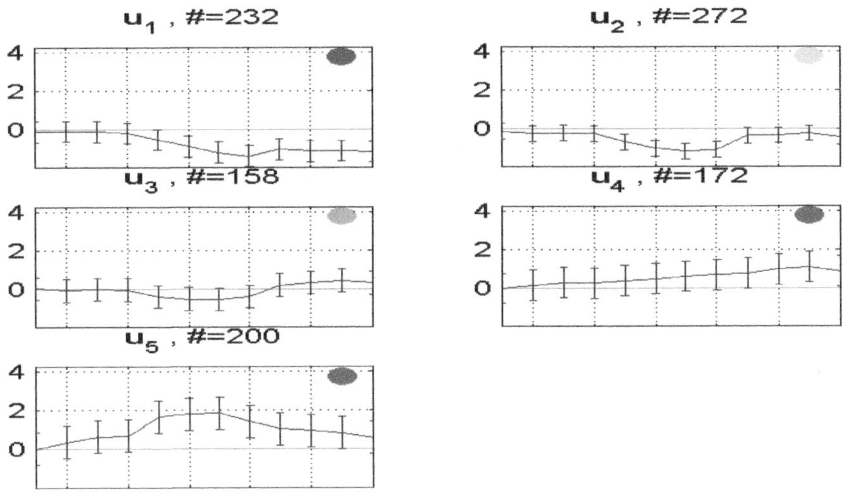

FIGURE 14.3 Seed means with standard-deviation bars for the five congruency classes.

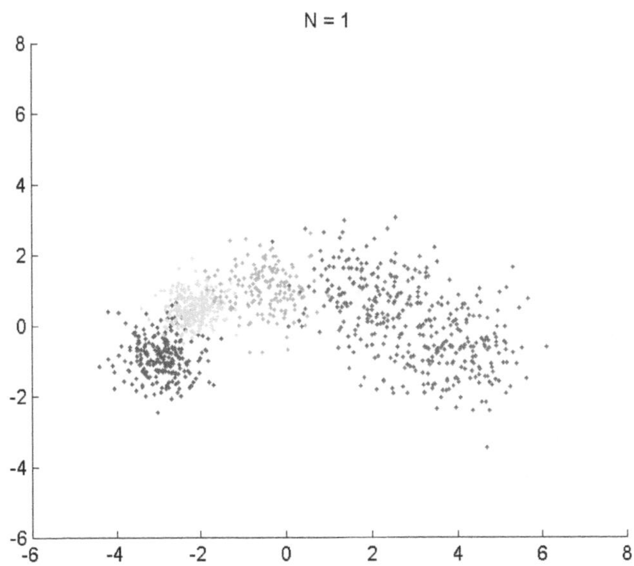

FIGURE 14.4 Two-dimensional principle-component plots of the generated data.

FIGURE 14.5 Results of fuzzy k-means clustering with SOM seed: (1) raw data; (2) clusters; (3) means of congruency classes assigned to clusters.

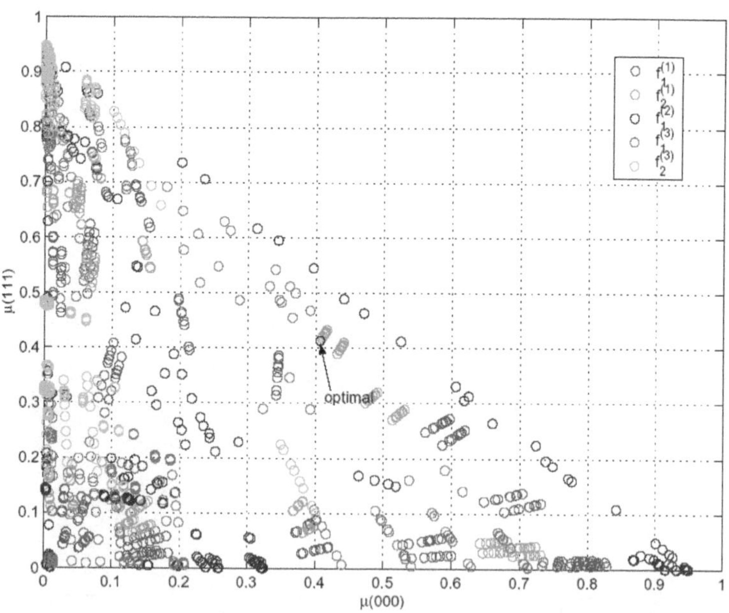

FIGURE 16.3 Each circle represents one of the 1280 possible alterations to the predictors. The x-axis is μ(000) and y-axis is μ(111). The optimal choice is shown with an arrow, as it comes closest to 0.4 for both stationary probabilities. The colors of the circles represent the predictor that is altered (see legend) [120].

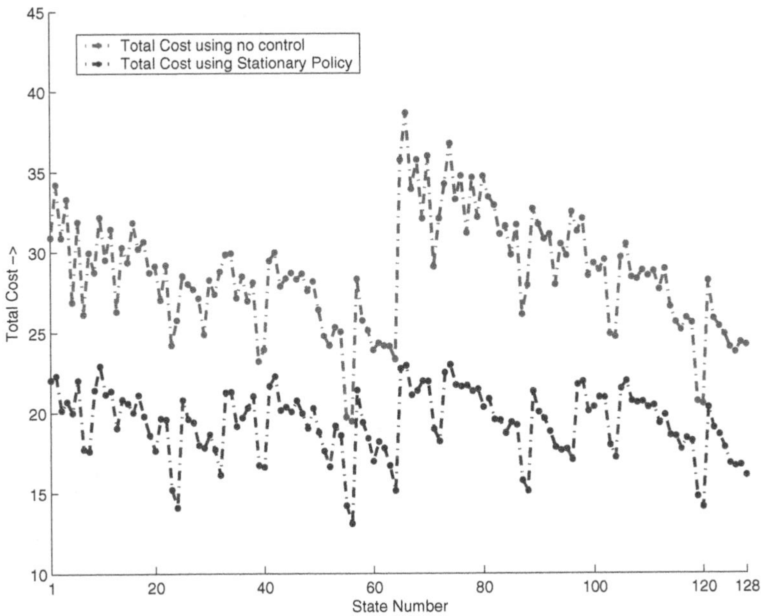

FIGURE 17.17 Total cost originating from different initial states [127] ©2006 IEEE.

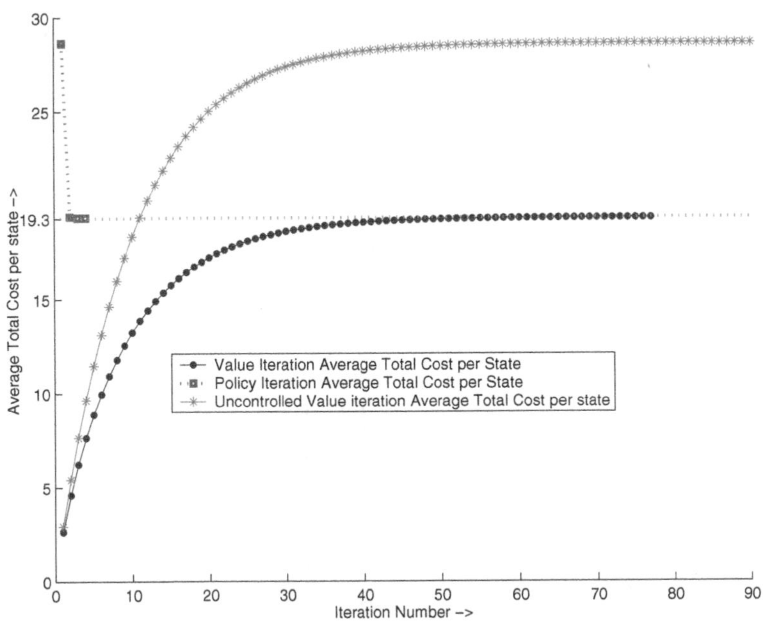

FIGURE 17.19 Average cost per state using discounted total cost [127] ©2006 IEEE.

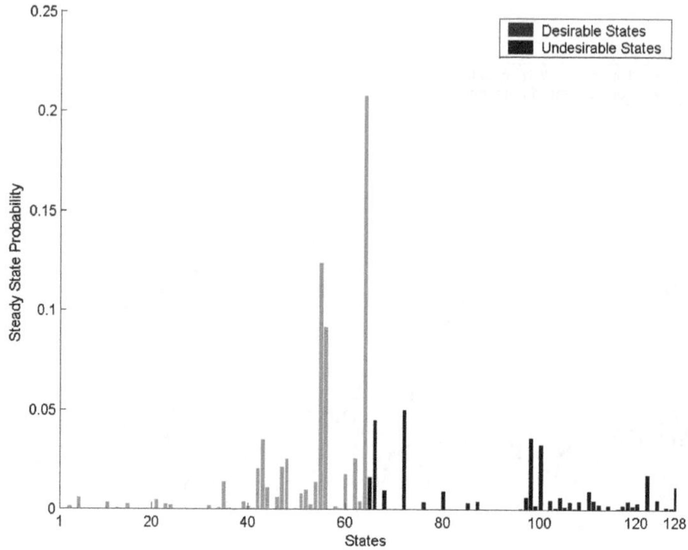

FIGURE 17.20 Steady state using discounted cost stationary policy [127] ©2006 IEEE.

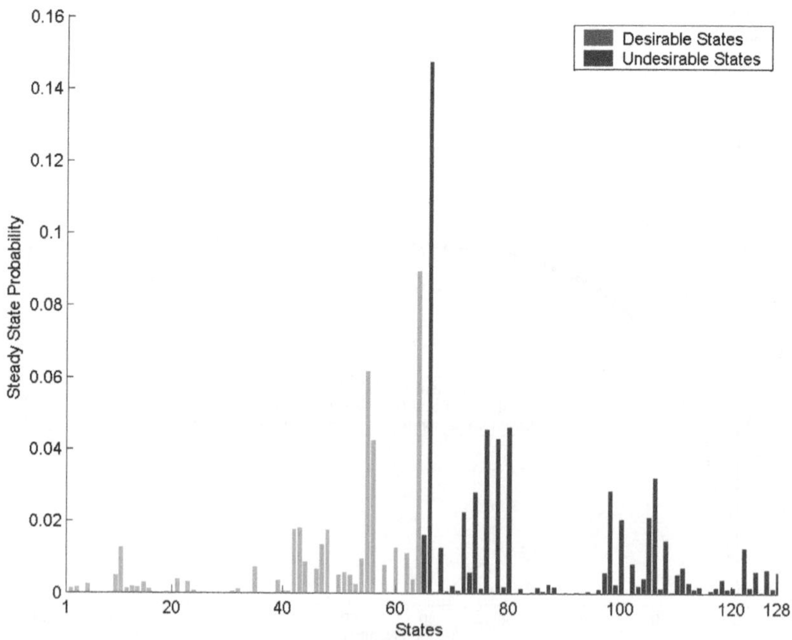

FIGURE 17.21 Original steady state [127] ©2006 IEEE.

12

Expression Microarrays

Cellular control results from multivariate activity among cohorts of genes and their products. Since all three levels in the central dogma — DNA, RNA, and protein — interact, it is not possible to fully separate them, and ultimately information from all realms must be combined for full understanding; nevertheless, the high level of interactivity between levels insures that a significant amount of the system information is available in each of the levels, so that focused studies provide useful insights. Much current effort is focused at the RNA level owing to measurement considerations. High-throughput technologies make it possible to simultaneously measure the RNA abundances of tens of thousands of mRNAs. In particular, expression microarrays result from a complex biochemical-optical system incorporating robotic spotting and computer image formation [3]. These arrays are grids of thousands of different single-stranded DNA molecules attached to a surface to serve as probes. Two major kinds are those using synthesized oligonucleotides and those using spotted cDNAs (complementary-DNA molecules). The basic procedure is to extract RNA from cells, convert the RNA to single-stranded cDNA, attach fluorescent labels to the different cDNAs, allow the single-stranded cDNAs to hybridize to their complementary probes on the microarray, and then detect the resulting fluor-tagged hybrids via excitation of the attached fluors and image formation using a scanning confocal microscope. Relative RNA abundance is measured via measurement of signal intensity from the attached fluors. This intensity is obtained by image processing and statistical analysis, with particular attention often paid to the detection of high- or low-expressing genes [4], and beyond that to expression-based phenotype classification [5] and the discovery of multivariate inter-gene predictive relationships [6]. This chapter briefly discusses microarrays, their hybridization-based foundation, normalization, and ratio analysis. At this point in time there is an extensive literature on microarray technology, signal extraction, and basic data analysis. We refer those interested to the various books on the subject [7, 8, 9, 10].

12.1 cDNA Microarrays

The principle behind cDNA spotted arrays is that an mRNA transcript to be detected is copied into a complementary DNA (cDNA) and this copied form of the transcript is immobilized on a glass microscope slide. The slides are usually coated with poly-lysine or poly-amine to immobilize the DNA molecules on the surface. For robotic spotting, the robot dips its pins into solutions that contain the cDNA and then the tiny amounts of solution that adhere to the pins are transferred to the surface. Each pin produces a printed spot containing cDNA. Rather than use robotic pins, another method uses ink-jet printing, in which the cDNA is expelled from a small nozzle equipped with a piezoelectric fitting by applying electric current. A complete cDNA microarray is prepared by printing thousands of cDNAs in an array format on the glass slide, and these provide gene-specific hybridization targets. Figure 12.1 provides a schematic representation of the preparation, hybridization, image acquisition, and analysis for cDNA microarrays.

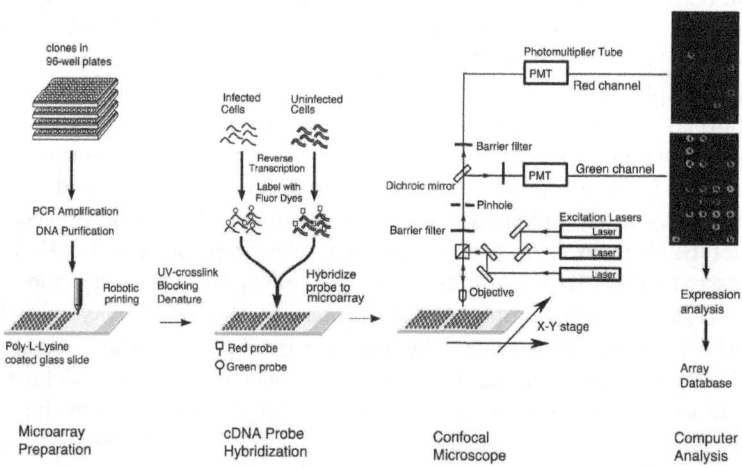

FIGURE 12.1 Microarray flow chart (See color figure following page 146).

A digital image reflecting the abundance of different mRNAs in a sample is formed in a number of steps. RNA is extracted from the cells of interest, converted to cDNA, and then amplified by a reverse transcriptase-polymerase chain reaction. During the process, fluorescent molecules are attached to the DNA. If a specific mRNA molecule is produced by the cells, then the generated fluorescently labeled cDNA molecule will hybridize to its complementary single-stranded microarray probe. The cDNA molecules that do not find their complementary single-stranded DNA sequences on the microarray are removed in a washing step. Since the fluorescent tags are attached to the cDNA strands that hybridize, the corresponding spots will fluoresce when provided fluorescence excitation energy and be detected at the level of emitted light. This yields a digital image whose intensities reflect levels of measured fluorescence, which in turn reflect mRNA abundances.

In practice, it is commonplace to label mRNA molecules from distinct sources with different fluorescent tags and then co-hybridize them onto each arrayed gene. Two monochrome images are obtained from laser excitations at two different wavelengths. Monochrome images of the intensity for each fluorescent label are combined by placing each image in the appropriate color channel of an RGB image. In this composite image, one can visualize the differential expression of genes in the two cell types, the test sample typically being placed in the red channel, with the reference sample in the green channel. Intense red fluorescence at a spot indicates a high level of expression of that gene in the test sample relative to the reference sample. Conversely, intense green fluorescence at a spot indicates relatively low expression of that gene in the test sample compared to the reference. When both test and reference samples express a gene at similar levels, the observed array spot is yellow. Assuming that specific DNA products from two samples have an equal probability of hybridizing to the specific target, the fluorescent intensity measurement is a function of the amount of specific RNA available within each sample, provided samples are well-mixed and there is sufficiently abundant cDNA deposited at each target location. Ratios or direct intensity measurements of gene-expression levels between the samples can be used to detect meaningfully different expression levels between the samples for a given gene.

When using cDNA microarrays, the signal must be extracted from the background. This requires image processing to extract signals, variability analysis, and measurement quality assessment [4]. The objective of the microarray image analysis is to extract probe intensities or ratios at each cDNA target location and then cross-link printed clone information so that biologists can easily interpret the outcomes and high-level analysis can be performed. A microarray image is first segmented into individual cDNA targets, either by manual interaction or an automated algorithm. For each target, the surrounding background fluorescent intensity is estimated, along with the exact target location, fluorescent intensity and expression ratios.

As with any random data, understanding variation is important to microarray data analysis. A major impediment to an effective understanding of variation is the large number of sources of variance inherent in microarray measurements. In many statistical analysis publications, the measured gene expression data are assumed to have multiple noise sources: noise due to sample preparation, labeling, hybridization, background fluorescence, different arrays, fluorescent dyes, and different printing locations. Various approaches have been taken to quantify and treat the noise levels in a set of experiments, including log-transformed signal-plus-noise ANOVA models [11, 12], mixture models [13], multiplicative models [14], ratio-distribution models [4, 15], rank-based models less sensitive to noise distributions [16], replicates using mixed models [17], and quantitative noise analysis [18]. In addition to the many studies on noise estimation, there is a large literature dealing with methods to isolate and eliminate the noise component from the measured signal. These studies suffer from the daunting complexity and inhomogeneity of the noise.

12.1.1 Normalization

Besides variation due to random effects, such as biochemical and scanner noise, simultaneous measurement of mRNA expression levels via cDNA microarrays involves variation owing to system sources, including labeling bias, imperfections due to spot extraction, and cross hybridization. Even with the development of good extraction algorithms and the use of control probes at the array printing stage to aid in accounting for cross hybridization, we are left with labeling bias resulting from the fluorescent tags as systemic error. Although different experimental designs target different profiling objectives, be it global cancer tissue profiling or a single induction experiment with one gene perturbed, normalization to correct labeling bias is a common preliminary step before further statistical or computational analysis is applied, its objective being to reduce the variation between arrays [19, 20]. Normalization is usually implemented for an individual array and is then called *intra-array* normalization, which is what we consider here. Assessment of the effectiveness of normalization has mainly been confined to the ability to detect differentially expressed genes. It has also been shown that normalization can benefit classification accuracy; the significance of this benefit depending on the bias properties [21].

The simplest and most commonly used normalization is the *offset method* [15] . To describe it, let the red and green channel intensities of the kth gene be r_k and g_k, respectively. In many cases these are background-subtracted intensities. In an ideal case where two identical biological samples are labeled and co-hybridized to the array, we expect the sum of the log-transformed ratios to be 0; however, due to various reasons (dye efficiency, scanner PMT control, etc.), this assumption may not be true. If we assume that the two channels are equivalent, except for a signal amplification factor, then the ratio of the kth gene, t_k, can be calculated by

$$\log t_k = \log\left(r_k/g_k\right) - \frac{1}{N_q}\sum_{k=1}^{N_q}\log\left(r_k/g_k\right) \qquad (12.1)$$

where the second term in Eq. 12.1 is a constant offset that simply shifts the

r_k vs. g_k scatter plot to a 45° diagonal line intersecting the origin and N_q is the number of probes that have measurement quality score of 1.0 (see [15] for a discussion of spot quality scores).

In some cases the scatter plot may not be perfectly at a 45° diagonal line due to the difference when the scanner's two channels may operate at different linear characteristic regions. In this case, full *linear regression*, instead of requiring the line to intersect at the origin, may be necessary. This is achieved by finding coefficients a and b to minimize the expectation $E[(g_k-(ar_k-b))^2]$.

Some microarray expression levels may have large dynamic range that will cause systematic scanner deviations such as nonlinear response at lower intensity range and saturation at higher intensity. Although data falling into these ranges are commonly discarded for further analysis, the transition range, without proper handling, may still cause some significant error in differential-expression gene detection. To account for this deviation, *locally weighted linear regression* (*lowess*) is regularly employed as a normalization method for such intensity-dependent effects [22, 23]. A newer method, *loess*, is essentially the same algorithm except that it uses a quadratic polynomial to fit the piecewise segments.

Figure 12.2 shows the effects of the three normalizations.

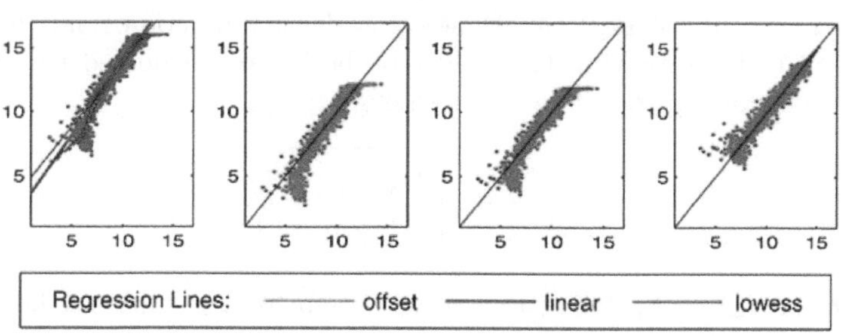

FIGURE 12.2 Effects of normalization: the left-hand scatter plot shows the regression lines and the red scatter plots show the normalized scatter plots, offset, linear, and lowess, from left to right (See color figure following page 146).

12.1.2 Ratio Analysis

It may be that the two channels in a cDNA microarray represent two sources of mRNA to be compared or that the red channel corresponds to the source of interest while the green channel serves as a reference channel. In either case, a basic question is whether the red intensity is significantly greater than or less than the green intensity. Such a question is naturally approached in the framework of hypothesis tests.

For a microarray having n genes, with red and green fluorescent expression values labeled by R_1, R_2, ..., R_n and G_1, G_2, ..., G_n, respectively, the issue is whether R_k is over- or under-expressed relative to G_k. Letting μ_{R_k} and σ_{R_k} denote the mean and standard deviation of R_k (similarly for G_k), the relevant hypothesis test is

$$\begin{aligned} H_0 &: \mu_{R_k} = \mu_{G_k} \\ H_1 &: \mu_{R_k} \neq \mu_{G_k} \end{aligned} \tag{12.2}$$

using the ratio test statistic $T_k = R_k/G_k$.

There is biological plausibility for assuming that the coefficient of variation is approximately constant across the microarray [4]. If we make the simplifying assumption of a constant coefficient of variation, meaning that

$$\begin{aligned} \sigma_{R_k} &= c\mu_{R_k} \\ \sigma_{G_k} &= c\mu_{G_k} \end{aligned} \tag{12.3}$$

where c denotes the common coefficient of variation (cv), then under the null hypothesis H_0, Eq.12.3 implies that $\sigma_{R_k} = \sigma_{G_k}$. Assuming R_k and G_k to be normally and identically distributed, T_k has the density function

$$f_{T_k}(t;c) = \frac{(1+t)\sqrt{1+t^2}}{c(1+t^2)^2\sqrt{2\pi}} \exp\left[\frac{-(t-1)^2}{2c^2(1+t)}\right] \tag{12.4}$$

Owing to the constant-cv assumption, the subscript k does not appear on the right-hand side of Eq.12.4. Hence, the density function holds for all genes and all ratios satisfying the null hypothesis can be pooled to estimate the parameter of Eq.12.4. The estimate is given by

$$\hat{c} = \sqrt{\frac{1}{n}\sum_{i=1}^{n}\frac{(t_i-1)^2}{(t_i^2+1)}} \tag{12.5}$$

where t_1, t_2, ..., t_n are ratio samples taken from a family of housekeeping* genes on a microarray.

---

*Housekeeping genes, encountered by us in Chapter 7, are continually transcribed at relatively constant levels, for sustenance of the cell. Since their expression is independent of the experimental conditions, they serve as an internal standard in quantitative expression analysis.

For a microarray derived from two identical mRNA samples co-hybridized on one slide (self-self experiment), the parameter c (cv of the fluorescent intensity) provides the *variation of assay.* However, for any experiment, to guarantee the null hypothesis condition in Eq.12.2 is not always possible. One alternative is to duplicate some or all clones where the same expression ratio is expected. For the ratio, T, of expression ratios, it can be shown that

$$\sigma_{\log T}^2 \approx 4c^2 \tag{12.6}$$

when the measurement of the log-transformed expression level is approximately normally distributed. For any given experiment with some duplicated clones, $\sigma_{\log T}^2$ is easily calculated, and along with it the coefficient of variation of the assay.

In practical applications, a constant amplification gain m may apply to one signal channel, in which case the null hypothesis in Eq.12.2 may become $\mu_{R_k} = m\mu_{G_k}$. Under this uncalibrated signal detection setting, the ratio density can be modified by

$$f_T(t; c, m) = \frac{1}{m} f_T(t/m; c, 1) \tag{12.7}$$

where $f_T(\bullet; c, 1)$ is given by Eq.12.4. In [4], an estimation procedure for the parameter m is proposed.

To decide whether expression levels of a gene from two samples are significantly different, we would like to find a confidence interval such that within the confidence interval, the null hypothesis given in Eq.12.2 cannot be rejected: the expression ratio, $T_k = R_k/G_k$, of the gene under consideration is not significantly deviated from 1.0 if the ratio is within the confidence interval. The confidence interval can be evaluated by integrating the ratio density function given by Eq.12.4. Since the confidence interval is determined by the parameter c, one can either use the parameter derived from preselected housekeeping genes (Eq.12.5), or a set of duplicated genes (Eq.12.6) if they are available in the array. The former confidence intervals contain some levels of variation from the fluctuation of the biological system that also affect the housekeeping genes, while the latter contains no variation of the biological fluctuation, but contains possible spot-to-spot variation. Spot-to-spot variation is not avoidable if one wishes to repeat the experiment. The confidence interval derived from the duplicated genes is termed as the *confidence interval of the assay.*

The constant-cv condition permits all ratios satisfying the null hypothesis to be pooled to estimate the parameter of Eq.12.5; however, studies indicate that the cv varies with the expression intensities [24, 25]. The constant-cv condition is made under the assumption that R_k and G_k are the expression levels detected from two fluorescent channels. The latter assumption is based on the assumption that image processing successfully extracts the true signal. This proves to be quite accurate for strong signals, but problems arise when the signal is weak in comparison to the background. Figure 12.3 illustrates how in many real-world situations the weaker expression signals (at the lower left corner of the scatter plot) produce a larger spread of gene placement.

Even with image processing, the actual expression intensity measurement is of the form

$$R_k = (SR_k + BR_k) - \mu_{BR_k} \tag{12.8}$$

where SR_k is the expression intensity measurement of gene k, BR_k is the fluorescent background level, and μ_{BR_k} is the mean background level. The measurable quantities are (1) signal with background, $SR_k + BR_k$, and (2) the surrounding background. The null hypothesis of interest is $\mu_{SR_k} = \mu_{SG_k}$. Taking the expectation in Eq.12.8 yields

$$\mu_{R_k} = E[R_k] = E[(SR_k + BR_k) - \mu_{BR_k}] = \mu_{SR_k} \tag{12.9}$$

Since $\mu_{G_k} = \mu_{SG_k}$, the hypothesis test of Eq.12.2 is still the one with which we are concerned, and we still apply the test statistic T_k. There is, however, a major difference. The assumption of a constant cv applies to SR_k and SG_k, not to R_k and G_k, and the density of Eq.12.4 is not applicable. The ratio analysis performed under the constant-cv condition does not apply in these new circumstances. We leave its analysis to the literature [15].

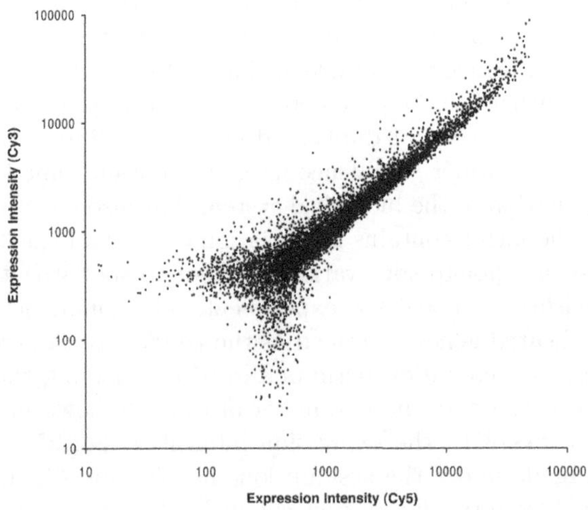

FIGURE 12.3 Dispersion owing to weak signals.

12.2 Synthetic Oligonucleotide Arrays

A different approach than that of immobilizing cDNA strands copied from particular mRNA transcripts onto a surface is to fabricate synthetic polynucleotides onto the surface. For the popular *Affymetrix* chip, photolithography and solid-phase DNA synthesis are used to synthesize desired polynucleotide probes at specified coordinates on the slide. Since these synthetic probes are much shorter than the genes on a cDNA microarray, they must be designed in such a way that their collective action provides accurate measurement of transcript targets. Here we are discussing probes consisting of only 25 bases (25-mers) and a single 25-mer may pair with a large number of mRNAs. For each mRNA of interest, there is a set of non-overlapping (almost non-overlapping) 25-mer oligonucleotides to provide sequence-specific detection that can lead to accurate abundance measurement at the next stage of the process. Probes must be designed so that the final measurement for a particular gene is not unduly affected by the fact that the short probe will hybridize to transcripts from other genes and other kinds of RNA in the sample.

The basic approach is to have, for each mRNA to be detected, a probe set for which different probes hybridize to different regions of the mRNA. In this way the measurement for a specific mRNA is obtained via averaging across a collection of probes, thereby mitigating the effects of noise, outliers, and cross-hybridization.

To further mitigate the effects of cross-hybridization, two sets of probes can be used. *Perfect-match* (*PM*) probes are the ones designed to perfectly match selected nucleotide sequences in the target mRNA, whereas *mismatch* (*MM*) probes are derived from the PM probes by changing one base at a central location. If a PM probe is aimed at identifying a particular mRNA by matching a string of bases in the mRNA, then it is more probable that the mRNA will hybridize to the PM probe rather than to the MM probe derived from it.

It is argued that subtraction of the intensity for the MM probe accomplishes a correction for both background noise and cross-hybridization. With this in mind, the expression level for a probe set on a microarray can be estimated by the *average difference*,

$$AvDiff = \frac{1}{|A|} \sum_{k \in A} PM_k - MM_k \tag{12.10}$$

where A is the subset of all probes k such that $PM_k - MM_k$ is within 3 standard deviations of the trimmed average of all differences $PM_j - MM_j$, the smallest and largest difference being left out of the average, and where $|A|$ is the number of probes in A. Alterations of this basic average difference have been proposed. Various problems have been pointed out regarding the average difference, including the fact that MM often exceeds PM. A number of other difference-based measures have been proposed, as have a number of alternative approaches [26, 27, 28, 29].

13

Classification

Pattern classification plays an important role in genomic signal processing. For instance, cDNA microarrays can provide expression measurements for thousands of genes at once, and a key goal is to perform classification via different expression patterns — between different kinds of cancer, different stages of tumor development, etc. [5, 16, 30, 31]. This requires designing a classifier (decision function) that takes a vector of gene expression levels as input, and outputs a class label that predicts the class containing the input vector. Classifiers are designed from a sample of expression vectors. This involves assessing expression levels from RNA obtained from the different tissues with microarrays, determining genes whose expression levels can be used as classifier features (variables), and then applying some rule to design the classifier from the sample microarray data. Design, performance evaluation, and application must take this randomness due to both biological and experimental variability. Three critical issues arise: (1) Given a set of features, how does one design a classifier from the sample data that provides good classification over the general population? (2) How does one estimate the error of a designed classifier when data are limited? (3) Given a large set of potential features, such as the large number of expression levels provided by each microarray, how does one select a set of features as the input to the classifier? Small samples (relative to the number of features) are ubiquitous in genomic signal processing and impact all three issues [32].

13.1 Classifier Design

Classification involves a *feature vector* $\mathbf{X} = (X_1, X_2, \cdots, X_d)$ on d-dimensional Euclidean space $\Re^d$ composed of random variables (*features*), a binary random variable Y, and a *classifier* $\psi : \Re^d \rightarrow \{0, 1\}$ to serve as a predictor of Y, which means that Y is to be predicted by $\psi(\mathbf{X})$. The values, 0 or 1, of Y are treated as class *labels*. The error, $\varepsilon[\psi]$, of ψ is the probability that the classification is erroneous, namely, $\varepsilon[\psi] = P(\psi(\mathbf{X}) \neq Y)$. It equals the expected (mean) absolute difference, $E[|Y - \psi(\mathbf{X})|]$, between the label and the classification. Owing to the binary nature of $\psi(\mathbf{X})$ and Y, it also equals the mean-square error.

13.1.1 Bayes Classifier

An optimal classifier, ψ_d, is one having minimal error, ε_d, among all binary functions on $\Re^d$, so that it is the minimal mean-absolute-error predictor of Y. ψ_d and ε_d are called the *Bayes classifier* and *Bayes error*, respectively. Classification accuracy, and thus the error, depends on the probability distribution of the feature-label pair $(\mathbf{X}, Y)$.

The posterior distribution for $\mathbf{X}$ is defined by $\eta(\mathbf{x}) = f_{\mathbf{X},Y}(\mathbf{x},1)/f_{\mathbf{X}}(\mathbf{x})$, where $f_{\mathbf{X},Y}(\mathbf{x},y)$ and $f_{\mathbf{X}}(\mathbf{x})$ are the densities for $(\mathbf{X},Y)$ and $\mathbf{X}$, respectively. $\eta(\mathbf{x})$ gives the probability that $Y = 1$, given $\mathbf{X} = \mathbf{x}$. In this binary setting, $\eta(\mathbf{x}) = E[Y|\mathbf{x}]$ is the conditional expectation of Y given $\mathbf{x}$. The error of an arbitrary classifier can be expressed as

$$\varepsilon[\psi] = \int_{\{\mathbf{x}:\psi(\mathbf{x})=0\}} \eta(\mathbf{x}) f_{\mathbf{X}}(\mathbf{x}) d\mathbf{x} + \int_{\{\mathbf{x}:\psi(\mathbf{x})=1\}} (1-\eta(\mathbf{x})) f_{\mathbf{X}}(\mathbf{x}) d\mathbf{x} \qquad (13.1)$$

The *class conditional probability* of Y given $\mathbf{x}$ is defined by $P(Y = 1|\mathbf{x}) = \eta(\mathbf{x})$. As a function of $\mathbf{X}$, $P(Y = 1|\mathbf{X})$ is a random variable dependent on $f_{\mathbf{X}}(\mathbf{x})$. Since $0 \leq \eta(\mathbf{x}) \leq 1$, the right-hand side of (13.1) is minimized by

$$\psi_d(\mathbf{x}) = \begin{cases} 0, \text{if } P(Y = 1|\mathbf{x}) \leq 1/2 \\ 1, \text{if } P(Y = 1|\mathbf{x}) > 1/2 \end{cases} \qquad (13.2)$$

$\psi_d(\mathbf{x})$ is defined to be 0 or 1 according to whether Y is less or more likely to be 1 given $\mathbf{x}$. It follows from Eqs. 13.1 and 13.2 that the Bayes error is given by

$$\varepsilon_d = \int_{\{\mathbf{x}:\eta(\mathbf{x})\leq 1/2\}} \eta(\mathbf{x}) f_{\mathbf{X}}(\mathbf{x}) d\mathbf{x} + \int_{\{\mathbf{x}:\eta(\mathbf{x})>1/2\}} (1-\eta(\mathbf{x})) f_{\mathbf{X}}(\mathbf{x}) d\mathbf{x} \qquad (13.3)$$

The problem with the Bayes classifier is that we typically do not know the class conditional probabilities, and therefore must design a classifier from sample data. An obvious approach would be to estimate the conditional probabilities from data, but often we do not have sufficient data to obtain good estimates. Moreover, good classifiers can be obtained even when we lack sufficient data for satisfactory density estimation.

13.1.2 Classification Rules

Design of a classifier ψ_n from a random sample $S_n = \{(\mathbf{X}_1, Y_1), (\mathbf{X}_2, Y_2), \cdots, (\mathbf{X}_n, Y_n)\}$ of vector-label pairs drawn from the feature-label distribution requires a *classification rule* that operates on random samples to yield a classifier. A *classification rule* is a mapping $\Psi_n : [\Re^d \times \{0,1\}]^n \rightarrow \mathcal{F}$, where $\mathcal{F}$ is the family of $\{0,1\}$-valued functions on $\Re^d$. Given a sample S_n, we obtain a designed classifier $\psi_n = \Psi_n(S_n)$ according to the rule Ψ_n. To be fully formal, one might write $\psi(S_n; \mathbf{X})$ rather than $\psi_n(\mathbf{X})$; however, we will use the simpler notation, keeping in mind that ψ_n derives from a classification rule applied to a feature-label sample. Note that a classification rule is really a sequence of classification rules depending on n.

The Bayes error ε_d is estimated by the error ε_n of ψ_n. There is a *design cost*

$$\Delta_n = \varepsilon_n - \varepsilon_d \tag{13.4}$$

ε_n and Δ_n being sample-dependent random variables. The expected design cost is $E[\Delta_n]$, the expectation being relative to all possible samples. The expected error of ψ_n is decomposed according to

$$E[\varepsilon_n] = \varepsilon_d + E[\Delta_n] \tag{13.5}$$

Asymptotic properties of a classification rule concern large samples (as $n \to \infty$). A rule is said to be *consistent* for a distribution of $(\mathbf{X}, Y)$ if $\Delta_n \to 0$ in the mean, meaning $E[\Delta_n] \to 0$ as $n \to \infty$. For a consistent rule, the expected design cost can be made arbitrarily small for a sufficiently large amount of data. Since rarely is there a good estimate of the distribution in pattern recognition, rules for which convergence is independent of the distribution are desirable. A classification rule is *universally consistent* if $\Delta_n \to 0$ in the mean for any distribution of $(\mathbf{X}, Y)$. Universal consistency is useful for large samples, but has little consequence for small samples.

There is a host of classification rules in the literature [33]. In this chapter we will discuss a few well-known and well-studied classification rules, our intent being to use them to illustrate methodology and issues, not to provide a tutorial on classification rules.

For the basic *nearest-neighbor* (NN) rule, ψ_n is defined for each $\mathbf{x} \in \mathcal{R}^d$ by letting $\psi_n(\mathbf{x})$ take the label of the sample point closest to $\mathbf{x}$. For the NN rule, no matter the distribution of $(\mathbf{X}, Y)$, $\varepsilon_d \leq \lim_{n\to\infty} \; E[\varepsilon_n] \leq 2\varepsilon_d$ [34]. It follows that $\lim_{n\to\infty} \; E[\Delta_n] \leq \varepsilon_d$. Hence, the asymptotic expected cost of design is small if the Bayes error is small; however, this result does not give consistency. More generally, for the *k-nearest-neighbor* (kNN) *rule*, k odd, the k points closest to $\mathbf{x}$ are selected and $\psi_n(\mathbf{x})$ is defined to be 0 or 1 according to which is the majority among the labels of these points. If $k = 1$, this gives the NN rule. We will not consider even k. The limit of $E[\varepsilon_n]$ as $n \to \infty$ can be expressed analytically and various upper bounds exist. In particular, $\lim_{n\to\infty} \; E[\Delta_n] \leq (ke)^{-1/2}$. This does not give consistency, but it does show that the design cost gets arbitrarily small for sufficiently large k as $n \to \infty$. The kNN rule is universally consistent if $k \to \infty$ and $k/n \to 0$ as $n \to \infty$ [35].

A classification rule can yield a classifier that makes very few, or no, errors on the sample data on which it is designed, but performs poorly on the distribution as a whole, and therefore on new data to which it is applied. This situation is exacerbated by complex classifiers and small samples. If the sample size is dictated by experimental conditions, such as cost or the availability of patient RNA for expression microarrays, then one only has control over classifier complexity. The situation with which we are concerned is typically referred to as *overfitting*. The basic point is that a classification rule should

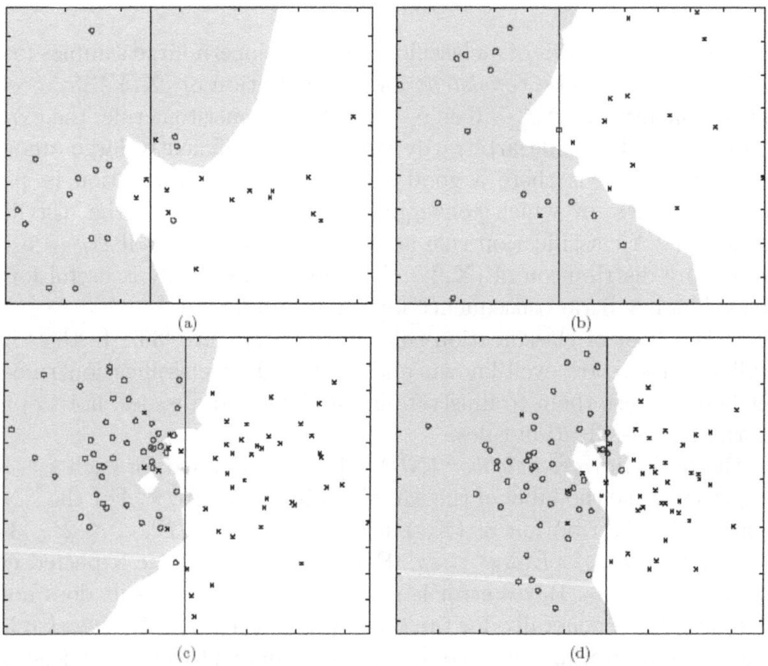

FIGURE 13.1 3NN classification applied to two equal-variance circular Gaussian class conditional distributions: (a) for a 30-point sample; (b) for a second 30-point sample; (c) for a 90-point sample; (d) for a second 90-point sample.

not cut up the feature space in a manner too complex for the amount of sample data available. The problem is not necessarily mitigated by applying an error-estimation rule to the designed classifier to see if it "actually" performs well, since when there is only a small amount of data available, error-estimation rules (as we will see subsequently) are very imprecise, and the imprecision tends to be worse for complex classification rules. Hence, a low error estimate is not sufficient to fully overcome our expectation of a large design error when using a complex classifier with a small data set.

Overfitting is illustrated in Fig. 13.1, in which the 3NN rule is applied to two equal-variance circular Gaussian class conditional distributions. Parts (a) and (b) show the 3NN classifier for two 30-point samples and parts (c) and (d) show the 3NN classifier for two 90-point samples. Note the greater overfitting of the data for the 30-point samples, in particular, the greater difference in the two 30-point designed classifiers as compared to the difference between the two 90-point classifiers, and the closer the latter are to the Bayes classifier given by the vertical line.

13.1.3 Constrained Classifier Design

To mitigate overfitting, we need to use constrained classification rules. Constraining classifier design means restricting the functions from which a classifier can be chosen to a class $\mathcal{C}$. This leads to trying to find an optimal *constrained* classifier, $\psi_{\mathcal{C}} \in \mathcal{C}$, having error $\varepsilon_{\mathcal{C}}$. Constraining the classifier can reduce the expected design error, but at the cost of increasing the error of the best possible classifier. Since optimization in $\mathcal{C}$ is over a subclass of classifiers, the error, $\varepsilon_{\mathcal{C}}$, of $\psi_{\mathcal{C}}$ will typically exceed the Bayes error, unless the Bayes classifier happens to be in $\mathcal{C}$. This *cost of constraint (approximation)* is

$$\Delta_{\mathcal{C}} = \varepsilon_{\mathcal{C}} - \varepsilon_d. \tag{13.6}$$

A classification rule yields a classifier $\psi_{n,\mathcal{C}} \in \mathcal{C}$ with error $\varepsilon_{n,\mathcal{C}}$, and $\varepsilon_{n,\mathcal{C}} \geq \varepsilon_{\mathcal{C}} \geq \varepsilon_d$. Design error for constrained classification is

$$\Delta_{n,\mathcal{C}} = \varepsilon_{n,\mathcal{C}} - \varepsilon_{\mathcal{C}}. \tag{13.7}$$

For small samples, this can be substantially less than Δ_n, depending on $\mathcal{C}$ and the rule. The error of the designed constrained classifier is decomposed as

$$\varepsilon_{n,\mathcal{C}} = \varepsilon_d + \Delta_{\mathcal{C}} + \Delta_{n,\mathcal{C}}. \tag{13.8}$$

The expected error of the designed classifier from $\mathcal{C}$ can be decomposed as

$$E[\varepsilon_{n,\mathcal{C}}] = \varepsilon_d + \Delta_{\mathcal{C}} + E[\Delta_{n,\mathcal{C}}]. \tag{13.9}$$

The constraint is beneficial if and only if $E[\varepsilon_{n,\mathcal{C}}] < E[\varepsilon_n]$, which means that

$$\Delta_{\mathcal{C}} < E[\Delta_n] - E[\Delta_{n,\mathcal{C}}]. \tag{13.10}$$

If the cost of constraint is less than the decrease in expected design cost, then the expected error of $\psi_{n,\mathcal{C}}$ is less than that of ψ_n. Choosing a complex classifier that can tightly fit the data reduces the constraint cost but increases the expected design error $E[\Delta_{n,\mathcal{C}}]$, whereas choosing a simple classifier that does not overfit the data increases the constraint cost but decreases the expected design error. In most microarray experiments the sample size is very small and design error is such a problem that simple classifiers are almost always preferable.

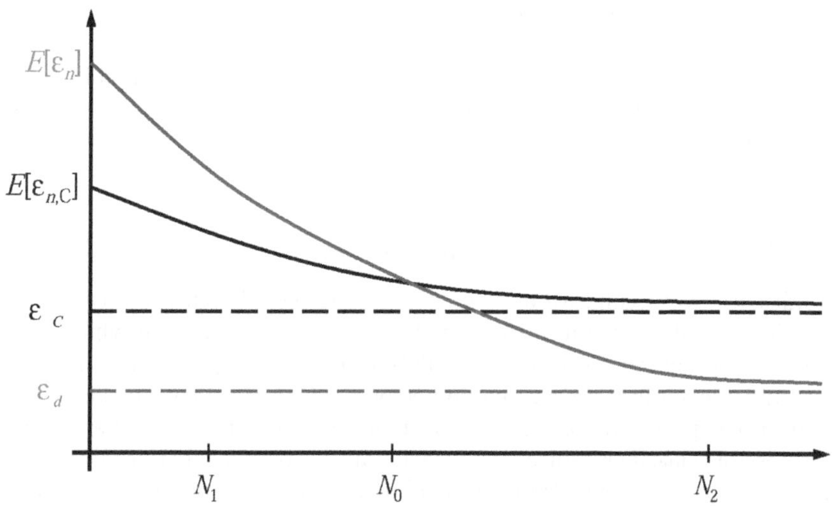

FIGURE 13.2 Errors for unconstrained and constrained classification as a function of sample size.

The matter can be graphically illustrated. For some rules, $E[\Delta_n]$ is non-increasing, meaning that $E[\Delta_{n+1}] \leq E[\Delta_n]$. This means that the expected design error never increases as sample sizes increase, and it holds for any feature-label distribution. Such classification rules are called *smart.* They fit our intuition about increasing sample sizes. The nearest-neighbor rule is not smart because there exist distributions for which $E[\Delta_{n+1}] \leq E[\Delta_n]$ does not hold for all n. Now consider a consistent rule, constraint, and distribution for which $E[\Delta_{n+1}] \leq E[\Delta_n]$ and $E[\Delta_{n+1,\mathcal{C}}] \leq E[\Delta_{n,\mathcal{C}}]$. Figure 13.2 illustrates the design problem. The axes correspond to sample size and error. The horizontal dashed lines represent $\varepsilon_{\mathcal{C}}$ and ε_d; the decreasing solid lines represent $E[\varepsilon_{n,\mathcal{C}}]$ and $E[\varepsilon_n]$. If n is sufficiently large, then $E[\varepsilon_n] < E[\varepsilon_{n,\mathcal{C}}]$; however, if n is sufficiently small, then $E[\varepsilon_n] > E[\varepsilon_{n,\mathcal{C}}]$. The point N_0 at which the decreasing lines cross is the cut-off: for $n > N_0$, the constraint is detrimental; for $n < N_0$, it is beneficial. If $n < N_0$, then the advantage of the constraint is the difference between the decreasing solid lines.

A classical example of a constrained classification rule is *quadratic discriminant analysis (QDA)*. To discuss it, let R_k denote the region in $\Re^d$ where the Bayes classifier has the value k, for $k = 0, 1$. According to Eq. 13.1, $\mathbf{x} \in R_k$ if $f_{\mathbf{X},Y}(\mathbf{x}, k) \geq f_{\mathbf{X},Y}(\mathbf{x}, j)$ for $j \neq k$. Since $f(\mathbf{x}, k) = f(\mathbf{x}|k)f(k)$, upon taking the logarithm, this is equivalent to $\mathbf{x} \in R_k$ if $d_k(\mathbf{x}) \geq d_j(\mathbf{x})$, where the *discriminant* $d_k(\mathbf{x})$ is defined by

$$d_k(\mathbf{x}) = \log f(\mathbf{x}|k) + \log f(k) \tag{13.11}$$

If the conditional densities $f(\mathbf{x}|0)$ and $f(\mathbf{x}|1)$ are normally distributed, then

$$f(\mathbf{x}|k) = \frac{1}{\sqrt{(2\pi)^n \det[\mathbf{K}_k]}} \exp\left[-\frac{1}{2}(\mathbf{x} - \mathbf{u}_k)'\mathbf{K}_k^{-1}(\mathbf{x} - \mathbf{u}_k)\right] \tag{13.12}$$

where $\mathbf{K}_k$ and $\mathbf{u}_k$ are the covariance matrix and mean vector, respectively. Dropping the constant terms and multiplying by the factor 2 (which has no effect on classification), the discriminant becomes

$$d_k(\mathbf{x}) = -(\mathbf{x} - \mathbf{u}_k)'\mathbf{K}_k^{-1}(\mathbf{x} - \mathbf{u}_k) - \log(\det[\mathbf{K}_k]) + 2\log f(k) \tag{13.13}$$

The form of this equation shows that the decision boundary $d_k(\mathbf{x}) = d_j(\mathbf{x})$ is quadratic — thus, QDA.

If both conditional densities possess the same covariance matrix $\mathbf{K}$, then $\log(\det[\mathbf{K}_k])$ can be dropped from $d_k(\mathbf{x})$ and

$$d_k(\mathbf{x}) = -(\mathbf{x} - \mathbf{u}_k)'\mathbf{K}_k^{-1}(\mathbf{x} - \mathbf{u}_k) + 2\log f(k) \tag{13.14}$$

which is a linear function of $\mathbf{x}$ and produces hyperplane decision boundaries. If we assume that the classes possess equal prior probabilities $f(k)$, then the logarithm term can be dropped. The discriminant characterizes *linear discriminant analysis (LDA)*. This classifier is of the form

$$\psi(\mathbf{x}) = \mathrm{T}\left[a_0 + \sum_{i=1}^{d} a_i x_i\right] \tag{13.15}$$

where $\mathbf{x} = (x_1, x_2, \cdots, x_d)$ and T thresholds at 0 and yields -1 or 1. It divides the space into two half-spaces determined by the hyperplane defined by the parameters $a_0, a_1, ., a_d$. The hyperplane is determined by the equation formed from setting the linear combination equal to 0.

Equations 13.13 and 13.14 for quadratic and linear discriminant analysis are derived under the Gaussian assumption, but in practice can perform well so long as the class conditional densities are approximately Gaussian — and one can obtain good estimates of the relevant covariance matrices. Owing to the greater number of parameters to be estimated for QDA as opposed to LDA, one can proceed with smaller samples with LDA than with QDA. Although QDA and LDA are derived under certain assumptions, they are often applied in situations where the assumptions are violated in order to achieve parametric quadratic or linear classifiers. In such circumstances, their performance depends on the actual feature-label distribution and the sample size. LDA appears to be more robust relative to the underlying assumptions than QDA [36].

A popular classification rule to construct linear classifiers is the *support vector machine* (*SVM*) [37]. Figure 13.3 shows a linearly separable data set and three hyperplanes (lines). The outer lines pass through points in the sample data, and the third, called the *maximal-margin hyperplane* (*MMH*) is equidistant between the outer lines. It has the property that the distance from it to the nearest −1-labeled vector is equal to the distance from it to the nearest 1-labeled vector. The vectors closest to it are called *support vectors.* The distance from the MMH to any support vector is called the *margin.* The matter is formalized by recognizing that differently labeled sets are separable by the hyperplane $\mathbf{u} \bullet \mathbf{x} = c$, where $\mathbf{u}$ is a unit vector and c is a constant, if $\mathbf{u} \bullet \mathbf{x}_k > c$ for $y_k = 1$ and $\mathbf{u} \bullet \mathbf{x}_k < c$ for $y_k = -1$. For any unit vector $\mathbf{u}$, the margin is defined by

$$\rho(\mathbf{u}) = \frac{1}{2}\left(\min_{\{\mathbf{x}_k : y_k = 1\}} \mathbf{u} \bullet \mathbf{x}_k - \max_{\{\mathbf{x}_k : y_k = -1\}} \mathbf{u} \bullet \mathbf{x}_k\right) \tag{13.16}$$

The MMH, which is unique, can be found by solving the following quadratic optimization problem: among the set of all vectors $\mathbf{v}$ for which there exists a constant b such that

$$\begin{aligned} \mathbf{v} \bullet \mathbf{x}_k + b \geq 1, &\text{ if } y_k = 1 \\ \mathbf{v} \bullet \mathbf{x}_k + b \leq -1, &\text{ if } y_k = -1 \end{aligned} \tag{13.17}$$

find the vector of minimum norm, $||\mathbf{v}||$. If $\mathbf{v}_0$ satisfies this optimization problem, then the vector defining the MMH and the margin are given by $\mathbf{u}_0 = \mathbf{v}_0/||\ \mathbf{v}_0||$ and $\rho(\mathbf{u}_0) = ||\mathbf{v}_0||^{-1}$, respectively.

If the sample is not linearly separable, then one has two choices: try to find a good linear classifier or to find a nonlinear classifier. In the first case, the preceding method can be modified [37]. A second approach is to map the sample points into a higher dimensional space, find a hyperplane in that space, and then map back into the original space.

13.1.4 Regularization for Quadratic Discriminant Analysis

Rather than design a classifier precisely according to some classification rule when the sample is small, it can be beneficial to *regularize* the sample data or the parameters estimated from the data, where by regularization we mean some alteration of the data or modification of the estimation rule for the parameters.

Relative to QDA, a simple regularization is to apply LDA even though the covariance matrices are not equal. This means estimating a single covariance

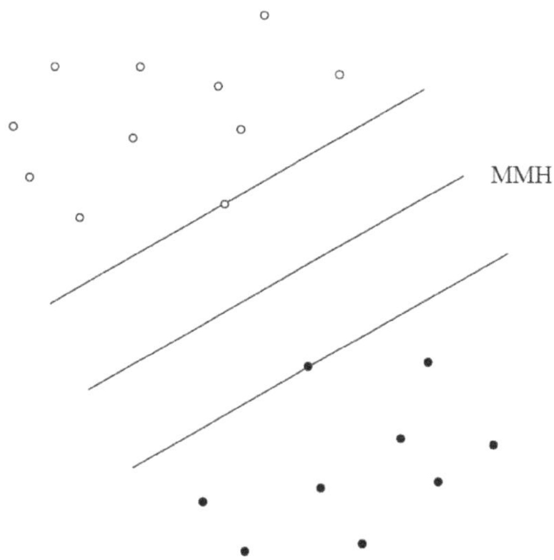

FIGURE 13.3 Linear support vector machine.

matrix by pooling the data. It reduces the number of parameters to be estimated and increases the sample size relative to the smaller set of parameters. The manner in which it can be beneficial to use LDA when QDA (and not LDA) is optimal relative to the feature-label distribution is illustrated in Fig. 13.4, where the curved dark line indicates the decision boundary of the optimal classifier relative to the feature-label distribution. In all parts of the figure, the sample is drawn from two circular Gaussian class conditional distributions, the one on the left possessing a smaller variance. The sample in parts a and b of the figure is of size 30 and we see that LDA results in less error than does QDA, even though the latter is optimal for the full class conditional distributions. With the sample being increased to 90 points in parts c and d, there is little change in the LDA classifier but the QDA decision boundary for the sample much more tightly fits the QDA decision boundary for the full distribution.

The small-sample problem for QDA can be appreciated by considering the spectral decompositions of the covariance matrices,

$$\mathbf{K}_k = \sum_{j=1}^{d} \lambda_{kj} \mathbf{v}_{kj} \mathbf{v}_{kj}^t \tag{13.18}$$

where λ_{k1}, $\lambda_{k2}, \ldots,$ λ_{kd} are the eigenvalues of $\mathbf{K}_k$ in decreasing order and $\mathbf{v}_{kj}$ is the eigenvector corresponding to λ_{kj}. Then it can be shown that the

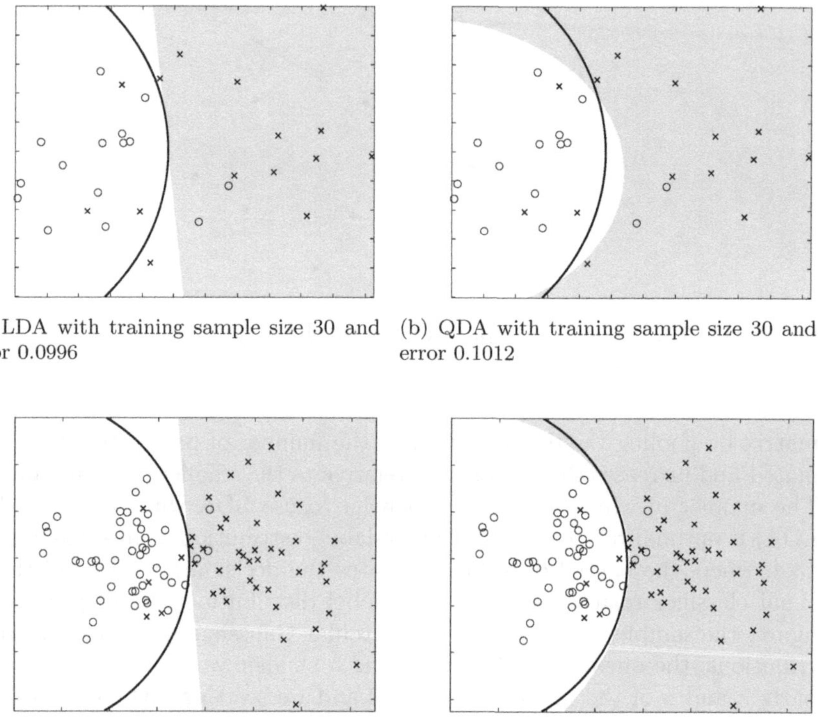

(a) LDA with training sample size 30 and error 0.0996

(b) QDA with training sample size 30 and error 0.1012

(c) LDA with training sample size 90 and error 0.1036

(d) QDA with training sample size 90 and error 0.0876

FIGURE 13.4 LDA and QDA classification applied to two unequal-variance circular Gaussian class conditional distributions: (a) LDA for a 30-point sample; (b) QDA for the same 30-point sample; (c) LDA for a 90-point sample; (d) QDA for the same 90-point sample.

quadratic discriminant takes the form

$$d_k(\mathbf{x}) = -\sum_{j=1}^{d} \frac{[\mathbf{v}_{kj}(\mathbf{x} - \mathbf{u}_k)]^2}{\lambda_{kj}} - \sum_{j=1}^{d} \log \lambda_{kj} + 2logf(k) \tag{13.19}$$

The discriminant is strongly influenced by the smallest eigenvalues. This creates a difficulty because the large eigenvalues of the sample covariance matrix are biased high and the small eigenvalues are biased low — and this phenomenon is accentuated for small samples.

A softer approach than strictly going from QDA to LDA is to shrink the individual covariance estimates in the direction of the pooled estimate. This can be accomplished by introducing a parameter α between 0 and 1 and using the estimates

$$\hat{\mathbf{K}}_k(\alpha) = \frac{n_k(1-\alpha)\hat{\mathbf{K}}_k + n\alpha\hat{\mathbf{K}}}{n_k(1-\alpha) + \alpha n} \tag{13.20}$$

where n_k is the number of points corresponding to $Y = k$, n is the total number of points, $\hat{\mathbf{K}}_k$ is the sample covariance matrix, and $\hat{\mathbf{K}}$ is the pooled estimate of the covariance matrix. QDA results from $\alpha = 0$ and LDA from $\alpha = 1$, with different amounts of shrinkage occurring for $0 < \alpha < 1$ [38]. While reducing variance, one must be prudent in choosing α, especially when the covariance matrices are very different.

To get more regularization while not overly increasing bias, one can shrink the regularized sample covariance matrix $\hat{\mathbf{K}}_k(\alpha)$ towards the identity multiplied by the average eigenvalue of $\hat{\mathbf{K}}_k(\alpha)$. This has the effect of decreasing large eigenvalues and increasing small eigenvalues, thereby offsetting the biasing effect seen in Eq. 13.19 [39]. Thus, we consider the estimate

$$\hat{\mathbf{K}}_k(\alpha, \beta) = (1-\beta)\hat{\mathbf{K}}_k(\alpha) + \frac{\beta}{n}\mathrm{tr}[\hat{\mathbf{K}}_k(\alpha)]\mathbf{I} \tag{13.21}$$

where $\mathrm{tr}[\hat{\mathbf{K}}_k(\alpha)]$ is the trace of $\hat{\mathbf{K}}_k(\alpha)$, $\mathbf{I}$ is the identity, and $0 \leq \beta \leq 1$. To apply this *regularized discriminant analysis* using $\hat{\mathbf{K}}_k(\alpha, \beta)$ requires selecting two model parameters. Model selection is critical to advantageous regularization and typically is problematic; nevertheless, simulation results for Gaussian conditional distributions indicate significant benefit of regularization for various covariance models and very little increase in error even in models where it does not appear to help.

13.1.5 Regularization by Noise Injection

A general procedure is to regularize the data itself by *noise injection* [40, 41, 42, 43]. This can be done by "spreading" the sample data by generating synthetic data about each sample point, thereby creating a large synthetic sample from which to design the classifier while at the same time making the designed classifier less dependent on the specific points in the small data set.

For instance, one may place a circular Gaussian distribution at each sample point, randomly generate points from each such distribution, and then apply a classification rule. This approach has been extensively examined relative to LDA [44]. An immediate advantage is that noise injection can be used in strictly data-driven iterative classifier designs. A spherical distribution need not be employed; indeed, it has been demonstrated that it can be advantageous to base noise injection at a sample point based on the nearest neighbors of the point [44]. Noise injection has been extensively examined relative to the classification rule and the amount of noise injected [45]. Figure 13.5 illustrates noise injection with different spreads for LDA.

Noise injection can be posed analytically in terms of matrix operations for linear classification [46]. This is important in situations where a large number of feature sets must be examined, as is often the case for microarray-based classification. We illustrate some results based on this form of regularized linear classification in the case of breast tumors from patients carrying mutations in the predisposing genes, BRCA1 or BRCA2, or from patients not expected to carry a hereditary predisposing mutation. Pathological and genetic differences appear to imply different but overlapping functions for BRCA1 and BRCA2, and in an early study involving expression-based classification, cDNA microarrays are used to show the feasibility of using differences in global gene expression profiles to separate BRCA1 and BRCA2 mutation-positive breast cancers [30]. Using data from that study, the analytic noise-injection method has been applied to derive a linear classifier to discriminate BRCA1 tumors from BRCA2 and sporadic tumors. Figure 13.6 shows the designed classifier based on genes KRT8 and DRPLA, along with its relation to the sample data pairs for these two genes. What can be inferred from this separation relative to population classification between the tumor classes depends on the method used to select the features KRT8 and DRPLA, the classification rule to construct the separating line, and the relationship between the estimated error based on the sample data and the true classifier error relative to the feature-label distribution.

13.2 Feature Selection

The Bayes error is monotone, meaning that if A and B are feature sets for which $A \subset B$, then $\varepsilon_B \leq \varepsilon_A$, where ε_A and ε_B are the Bayes errors corresponding to A and B, respectively. Thus, relative to the Bayes error, if one has a large set of potential features, it is statistically safe to use all of the features when forming a classifier. However, if $\varepsilon_{A,n}$ and $\varepsilon_{B,n}$ are the corresponding errors resulting from designed classifiers on a sample of size n, then it cannot be asserted that $E[\varepsilon_{B,n}] \leq E[\varepsilon_{A,n}]$. Monotonicity does not apply to

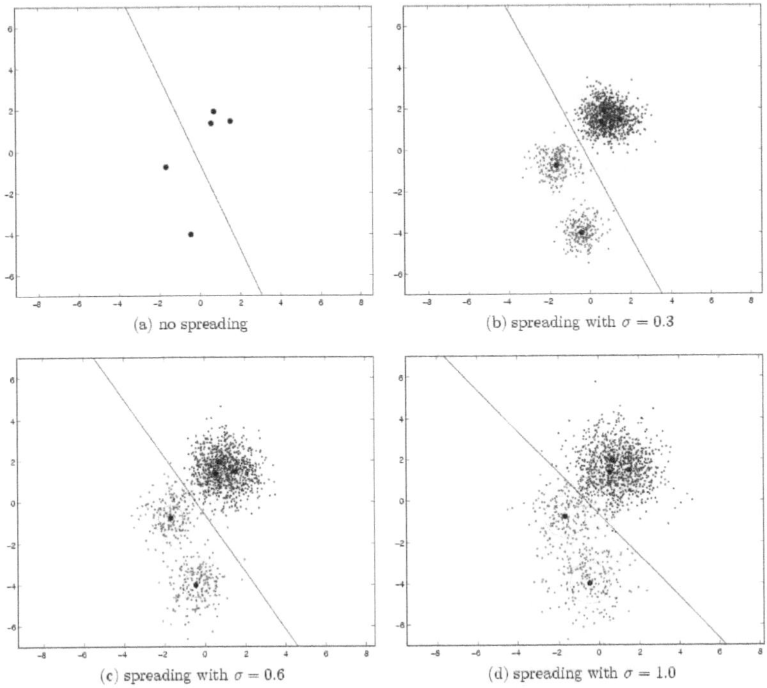

FIGURE 13.5 Noise injection for LDA with different spreads (See color figure following page 146).

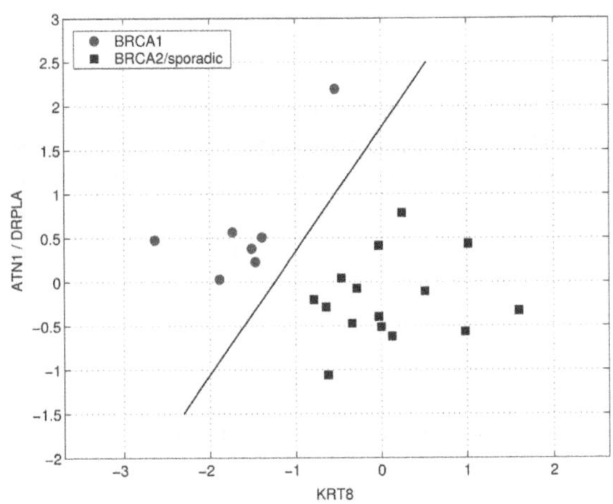

FIGURE 13.6 Linear classification of BRCA1 versus BRCA2 and sporadic patients.

the expected errors of designed classifiers. Regarding the lack of monotonicity for the errors of designed classifiers, it is commonplace for the expected error of designed classifiers to decrease and then increase for increasingly large feature sets. This is called the *peaking phenomenon* [47, 48]. It is illustrated in Fig. 13.7, where the horizontal axis corresponds to a sequence of features, x_1, $x_2, \ldots, x_d, \ldots$, and the vertical axis gives the error. For d features, the Bayes error ε_d continues to decline but the expected error, $E[\varepsilon_{d,n}]$, of the designed classifier decreases and then increases.

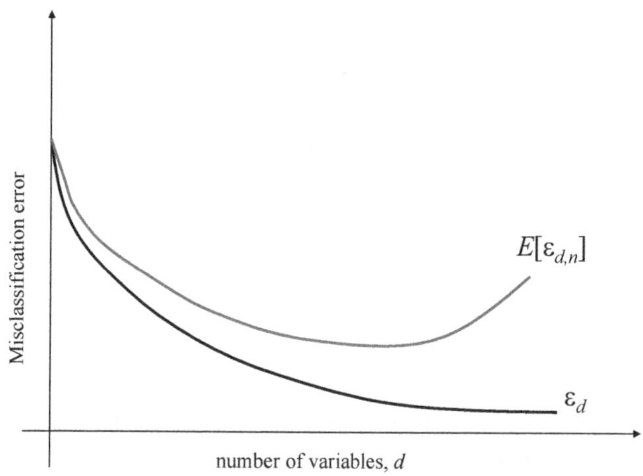

FIGURE 13.7 Peaking phenomenon.

One might hope that there is some way of avoiding checking all possible feature sets to find the best. Unfortunately, a classical result states that to be assured of finding the optimal k-element feature set from among n features, one must check all k-element feature sets, unless one has some mitigating prior distributional knowledge, which is generally not the case in practical situations [49]. A further problem is that, even were one able to check all feature sets, the need for error estimation can greatly impact finding the best one [50].

The most obvious approach to suboptimal feature selection is to consider each feature by itself and choose the k features that perform individually the best. While easy, this method is subject to choosing a feature set with a large number of redundant features, thereby obtaining a feature set that is much worse than the optimal. Moreover, features that perform poorly individually may do well in combination with other features.

A common approach to suboptimal feature selection is sequential selection, either forward or backward, and their variants. *Sequential forward selection* (*SFS*) begins with a small set of features, perhaps one, and iteratively builds the feature set. When there are k features, x_1, x_2,..., x_k, in the growing feature set, all feature sets of the form $\{x_1, x_2, \ldots, x_k, w\}$ are compared and the best one chosen to form the feature set of size $k + 1$. A problem with SFS is that there is no way to delete a feature adjoined early in the iteration that may not perform as well in combination as other features. The SFS look-back algorithm aims to mitigate this problem by allowing deletion. For it, when there are k features, x_1, x_2,..., x_k, in the growing feature set, all feature sets of the form $\{x_1, x_2, \ldots, x_k, w, z\}$ are compared and the best one chosen. Then all $(k + 1)$-element subsets are checked to allow the possibility of one of the earlier chosen features to be deleted, the result being the $k + 1$ features that will form the basis for the next stage of the algorithm. Flexibility can be added by considering *sequential forward floating selection* (*SFFS*), where the number of features to be adjoined and deleted is not fixed, but is allowed to "float" [51].

When selecting features via an algorithm like SFFS that employs error estimation within it, one should expect the choice of error estimator to impact feature selection, the degree depending on the classification rule and feature-label distribution [52].

In general, feature selection is an important and difficult problem, with many proposed algorithms but little theoretical support. For the most part, comparison of feature selection algorithms is via simulation in which algorithms are applied to particular feature-label distributions [53, 54, 55]. Even if good feature sets exists, for small samples, the likelihood of finding one whose performance is close to optimal may be small [56, 57]. This is due to the difficulty of classifier design and inadequate error estimation. The problem is exacerbated for counting-based estimators like cross-validation on account of a multitude of ties among feature sets [58].

13.3 Error Estimation

Error estimation is a key aspect of classification. If a classifier ψ_n is designed from a random sample S_n, then the error of the classifier relative to a particular sample is given by

$$\varepsilon_n = E_F[|Y - \psi_n(\mathbf{X})|] \tag{13.22}$$

where the expectation is taken relative to the feature-label distribution F (as indicated by the notation E_F). The expected error of the classifier over all samples of size n is given by

$$E[\varepsilon_n] = E_{F_n} E_F[|Y - \psi_n(\mathbf{X})|] \tag{13.23}$$

where the outer expectation is with respect to the joint distribution of the sample S_n. In practice the feature-label distribution is unknown and the expected error must be estimated.

If there is an abundance of sample data, then it can be split into *training* and *test* data. A classifier is designed on the training data, and its estimated error is the proportion of errors it makes on the test data. The problem with using both training and test data is that one would like to use all the data for design because increased data means decreased design cost. This is especially the case with small samples.

13.3.1 Error Estimation Using the Training Data

One approach is to use all sample data to design a classifier ψ_n, and estimate ε_n by applying ψ_n to the same data. The *resubstitution estimate*, $\bar{\varepsilon}_n$, is the fraction of errors made by ψ_n on the sample (training) data. Resubstitution is usually biased low, meaning $E[\bar{\varepsilon}_n] \leq E[\varepsilon_n]$, and it is often severely low-biased when samples are small. As with all error estimators, performance depends on the classification rule. An extreme case is that the resubstitution estimate is always 0 for the nearest-neighbor classifier.

A common way of performing error estimation using the same data on which the classifier is designed is to apply a re-sampling strategy. *Cross-validation* is a re-sampling strategy in which classifiers are designed from parts of the sample, each is tested on the remaining data, and ε_n is estimated by averaging the errors. In *k-fold cross-validation*, the sample S_n is partitioned into k folds $S_{(i)}$, for i= 1, 2,..., k. Each fold is left out of the design process and used as a test set, and the estimate, $\varepsilon_n^{cv(k)}$, is the average error committed on all folds. A k-fold cross-validation estimator is unbiased as an estimator of $E[\varepsilon_{n-n/k}]$, meaning $E[\varepsilon_n^{cv(k)}] = E[\varepsilon_{n-n/k}]$. The special case of n-fold cross-validation yields the *leave-one-out estimator*, $\hat{\varepsilon}_n^{loo}$, which is an unbiased estimator of $E[\varepsilon_{n-1}]$.

While not suffering from severe low bias like resubstitution, cross-validation has large variance in small-sample settings and therefore its use is problematic [59]. Focusing on the unbiasedness of leave-one-out estimation, $E[\hat{\varepsilon}_n^{loo}] = E[\varepsilon_{n-1}]$, so that $E[\hat{\varepsilon}_n^{loo} - \varepsilon_n] \approx 0$. Thus, the expected difference between the error estimator and the error is approximately 0. But we are not interested in the expected difference between the error estimator and the error; rather, we are interested in the precision of the error estimator in estimating the error. Our concern is the expected deviation, $E[|\hat{\varepsilon}_n^{loo} - \varepsilon_n|]$, and unless the cross-validation variance is small, which it is not for small samples, this expected deviation will not be small.

Bootstrap is a general re-sampling strategy that can be applied to error estimation [60]. A *bootstrap sample* consists of n equally-likely draws with replacement from the original sample S_n. Some points may appear multiple times, whereas others may not appear at all. For the basic bootstrap estimator, $\hat{\varepsilon}_n^b$, the classifier is designed on the bootstrap sample and tested on the points left out, this is done repeatedly, and the bootstrap estimate is the average error made on the left-out points. $\hat{\varepsilon}_n^b$ tends to be a high-biased estimator of $E[\varepsilon_n]$, since the number of points available for design is on average only $0.632n$. The *.632 bootstrap estimator* tries to correct this bias via a weighted average of $\hat{\varepsilon}_n^b$ and resubstitution [61],

$$\hat{\varepsilon}_n^{b632} = 0.368\varepsilon_n^{res} + 0.632\hat{\varepsilon}_n^b \qquad (13.24)$$

In resubstitution there is no distinction between points near and far from the decision boundary; the *bolstered-resubstitution* estimator is based on the heuristic that, relative to making an error, more confidence should be attributed to points far from the decision boundary than points near it [62]. This is achieved by placing a distribution, called a *bolstering kernel*, at each point and estimating the error by integrating the bolstering kernel for each point over the decision region that the point should not belong to and then summing the integrals (rather than simply counting the misclassified points as in resubstitution). The procedure is illustrated in Fig. 13.8 for a linear classifier. A key issue is the amount of bolstering (spread of the bolstering kernels). Since the purpose of bolstering is to improve error estimation in the small-sample setting, we need to be cautious about using bolstering kernels that require complicated inferences. Hence, zero-mean, spherical bolstering kernels with covariance matrices of the form $\sigma_i\mathbf{I}$ are commonly employed. The choice of the parameters σ_1, σ_2,..., σ_n determines the variance and bias properties of the corresponding bolstered estimator. If $\sigma_1 = \sigma_2 = \cdots = \sigma_n = 0$, then there is no bolstering and the bolstered estimator reduces to the original estimator. As a general rule, larger σ_i lead to lower-variance estimators, but after a certain point this advantage is offset by increasing bias. A method has been proposed to compute the amount of bolstering based on the data [62]. In situations where resubstitution is very low-biased, it may not be a good idea to spread the incorrectly classified points. This approach yields the *semi-bolstered resubstitution* estimator.

13.3.2 Performance Issues

Large simulation studies involving the empirical distribution of $\hat{\varepsilon}_n - \varepsilon_n$ have been conducted to examine small-sample error-estimator performance [63, 64]. In the overall set of experiments, resubstitution, leave-one-out, and even 10-fold cross-validation are generally outperformed by the bootstrap and bolstered estimators. Bolstered resubstitution is very competitive with the bootstrap, in some cases beating it. For LDA, the best estimator overall is the bolstered resubstitution. An exception for the .632 bootstrap is CART (Classification and Regression Trees), where the bootstrap estimator is affected by the extreme low-biased behavior of CART resubstitution. In this case, bolstered resubstitution performs quite well, but the best overall estimator is

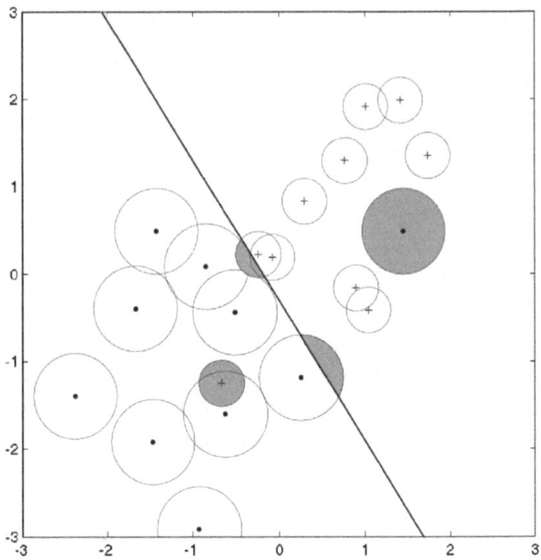

FIGURE 13.8 Bolstering error for linear classification.

semi-bolstered resubstitution. Increasing the sample size improves the performance of all the estimators considerably, but the general relative trends still hold, with few exceptions. When considering these kinds of results, it must be kept in mind that specific performance advantages depend on the classification rule.

To highlight the need for caution when comparing error estimators, consider the oft held opinion that cross-validation is "always" superior to resubstitution. In fact, this need not be true. In the case of discrete classification in which a point is classified according to the majority label observed for the point in the sample data, there exist exact analytic formulations for the mean-square error $E[|\hat{\varepsilon}_n - \varepsilon_n|^2]$ for both resubstitution and leave-one-out cross-validation, and these exact formulations are used to demonstrate the better performance of resubstitution when the number of points to be classified is small [65].

A key issue for error estimation is its effect on choosing feature sets. For instance, when choosing among a collection of potential feature sets for classification, it is natural to order the potential feature sets according to the error rates of their corresponding classifiers. Hence, it is important to apply error estimators that provide rankings that better correspond to rankings produced by the true errors. This is especially important in small-sample settings, where all feature-selection algorithms are subject to significant errors and feature selection may be viewed as finding a list of potential feature sets, and not as trying to find a best feature set. For instance, in phenotype classification based on gene expression, feature selection can be viewed as gene selection, finding sets of genes whose expressions can be used for phenotypic discrimination; indeed, gene selection in the context of small samples can be viewed as an exploratory methodology. Regarding feature-set ranking, in an extensive simulation study, it has been observed that .632 bootstrap tends to significantly outperform cross-validation methodologies and bolstering generally outperforms .632 bootstrap [50].

A related issue to feature-set ranking is the manner in which error estimation affects selecting features via an algorithm like SFFS that employs error estimation within it. In a large simulation study, it has been seen that the effect of error estimation is enormous in the case of small samples [52]. Indeed, the choice of error estimator can make a greater difference than the manner of feature selection. For instance, in that study, for LDA an exhaustive search using leave-one-out resulted in average true error 0.2224, whereas SFFS using bolstered resubstitution yielded an average true error of only 0.1918. SFFS using semi-bolstered resubstitution (0.2016) or bootstrap (0.2129) was also superior to exhaustive search using leave-one-out, although not as good as bolstered resubstitution. In the case of 3NN, once again SFFS with either bolstered resubstitution, semi-bolstered resubstitution, or bootstrap outperformed a full search using leave-one out. We reiterate that these are small-sample results and relate to particular error estimators and feature-selection algorithms; nevertheless, they demonstrate the salient effect of error estimation on feature selection.

Computation time can be critical for error estimation, in particular, when examining tens of thousands of potential feature sets. Resubstitution is the fastest estimator. Leave-one-out is fast for small number of samples, but its performance quickly degrades with increasing number of samples. The 10-fold cross-validation and the bootstrap estimator are the slowest estimators. Bolstered resubstitution can be hundreds of times faster than the bootstrap estimator.

14

Clustering

A cluster operator takes a set of data points and partitions the points into clusters (subsets). Clustering has become a popular data-analysis technique in genomic studies using gene-expression microarrays. Time-series clustering groups together genes whose expression levels exhibit similar behavior through time. Similarity indicates possible co-regulation. Another way to use expression data is to take expression profiles over various tissue samples, and then cluster these samples based on the expression levels for each sample. This approach offers the potential to discriminate pathologies based on their differential patterns of gene expression.

Classification exhibits two fundamental characteristics: (1) classifier error can be estimated under the assumption that the sample data arise from an underlying feature-label distribution; and (2) given a family of classifiers, sample data can be used to learn the optimal classifier in the family. Once designed, the classifier represents a mathematical model that provides a decision mechanism relative to real-world measurements. The model represents scientific knowledge to the extent that it has predictive capability [66]. The purpose of testing (error estimation) is to quantify the worth of the model. Clustering has historically lacked both fundamental characteristics of classification. Many validation techniques have been proposed for evaluating clustering results. These are generally based on the degree to which clusters derived from a set of sample data satisfy certain heuristic criteria. This is significantly different than classification, where the error of a classifier is given by the probability of an erroneous decision. The problem is illustrated in Fig. 14.1. Both cluster results appear "reasonable," but the performance of a clustering algorithm cannot be measured by observing the results of a single application; rather, as with classification, its performance must be measured relative to predictive results on a distribution.

For clustering, error estimation must assume that clusters resulting from a cluster algorithm can be compared to the correct clusters for the data set in the context of a probability distribution, thereby providing an error measure. The key to a general probabilistic theory of clustering, including both error estimation and learning, is to recognize that classification theory is based on operators on random variables, and that the theory of clustering needs to be based on operators on random sets [67]. While we will not discuss a general probabilistic theory of clustering, we will examine clustering algorithms relative to the correctness of their results in the context of model-based random

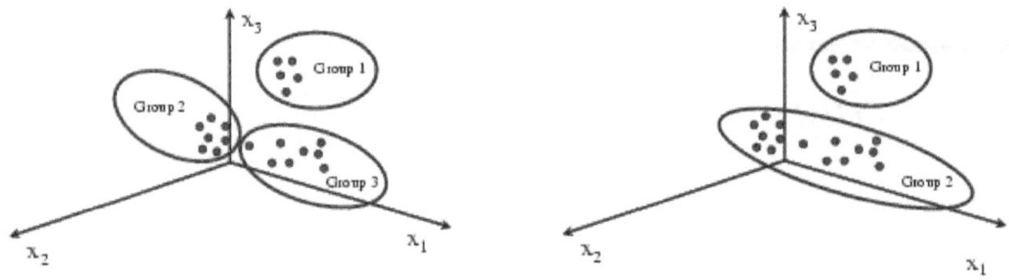

FIGURE 14.1 Two "reasonable" cluster results.

points.

14.1 Examples of Clustering Algorithms

A host of clustering algorithms has been proposed in the literature. We will discuss a few standard algorithms.

14.1.1 *k*-means

If we envision clusters formed by points $\mathbf{x}$ generated by a random sample S from a mixture of m circular Gaussian conditional distributions, then the points $\mathbf{a}_0^S, \mathbf{a}_1^S, \cdots, \mathbf{a}_{m-1}^S$ that minimize

$$\rho_S(\mathbf{a}_0, \mathbf{a}_1, \cdots, \mathbf{a}_{m-1}) = \frac{1}{|S|} \sum_{\mathbf{x} \in S} \min_{0 \leq j \leq m-1} \|\mathbf{x} - \mathbf{a}_j\|^2 \tag{14.1}$$

are a reasonable choice for the centroids of the m samples arising from the conditional distributions. Let $\mathcal{V} = \{V_1, V_2, \cdots, V_m\}$ be the Voronoi partition of $\Re^d$ induced by $\mathbf{a}_0^S, \mathbf{a}_1^S, \cdots, \mathbf{a}_{m-1}^S$: a point lies in V_k if its distance to $\mathbf{a}_k^S$ is no more than its distance to any other of the points $\mathbf{a}_0^S, \mathbf{a}_1^S, \cdots, \mathbf{a}_{m-1}^S$. For *Euclidean-distance clustering*, the sample points are clustered according to how they fall into the Voronoi partition.

Direct implementation of Euclidean-distance clustering is computationally prohibitive. A classical iterative approximation is given by the *k-means algorithm*, where k refers to the number of clusters provided by the algorithm. Each sample point is placed into a unique cluster during each iteration and the means are updated based on the classified samples. Given a sample S with n points to be placed into k clusters, initialize the algorithm with k means, $\mathbf{m}_1, \mathbf{m}_2, \cdots, \mathbf{m}_k$, among the points; for each point $\mathbf{x} \in S$, calculate the distance $\|\mathbf{x} - \mathbf{m}_i\|$, for $i = 1, 2, \cdots, k$; form clusters $C_1, C_1, \cdots, C_k$ by placing $\mathbf{x}$

into C_i if $\|\mathbf{x} - \mathbf{m}_i\| \leq \|\mathbf{x} - \mathbf{m}_j\|$ for $j = 1, 2, \cdots, k$; update $\mathbf{m}_1, \mathbf{m}_2, \cdots, \mathbf{m}_k$ as the means of $C_1, C_1, \cdots, C_k$, respectively; and repeat until the means do not change. At each stage of the algorithm, the clusters are determined by the Voronoi diagram associated with $\mathbf{m}_1, \mathbf{m}_2, \cdots, \mathbf{m}_k$. Two evident problems with the k-means algorithm are the prior assumption on the number of means and the choice of means to seed the algorithm.

14.1.2 Fuzzy k-means

Equation 14.1 can be rewritten in the following form

$$\rho_S(\mathbf{a}_0, \mathbf{a}_1, \cdots, \mathbf{a}_{m-1}) = \frac{1}{|S|} \sum_{i=1}^{n} \sum_{j=0}^{m-1} P(C_j|\mathbf{x}_i)^b \|\mathbf{x}_i - \mathbf{a}_j\|^2 \tag{14.2}$$

where $P(C_j|\mathbf{x}_i)$ is the probability that $\mathbf{x}_i \in C_j$, which is either 0 or 1, and is 1 only for the minimizing j of Eq. 14.1. A fuzzy approach results from letting the conditional probabilities reflect uncertainty, so that cluster inclusion is not crisp, and letting $b > 0$ be a parameter affecting the degree to which a point can belong to more than a single cluster. The conditional probabilities are constrained by the requirement that their sum is 1 for any fixed $\mathbf{x}_i$,

$$\sum_{j=0}^{m-1} P(C_j|\mathbf{x}_i) = 1 \tag{14.3}$$

Let p_j denote the prior probability of C_j. Since the conditional probabilities $P(C_j|\mathbf{x}_i)$ are not estimable and are heuristically set, we view them as fuzzy membership functions. In this case, for the minimizing values of $\rho_S(\mathbf{a}_0, \mathbf{a}_1, \cdots, \mathbf{a}_{m-1})$, the partial derivatives with respect to $\mathbf{a}_j$ and p_i satisfy $\partial\rho_S/\partial a_j = 0$ and $\partial\rho_S/\partial p_j = 0$. These partial-derivative identities yield

$$\mathbf{m}_j = \frac{\sum_{i=1}^{n} P(C_j|\mathbf{x}_i)^b \mathbf{x}_i}{\sum_{i=1}^{n} P(C_j|\mathbf{x}_i)^b} \tag{14.4}$$

$$P(C_j|\mathbf{x}_i) = \frac{\|\mathbf{x}_i - \mathbf{m}_j\|^{-1/(b-1)}}{\sum_{l=1}^{k} \|\mathbf{x}_i - \mathbf{m}_l\|^{-1/(b-1)}} \tag{14.5}$$

These lead to the *fuzzy k-means* iterative algorithm. Initialize the algorithm with b, k means $\mathbf{m}_1, \mathbf{m}_2, \cdots, \mathbf{m}_k$, and the membership functions $P(C_j|\mathbf{x}_i)$ for $j = 1, 2, \cdots, k$ and $i = 1, 2, \cdots, n$, where the membership functions must be normalized so that their sum is 1 for any fixed $\mathbf{x}_i$; re-compute $\mathbf{m}_j$ and $P(C_j|\mathbf{x}_i)$ by Eqs. 14.4 and 14.5; and repeat until there are only small pre-specified changes in the means and membership functions. The intent of fuzzifying the k-means algorithm is to keep the means from getting "stuck" during the iterative procedure.

14.1.3 Self-Organizing Maps

Self-organizing maps provide a different extension of the k-means concept [68, 69]. The idea is to map high-dimensional vectors in Euclidean space to a low-dimensional grid in a neighborhood-preserving manner, whereby we mean that vectors that are close in the high-dimensional space have close representations in the low-dimensional grid. To describe self-organizing maps, we begin with an ordered lattice, typically a one- or two-dimensional grid, that we index by $I = \{1, 2, \cdots, k\}$. Associated with I is a *neighborhood function* η_t, defined on $I \times I$ and parameterized by $t = 0, 1, \cdots$, satisfying three properties: (1) η_t depends only on the distance between points in I, meaning that $\eta_t(i, j) = \eta_t(\|i - j\|)$; (2) η is non-increasing relative to distance, meaning $\eta_t(\|i-j\|) \leq \eta_t(\|u-v\|)$ if $\|u-v\| \leq \|i-j\|$; and (3) the domain of η is non-increasing relative to t, meaning that for each t there exists a nonnegative integer parameter $\alpha(t)$ such that $\eta_t(\|i-j\|) = 0$ if $\|i-j\| > \alpha(t)$ and $\alpha(t+1) \leq \alpha(t)$. The properties of the neighborhood function have been chosen so that points nearby a chosen point in I will be updated in conjunction with the chosen point, with less adjustment for points further away from the chosen point, and this update neighborhood will decrease in size as the algorithm proceeds through its iterations, usually with the neighborhood reduced to a single point $[\alpha(t) = 0]$ during the last steps of the algorithm. It is assumed that the input vectors to the algorithm are derived from a random sample and they lie in some bounded convex set in $\Re^d$.

Each iteration of a self-organizing-map algorithm is characterized by a vector-valued state vector

$$\mathbf{m}(t) = (\mathbf{m}_1(t), \mathbf{m}_2(t), \cdots, \mathbf{m}_k(t)) \tag{14.6}$$

where $\mathbf{m}_i(t) \in \Re^d$ for $i = 1, 2, \cdots, k$. The algorithm is initialized at $\mathbf{m}(0)$. The algorithm proceeds in the following recursive fashion. Given an input vector $\mathbf{x}$ at time t, the index, ι_t, of the component state closest to $\mathbf{x}$ is selected, namely,

$$\iota_t = \arg\min_{i \in I} \|\mathbf{x} - \mathbf{m}_i(t)\| \tag{14.7}$$

The vector $\mathbf{m}(t)$ is then updated according to

$$\mathbf{m}_i(t+1) = \mathbf{m}_i(t) + \beta_t \eta_t(\iota_t, i)[\mathbf{m}_i(t) - \mathbf{x}] \tag{14.8}$$

for $i = 1, 2, \cdots, k$, where $\beta_t > 0$ is a parameter that can be lowered over time to lessen the adjustment of $\mathbf{m}(t)$. A critical role is played by the neighborhood function in achieving the preservation of neighborhood relations. The algorithm is very popular and used for various purposes; however, many theoretical questions remain regarding the organizational nature of the algorithm and convergence [70]. Clustering is achieved in the same manner as the k-means algorithm, with the clusters determined by the Voronoi diagram associated with $\mathbf{m}_1(t), \mathbf{m}_2(t), \cdots, \mathbf{m}_k(t)$ [71].

14.1.4 Hierarchical Clustering

Both k-means and fuzzy k-means are based on Euclidean-distance clustering. Another approach (albeit, one also often related to Euclidean distance) is to iteratively join clusters according to a similarity measure. The general *hierarchical clustering* algorithm is given by the following procedure: initialize the clusters by $C_i = \{\mathbf{x}_i\}$ for $i = 1, 2, \cdots, n$ and by a desired final number k of clusters; and then proceed to iteratively merge the nearest clusters according to the similarity measure until there are only k clusters. An alternative is to continue to merge until the similarity measure satisfies some criterion. The merging process can be pictorially represented in the form of a dendrogram, where joined arcs represent merging. As stated, the hierarchical clustering is *agglomerative*, in the sense that points are agglomerated into growing clusters. One can also consider *divisive* clustering, in which, beginning with a single cluster, the algorithm proceeds to iteratively split clusters. Various similarity measures have been proposed. Three popular ones are the minimum, maximum, and average measures given by

$$d_{min}(C_i, C_j) = \min_{\mathbf{x} \in C_i, \mathbf{x}' \in C_j} \|\mathbf{x} - \mathbf{x}'\| \tag{14.9}$$

$$d_{max}(C_i, C_j) = \max_{\mathbf{x} \in C_i, \mathbf{x}' \in C_j} \|\mathbf{x} - \mathbf{x}'\| \tag{14.10}$$

$$d_{av}(C_i, C_j) = \frac{1}{|C_i| \cdot |C_j|} \sum_{\mathbf{x} \in C_i, \mathbf{x}' \in C_j} \|\mathbf{x} - \mathbf{x}'\|. \tag{14.11}$$

Hierarchical clustering using the minimum distance is called *nearest-neighbor* clustering. If it halts when the distance between nearest clusters exceeds a pre-specified threshold, then it is called *single-linkage clustering*. Given a set of clusters at any stage of the algorithm, it merges the clusters possessing the nearest points. If we view the points as the nodes of a graph, then when two clusters are merged, an edge is placed between the nearest nodes in the two clusters. Hence, there are no closed loops created and the resulting graph is a tree. If the algorithm is not stopped until there is a single cluster, the result is a spanning tree, and the spanning tree is minimal. While the algorithm is intuitively pleasing owing to the manner in which it generates a hierarchical sub-cluster structure based on nearest neighbors, it is extremely sensitive to noise and can produce strange results, such as elongated clusters. It is also very sensitive to early mergings, since once joined, points cannot be separated.

Farthest-neighbor clustering results from using the maximum distance. If it halts when the distance between nearest clusters exceeds a pre-specified threshold, then it is called *complete-linkage clustering*. Given a set of clusters at any stage of the algorithm, it merges the clusters for which the greatest distance between points in the two clusters is minimized. This approach counteracts the tendency toward elongation from which nearest-neighbor clustering suffers. Finally, if the algorithm halts when the average distance between nearest clusters exceeds a pre-specified threshold, then it is called *average-linkage clustering*.

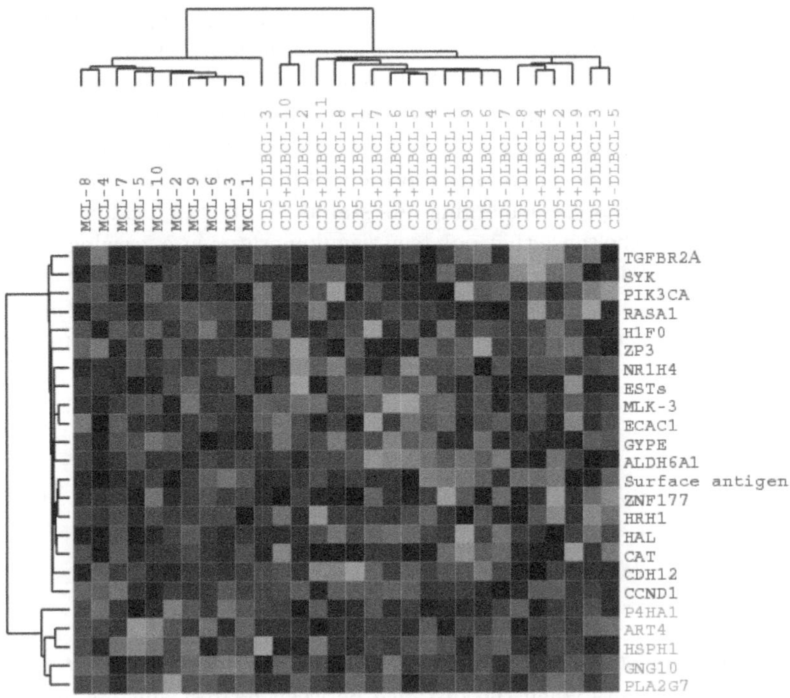

FIGURE 14.2 Hierarchical clustering for two types of lymphoma, DLBCL and MCL (See color figure following page 146).

Figure 14.2 shows a sample of 30 patients (columns) suffering from B-cell lymphoma and the expression profiles (rows) of 24 genes across the sample. The samples have been hierarchically clustered and the clusters correspond perfectly to two types of lymphoma, DLBCL and MCL [72]. The genes have also been hierarchically clustered and it appears from the figure that the red-labeled genes are up-regulated for MCL and down-regulated for DLBCL, whereas the green-labeled genes are up-regulated for DLBCL and down-regulated for MCL. Thus, the different gene clusters seem to "classify" the lymphomas. Perhaps more precisely, the clustering might provide feature selection in the sense that actual classification might be accomplished by a two-gene feature set, one red-labeled and one green-labeled.

14.2 Clustering Accuracy

A clustering algorithm operates on the point set as a whole by partitioning it. Its error is based on the accuracy of the partition. Essentially, if we assume the points come from different distributions, label them accordingly, and match the results from the clustering algorithm with the subsets of the partition in the best way possible, then the error is the number of mismatches. This idea can be formalized in the probabilistic theory of random sets [67].

14.2.1 Model-Based Clustering Error

Here we will not discuss the general theory, but will proceed with a model-based approach that is in concordance with the general theory [73]. Since our interest is genomics we place the discussion in the context of gene expression and we assume that each gene is sampled across time; however, there is nothing in the approach that requires gene-expression measurements, nor samples based on time. We assume that a congruency class is composed of a set of genes possessing the same mean, and that the expression profile of any gene is given by its mean plus a zero-mean random displacement that incorporates both the biologically inherent random variation of an expression together with experimental noise. Although the expression profiles of genes possessing identical means will differ upon observation, a clustering algorithm should place them in the same cluster.

For the model, there are m deterministic functions (*templates*) of the time points (sample indices), 1, 2,..., n. Each template corresponds to a mean function and is defined by an n-vector. There are m template vectors $\mathbf{u}_1$, $\mathbf{u}_2, \ldots, \mathbf{u}_m$, each of the form $\mathbf{u}_k = (u_{k1}, u_{k2}, \ldots, u_{kn})$, where u_{kj} is the value of the kth template at time j. Congruency classes are defined by randomizations of $\mathbf{u}_1, \mathbf{u}_2, \ldots, \mathbf{u}_m$. For $k = 1, 2, \ldots, m$ and $j = 1, 2, \ldots, n$, let N_{kj} be a random variable possessing a Gaussian distribution with mean 0 and variance σ_{kj}^2. Assume that the collection $\{N_{kj}\}$ is probabilistically independent. Letting $\mathbf{N}_k = (N_{k1}, N_{k2}, \ldots, N_{kn})$, the kth *congruency class*, U_k, is defined by the random vector

$$\mathbf{X}_k = \begin{pmatrix} X_{k1} \\ X_{k2} \\ \vdots \\ X_{kn} \end{pmatrix} = \begin{pmatrix} u_{k1} + N_{k1} \\ u_{k2} + N_{k2} \\ \vdots \\ u_{kn} + N_{kn} \end{pmatrix} = \mathbf{u}_k + \mathbf{N}_k \tag{14.12}$$

$\mathbf{X}_k$ is Gaussian, and has mean vector $\mathbf{u}_k$ and variance vector $\boldsymbol{\sigma}_k^2 = (\sigma_{k1}^2, \sigma_{k2}^2, \ldots, \sigma_{kn}^2)$. $\mathbf{X}_k$ and $\mathbf{X}_j$ are uncorrelated if $k \neq j$.

A random vector belongs to congruency class U_k if it is identically distributed to $\mathbf{X}_k$. In the present application, each such vector corresponds to an expression time series. Hence, we say that a gene is in congruency class U_k if its expression profile is modeled by $\mathbf{X}_k$. If there are T genes altogether, then there are m congruency classes $U_1, U_2, \ldots, U_m$ with r_k genes in class U_k, $T = r_1 + r_2 + \cdots + r_m$.

A single experiment produces a random sample of size r_k for each congruency class U_k. Each is a sample of $\mathbf{X}_k$. Because there are n time points, these correspond to r_k points $\mathbf{x}_k^1, \mathbf{x}_k^2, \ldots, \mathbf{x}_k^{r_k}$ in n-dimensional space. Each sample yields T points. The statistical model for the sampling is that there are r_k random vectors $\mathbf{X}_k^1, \mathbf{X}_k^2, \ldots, \mathbf{X}_k^{r_k}$ identically distributed to $\mathbf{X}_k$. A single sample produces the deterministic points $\mathbf{x}_k^1, \mathbf{x}_k^2, \ldots, \mathbf{x}_k^{r_k}$. A clustering algorithm is run on the T points $\mathbf{x}_1^1, \ldots \mathbf{x}_1^2, \ldots, \mathbf{x}_1^{r_1}, \mathbf{x}_2^1, \ldots, \mathbf{x}_m^{r_m}$. We assume the number of clusters is known beforehand, and therefore the algorithm is preset to have m clusters $C_1, C_2, \ldots, C_m$. To analyze clustering precision, we assign clusters to congruency classes. Cluster C_j is assigned to a congruency class by voting: C_j is assigned to congruency class U_k if the number of genes in C_j from U_k exceeds the number from any other congruency class. In case of ties, the congruency class is chosen randomly. If there are many misclassifications, then it is possible under this assignment scheme that different clusters may be assigned to the same congruency class, and some convention has to be employed to handle this situation. The number of misclassifications is the number of sample points assigned to the wrong congruency class. This number depends on the sample. The misclassification error, ρ_n, is the number of misclassifications divided by the number of sample points. We are mainly interested in the expected misclassification error, $E[\rho_n]$, as a measure of clustering precision.

14.2.2 Application to Real Data

The difficulty with application to real data, as opposed to synthetic data, is that the methodology requires the means and variances for the congruency classes, and the raw data does not include congruency classes. We proceed by applying a seed clustering algorithm to the raw data to form congruency classes with which to seed the algorithm. For instance, we might seed the model by applying SOM to form *seed* congruency classes and then apply clustering algorithms to the model based on those classes.

To be precise, suppose there are q clustering algorithms $A_1, A_2, \ldots, A_q$, and m congruency classes. A selected clustering algorithm A_k is used to form m clusters, which are then identified as seed congruency classes, $U_{k1}, U_{k2}, \ldots, U_{km}$. The model is seeded by computing the means and variances from $U_{k1}, U_{k2}, \ldots, U_{km}$. The simulation can then be run using the various clustering algorithms $A_1, A_2, \ldots, A_q$. This will produce expected error rates of $R_{k1}, R_{k2}, \ldots, R_{kq}$ corresponding to $A_1, A_2, \ldots, A_q$, respectively. These error rates correspond to seeding by algorithm A_k, and are dependent on this seeding.

Altogether, the error rates form an error matrix

$$\mathbf{R} = \begin{pmatrix} R_{11} & R_{12} & \cdots & R_{1q} \\ R_{21} & R_{22} & \cdots & R_{2q} \\ \vdots & \vdots & \ddots & \vdots \\ R_{q1} & R_{q2} & \cdots & R_{qq} \end{pmatrix} \tag{14.13}$$

The entry R_{kj} gives the error rate for algorithm A_j under model seeding by A_k.

It seems intuitive that the seeding algorithm should be favored when the clustering algorithms are applied to the model. Under fuzzy k-means seeding one might think that fuzzy k-means will outperform k-means. While this initialization advantage is often the case, it may not be if the seed algorithm has poor inference capability. To obtain results not so dependent on seeding, we can compute the various error rates for each seed and average the algorithm's performance over all seeds to obtain the global error rates $R_{\bullet 1}$, $R_{\bullet 2}, \ldots, R_{\bullet q}$, where

$$R_{\bullet j} = \frac{1}{q} \sum_{k=1}^{q} R_{kj} \tag{14.14}$$

is the average performance of algorithm A_j over all seeds. A slight modification occurs if one does not wish to average over all seeds, but only over a sub-collection of seeds. For instance, owing to its generally poor performance, one might not wish to include seeding by k-means. In this case the error rate corresponding to seeding by k-means is omitted from the average.

There are some considerations concerning variances for seed congruency classes. Consider the seed congruency class U_{ki}, the ith congruency class for the kth seeding algorithm. For the n time points, there are n means. Each of these is formed by the sample mean of the values at a time point of the profiles within the congruency class. Variances can similarly be formed from the sample variances at each time point. Alternatively, if some of the congruency classes are small, one can form the pooled variance over all time points, in which case, $\boldsymbol{\sigma}_k^2 = (\sigma_k^2, \sigma_k^2, \ldots, \sigma_k^2)$. This results in all time points having the same model variance, but it avoids poor variance estimates for small classes.

To illustrate application, we use data published in [74] from an experiment to see the response of human fibroblasts to serum. Figure 14.3 shows the five seed means generated from the data using SOM as a seed and Fig. 14.4 shows the corresponding two-dimensional principle-component plots of the synthetic data generated by the model. The first column of Fig. 14.5 provides a popular visualization of the time-series ratio data obtained by listing the genes vertically and time points horizontally, and using discrete pseudo-colored squares to indicate the ratio. Green indicates a ratio R/G less than 1, red a ratio greater than 1, and increasing color intensity reflects the degree to which the ratio is displaced from 1 (red labels for positive values of log R/G, and green labels for negatives values). The first (wide) column shows the expressions for the full gene set, in the initial order, with the true class of each gene indicated by a color in the associated tiny column. The second column shows the results of fuzzy k-means clustering and each block of the third column shows the mean of the congruency class to which a cluster has been assigned. The tiny column associated with the second column shows the true class of the gene in the cluster, and mismatches between this tiny column and the tiny column next to the third column indicate errors. Although the visualization looks like there has been pretty good clustering, there are 172 errors for an error rate of 16.6%. The results of the second column appear decent to the eye, but such visualizations can be misleading because the human visual system performs outstanding image restoration.

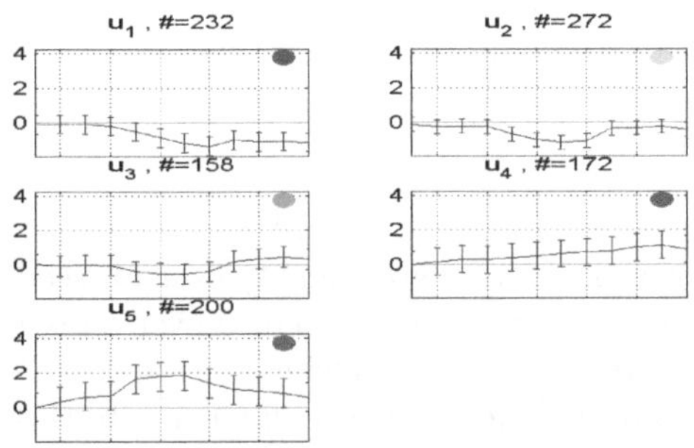

FIGURE 14.3 Seed means with standard-deviation bars for the five congruency classes (See color figure following page 146).

14.3 Cluster Validation

The error of a cluster operator can be defined in the context of random point sets in a manner analogous to classification error. One might loosely define a "valid" cluster operator as one possessing small error — or even more loosely as one that produces "good" clusters. Of course, such a definition is vacuous unless it is supported by a definition of goodness. What if one tries to evaluate a cluster operator in the absence of a probability distribution (labeled point process)? Then measures of validity (goodness) need to be defined. They will not apply to the cluster operator as an operator on random sets, but will depend on heuristic criteria. Their relation to future application of the operator will not be understood, for the whole notion of prediction depends

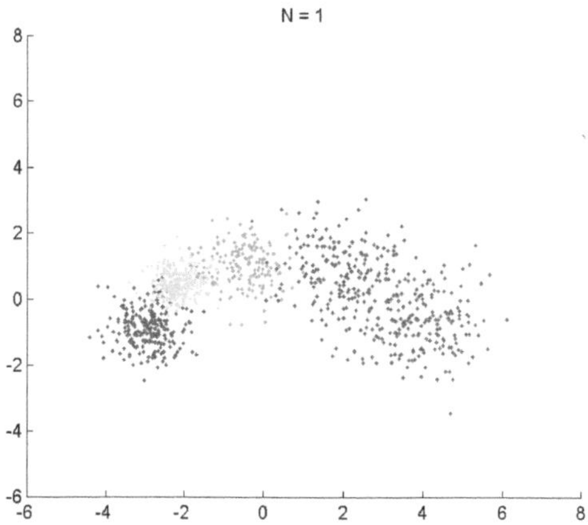

FIGURE 14.4 Two-dimensional principle-component plots of the generated data (See color figure following page 146).

on the existence of a random process. Nevertheless, a heuristic criterion can serve to give some indication of how the cluster operator is performing on the data at hand relative to some criterion. We will consider some methods that address the validity of a clustering output, or compare clustering outputs, based on heuristic criteria. These can be roughly divided into two categories. *Internal validation* methods evaluate the resulting clusters based solely on the data. *External validation* methods evaluate the resulting clusters based on pre-specified information.

External validation involves a criterion based on the comparison of the clusters produced by the algorithm and a partition chosen by some other means, say one produced by the investigator's understanding of the data. One criterion can simply be the number of mismatches, which provides an empirical error analogous to the clustering error discussed previously. If we were to randomly generate a set S of points and partition according to an underlying distributional model, then we would have a single-sample estimate of the error of the clustering algorithm relative to the distributional model; however, in the present situation no random process has been assumed as the generator of the point set S.

Rather than compute an empirical error directly, we can consider how pairs of points are commonly and uncommonly clustered by a cluster operator ζ and a heuristic partitioning. Suppose that P_S and P_ζ are the heuristic and ζ partitions, respectively. Define four quantities: a is the number of pairs of points in S such that the pair belongs to the same class in P_S and the same class in P_ζ; b is the number of pairs such that the pair belongs to the same class in P_S and different classes in P_ζ; c is the number of pairs such that the pair belongs to different classes in P_S and the same class in P_ζ; and d is the

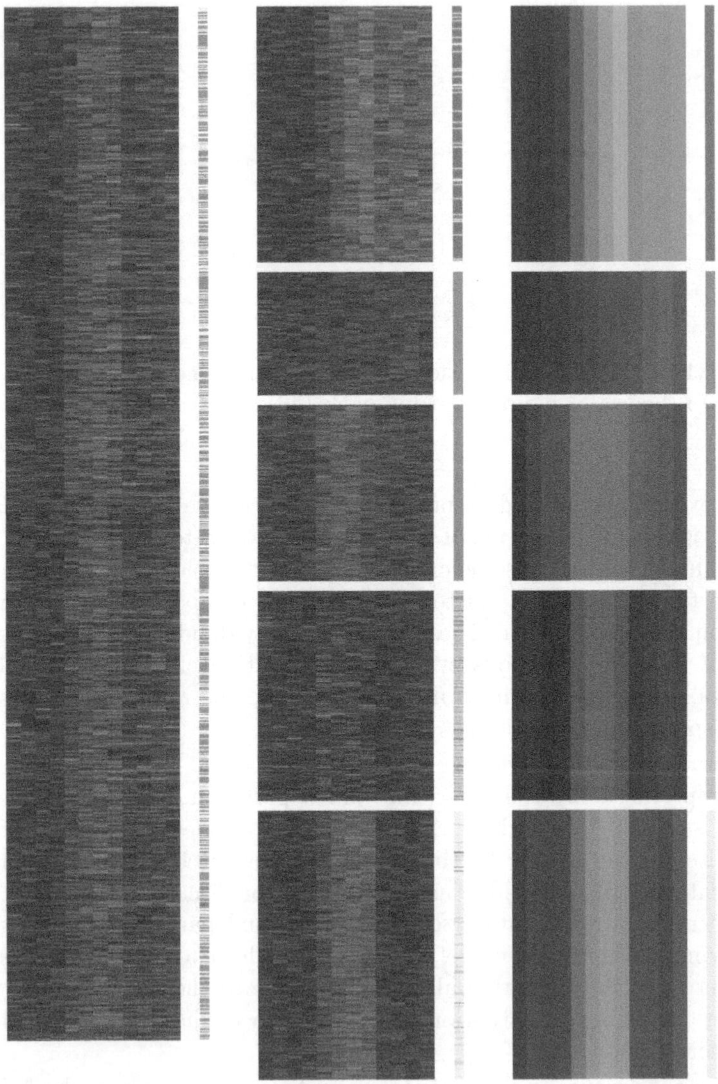

FIGURE 14.5 Results of fuzzy k-means clustering with SOM seed: (1) raw data; (2) clusters; (3) means of congruency classes assigned to clusters (See color figure following page 146).

number of pairs in S such that the pair belongs to different classes in P_S and different classes in P_ζ. If the partitions match exactly, then all pairs are either in the a or d classes. The *Rand index* is defined by $(a + d)/(a + b + c + d)$ [75]. The Rand index lies between 0 and 1. Numerous external validation procedures have been proposed [76].

Internal validation methods evaluate the clusters based solely on the data, without external information. Typically, a heuristic measure is defined to indicate the goodness of the clustering. It is important to keep in mind that the measure applies only to the data at hand and therefore is not predictive of the worth of a clustering algorithm — even with respect to the measure itself.

A common heuristic for spatial clustering is that, if the algorithm produces tight clusters and cleanly separated clusters, then it has done a good job clustering. We consider two indices based on this heuristic. Let $\mathcal{P} = \{S_1, S_2, ..., S_m\}$ be a partition of S, $\delta(S_i, S_j)$ be a between-cluster distance, and $\sigma(S_i)$ be a measure of cluster dispersion. The *Davies-Bouldin index* is defined by

$$\alpha(\mathcal{P}) = \frac{1}{k}\sum_{i=1}^{k} \max_{j \neq i} \left\{ \frac{\sigma(S_i) + \sigma(S_j)}{\delta(S_i, S_j)} \right\} \tag{14.15}$$

[77] and the Dunn index is defined by

$$\beta(\mathcal{P}) = \min_i \min_{j \neq i} \frac{\delta(S_i, S_j)}{\max_l \sigma(S_l)} \tag{14.16}$$

[78]. Low and high values are favorable for $\alpha(\mathcal{P})$ and $\beta(\mathcal{P})$, respectively. As defined, the indices leave open the distance and dispersion measures, and different ones have been employed. If we want the classes far apart, an obvious choice for δ is

$$\delta(S_i, S_j) = \min_{\mathbf{x} \in S_i, \mathbf{z} \in S_j} \|\mathbf{x} - \mathbf{z}\| \tag{14.17}$$

For tight classes, an evident choice is the diameter of the class,

$$\sigma(S_i) = \max_{\mathbf{x}, \mathbf{z} \in S_i} \|\mathbf{x} - \mathbf{z}\| \tag{14.18}$$

Since these kinds of measures do not possess predictive capability, it appears difficult to assess their worth — even what it means to be "worthy." But there have been simulation studies to observe how they behave [79]. The danger of relying on heuristic validation indices has been demonstrated in a study that has shown weak correlation between many validation indices and clustering error across various clustering algorithms and random point processes [80].

15

Genetic Regulatory Networks

A central focus of genomic research concerns understanding the manner in which cells execute and control the enormous number of operations required for normal function and the ways in which cellular systems fail in disease. In biological systems, decisions are reached by methods that are exceedingly parallel and extraordinarily integrated, as even a cursory examination of the wealth of controls associated with the intermediary metabolism network demonstrates. Feedback and damping are routine even for the most common of activities, cell cycling, where it seems that most proliferative signals are also apoptosis priming signals as well, and the final response to the signal results from successful negotiation of a large number of checkpoints, which themselves involve further extensive cross checks of cellular conditions.

Traditional biochemical and genetic characterizations of genes do not facilitate rapid sifting of these possibilities to identify the genes involved in different processes or the control mechanisms employed. Of course, when methods do exist to focus genetic and biochemical characterization procedures on a smaller number of genes likely to be involved in a process, progress in finding the relevant interactions and controls can be substantial. The earliest understandings of the mechanics of cellular gene control were derived in large measure from studies of just such a case, metabolism in simple cells. In metabolism, it is possible to use biochemistry to identify stepwise modifications of the metabolic intermediates and genetic complementation tests to identify the genes responsible for catalysis of these steps, and those genes and cis-regulator* elements involved in control of their expression. Standard methods of characterization guided by some knowledge of the connections could thus be used to identify process components and controls. Starting from the basic outline of the process, molecular biologists and biochemists have been able to build up a very detailed view of the processes and regulatory interactions operating within the metabolic domain.

In contrast, for most cellular processes, general methods to implicate likely participants and to suggest control relationships have not emerged from classical (often correlation-based) approaches. The resulting inability to produce overall schemata for most cellular processes has meant that gene function has been, for the most part, determined in a piecemeal fashion. Once a gene

*A *cis-regulator* is a DNA sequence that controls the transcription of a related gene.

is suspected of involvement in a particular process, research focuses on the role of that gene in a very narrow context. This typically results in the full breadth of important roles for well-known, highly characterized genes being slowly discovered. A particularly good example of this is the relatively recent appreciation that oncogenes such as Myc can stimulate apoptosis in addition to proliferation. Because transcriptional control is accomplished by a complex method that interprets a variety of inputs, the development of analytical tools that detect multivariate influences on decision-making present in complex genetic networks is essential. Modeling and analysis of gene regulation can substantially help to unravel the mechanisms underlying gene regulation and to understand gene function [81, 82, 83]. This, in turn, can have a profound effect on developing techniques for drug testing and therapeutic intervention for effective treatment of disease [84].

15.1 Nonlinear Dynamical Modeling of Gene Networks

Two salient aspects of a genetic regulatory system must be modeled and analyzed. One is the topology (connectivity structure) and the other is the set of interactions between the elements, the latter determining the dynamical behavior of the system. Exploration of the relationship between topology and dynamics may lead to valuable conclusions about the structure, behavior, and properties of genetic regulatory systems [85, 86].

Numerous mathematical and computational methods have been proposed for construction of formal models of genetic interactions. Generally, these models share certain characteristics: (1) they represent *systems* in that they characterize an interacting group of components forming a whole, can be viewed as a process that results in a transformation of signals, and generate outputs in response to input stimuli; (2) they are *dynamical* in that they capture the time-varying quality of the physical process under study and can change their own behavior over time; and (3) they can be considered to be generally *nonlinear*, in that the interactions within the system yield behavior that is more complicated than the sum of the behaviors of the agents.

The preceding characteristics are representative of nonlinear dynamical systems. These are composed of states, input and output signals, transition operators between states, and output operators. In their abstract form, they are very general. More mathematical structure is provided for particular application settings. For instance, in computer science they can be structured into the form of dataflow graphical networks that model asynchronous distributed computation, a model that is very close to genomic regulatory models. Indeed, most attempts to model gene regulatory networks fall within the scope of nonlinear dynamical systems, including probabilistic graphical models, such as Bayesian networks [87, 88, 89]; neural networks [90, 91]; and differential equations [92] — see [93] for a review. In particular, we mention an elaborate differential-equation-based discrete regulatory model involving cis-regulatory functionals that includes degradation parameters, transcription and translation rates, and delays [94]. Based on long experience in electrical and computer engineering, and more recent evidence from genomics itself, nonlinear dynamical systems appear to provide the appropriate framework to support the modeling of genomic systems. To build a model for a specific application requires abstracting from the specifics of the problem, and the breadth of nonlinear dynamical systems facilitates modeling within their framework.

Many concepts relevant to genomic regulation have been characterized from the perspectives of mathematical theory, estimation of model parameters, and application paradigms. We mention a few. *Structural stability* concerns the persistent behavior of a system under perturbation. It captures the idea of behavior that is not destroyed by small changes to the system. This is certainly a property of real genetic networks, since the cell must be able to maintain homeostasis[†] in the face of external perturbations and stimuli. *Uncertainty* relative to model behavior and knowledge acquisition has been extensively explored. Information theory, traditionally used for communications technology applications, is well suited to study uncertainty measures, quantified through the use of probability theory. *Distributed control* is common for complex systems, which have the property that no single agent is singularly in control of the system behavior; rather, control is dispersed among all agents, with varying levels of influence. This is the current view of genetic regulatory networks. To significantly change the global behavior of a system in a desired manner via external control, it is necessary to consider the effects holistically. This property is consistent with the inherent global stability of genetic networks in the presence of small changes to the system. This issue is addressed within control theory, where a central problem is *controllability*: how to select inputs so that the state of the system takes a desired value after some period of time. This is precisely the kind of issue that must be addressed for treatment of cancer and other genetically related diseases. In sum, nonlinear dynamical systems provide a framework for modeling and studying gene regulatory networks.

A key question concerns which model one should use. Model selection depends on the kind and amount of data available and the goals of the modeling and analysis. This choice involves classical engineering trade-offs. Should a model be *fine*, with many parameters to capture detailed low-level phenomena, such as protein concentrations and kinetics of reactions, but thereby requiring a great deal of data for inference; or should it be *coarse*, with fewer parameters and lower complexity, thus being limited to capturing high-level phenomena,

[†]Homeostatis is the ability of living systems to maintain internal equilibrium by adapting their physiology.

such as whether a gene is ON or OFF at a given time, but thereby having the advantage of requiring much smaller amounts of data [95]. Ultimately, model selection needs to obey the principle of Occam's razor; model complexity should be sufficient to faithfully explain the data but not be greater. From a pragmatic engineering perspective, this is interpreted to mean that the model should be as simple as possible to sufficiently solve the problem at hand. In the context of a functional network, complexity is determined by the number of nodes, the connectivity between the nodes, the complexity of the functional relations, and the quantization.

15.2 Boolean Networks

This section focuses on the original deterministic version of the Boolean model. The more recently proposed stochastic extension will be presented in Section 15.3. The Boolean model is archetypical of logical functional models and many of the issues that arise with it arise in other regulatory network models. A key issue in this text is intervention in gene regulatory networks and this has mainly been considered in the context of a probabilistic generalization of the Boolean model.

15.2.1 Boolean Model

The regulatory model that has perhaps received the most attention is the Boolean network model [96, 97, 98]. The model has been studied both in biology and physics. In the Boolean model, gene expression is quantized to two levels: ON and OFF. The expression level (state) of each gene is functionally related to the expression states of other genes using logical rules. Although binarization provides very coarse quantization, we note that it is commonplace to describe genetic behavior in binary logical language, such as *on* and *off*, *up-regulated* and *down-regulated*, and *responsive* and *non-responsive*. In the context of expression microarrays, consideration of differential expression leads to the categories of *low-expressed* and *high-expressed*, thereby leading to binary networks, or to the categories of *low-expressed*, *high-expressed*, and *invariant*, thereby leading to ternary valued networks that are treated in much the same way as binary networks and often referred to as Boolean networks.

Successful application of the Boolean model requires the inclusion of genes whose behavior is essentially binary (bi-modal). It has been demonstrated in the context of microarrays that there can be sufficiently many switch-like genes so that binary quantization can be successfully utilized for clustering [99] and classification [100]. From the perspective of logical prediction, numerous Boolean relations have been observed in the NCI 60 Anti-Cancer Drug Screen cell lines [101]. Some examples are

$$\begin{aligned} MRC1 &= VSNL1 \vee HTR2C \\ SCYA7 &= CASR \wedge MU5SAC \end{aligned} \qquad (15.1)$$

Moreover, using classical methods there is ample evidence demonstrating inherent logical genomic decision making [102, 103]. Figure 15.1 shows a biologically studied regulatory pathway and its corresponding Boolean representation. A full description of the biological model is given in [104]; here we restrict ourselves to noting that for cells to move into the S phase, cdk2 and cyclin E work together to phosphorylate the Rb protein and inactivate it, thereby releasing cells into the S phase, and that misregulation can result in unregulated cell growth.

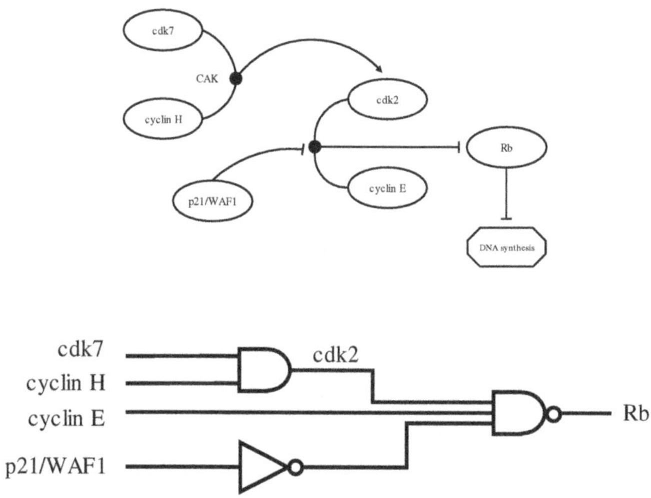

FIGURE 15.1 Regulation of the Rb protein in the cell cycle:(a) biological model; (b) Boolean representation [104].

A *Boolean network* is defined by a set of nodes, $V = \{x_1, x_2, \ldots, x_n\}$ and a list of Boolean functions, $F = \{f_1, f_2, \ldots, f_n\}$. Each x_k represents the state (expression) of a gene, g_k, where $x_k = 1$ or $x_k = 0$, depending on whether the gene is expressed or not expressed. The Boolean functions represent the rules of regulatory interaction between the genes. Network dynamics result from a synchronous clock with times $t = 0, 1, 2, \ldots$. The value of gene g_k at time $t + 1$ is determined by

$$x_k(t+1) = f_k(x_{k1}, x_{k2}, \ldots, x_{k,m(k)}) \qquad (15.2)$$

where the nodes in the argument of f_k form the *regulatory set* for x_k (gene g_k). The numbers of genes in the regulatory sets define the *connectivity* of the network, with maximum connectivity often limited. At time point t, the state vector

$$x(t) = (x_1(t), x_2(t), \ldots, x_n(t)) \tag{15.3}$$

is called the *gene activity profile* (*GAP*). The functions together with the regulatory sets determine the network wiring. A Boolean network is a very coarse model; nonetheless, it facilitates understanding of the generic properties of global network dynamics [105, 106], and its simplicity mitigates data requirements for inference.

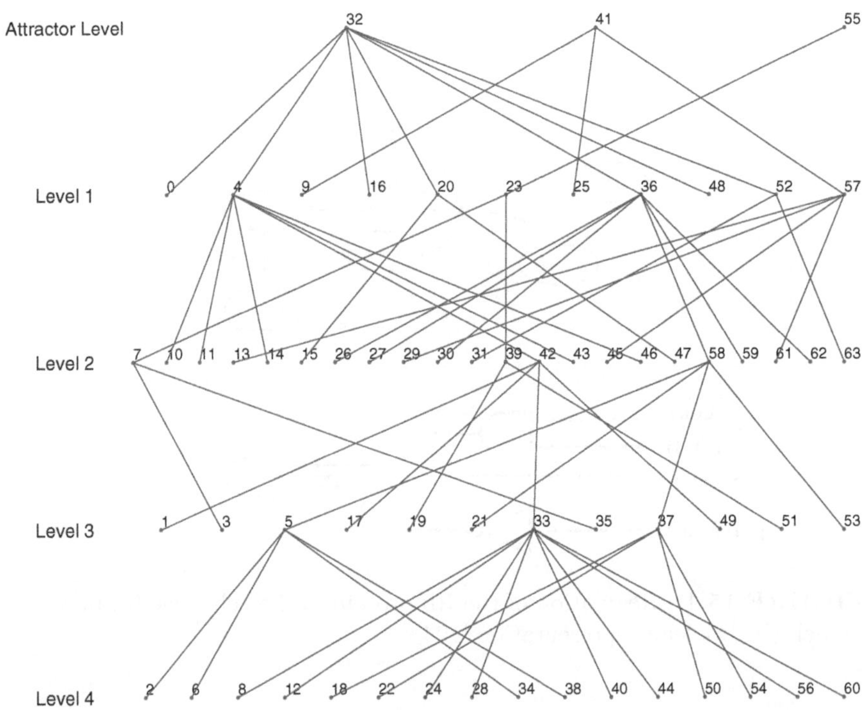

FIGURE 15.2 A Boolean network with three singleton attractors and four transient levels.

Attractors play a key role in Boolean networks. Given a starting state, within a finite number of steps, the network will transition into a cycle of states, called an *attractor*, and absent perturbation will continue to cycle thereafter. Each attractor is a subset of a *basin* composed of those states that lead to the attractor if chosen as starting states. The basins form a partition of the state space for the network. Non-attractor states are transient. They are visited at most once on any network trajectory. Figure 15.2 provides a transition-flow schematic for a Boolean network containing six genes, with states $0 = 000000$, $1 = 000001, \ldots, 63 = 111111$. There are three singleton attractors, 32, 41, and 55. There are four transient levels, where a state in level k transitions to an attractor in k time points.

The attractors of a Boolean network characterize the long-run behavior of the network and have been conjectured by Kauffman to be indicative of the cell type and phenotypic behavior of the cell [105]. Real biological systems are typically assumed to have short attractor cycles, with singleton attractors being of special import. For instance, it has been suggested that apoptosis and cell differentiation correspond to some singleton attractors and their basins, while cell proliferation corresponds to a cyclic attractor along with its associated basin [106]. Changes in the Boolean functions, via mutations or rearrangements, can lead to a rewiring in which attractors appear that are associated with tumorigenesis. This is likely to lead to a cancerous phenotype unless the corresponding basins are shrunk via new-rewiring, so that the cellular state is not driven to a tumorigenic phenotype, or, if already in a tumorigenic attractor, the cell is forced to a different state by flipping one or more genes. The objective of cancer therapy would be to use drugs to do one or both of the above.

15.2.2 Coefficient of Determination

By viewing gene status across different conditions, say, via microarrays, it is possible to establish relationships between genes that show variable status across the conditions. Owing to limited replications, we assume that gene expression data are quantized based on some statistical analysis of the raw data. One way to establish multivariate relationships among genes is to quantify how the estimate for the expression status of a particular *target gene* can be improved by knowledge of the status of some other *predictor genes*. This is formalized via the *coefficient of determination* (*CoD*) [6], which is defined by

$$CoD = \frac{\varepsilon_0 - \varepsilon_{opt}}{\varepsilon_0} \tag{15.4}$$

where ε_0 is the error of the best numerical predictor of the target gene in the absence of observation and ε_{opt} is the error of the optimal predictor of the target gene based on the predictor genes. This nonlinear form of the CoD is essentially a nonlinear, multivariate generalization of the familiar goodness of fit measure in linear regression. The CoD measures the degree to which the best estimate for the transcriptional activity of a target gene can be improved using the knowledge of the transcriptional activity of some predictor genes, relative to the best estimate in the absence of any knowledge of the transcriptional activity of the predictors. The CoD is a number between 0 and 1, a higher value indicating a tighter relationship.

Figure 15.3 shows a CoD diagram for the target gene p53 and predictor genes p21 and MDM2, in which the CoDs have been estimated in the context of a study involving stress response [107]. We see that the individual CoDs for p21 and MDM2 are 0.227 and 0.259, respectively, but when used jointly, the CoD for the predictor set {p21, MDM2} increases to 0.452. Biologically, it is known that p53 is influential but not determinative of the up regulation of both p21 and MDM2, and hence it is not surprising that some level of prediction of p53 should be possible by a combination of these two genes. Note that the prediction of p53 by p21 and MDM2 apparently results from the regulation of p53 on them, not the other way around. Going the other way, the same study found the CoD for p53 predicting p21 to be 0.473. The increased predictability of p53 using both MDM2 and p21 is expected because increasing the size of the predictor set cannot result in a decrease in CoD. The extent of the increase can be revealing. In Fig. 15.3, MDM2 and p21 have very similar CoDs relative to p53 and there is a significant increase when they are used in combination. On the other hand, it may be that very little, if any, predictability is gained by using predictors in combination. Moreover, it may be that the individual predictors have CoDs very close (or equal) to 0, but when used in combination the joint CoD is 1. This kind of situation shows that it is risky to assume that a predictor g_1 and target g_0 are unrelated because the CoD of g_1 predicting g_0 is very low. This situation is akin to that in classification where a feature may be poor if used alone but may be good if used in combination with other features. The issue in both settings is the danger of *marginal analysis* – drawing conclusions about variables from marginal relations instead of joint (multivariate) relations. The complex non-linear distributed regulation ubiquitous in biological systems makes marginal analysis highly risky.

15.3 Probabilistic Boolean Networks

Given a target gene, several predictor sets may provide equally good estimates of its transcriptional activity, as measured by the CoD. Moreover, one may rank several predictor sets via their CoDs. Such a ranking provides a quantitative measure to determine the relative ability of each predictor set to improve the estimate of the transcriptional activity of the particular target gene. While attempting to infer inter-gene relationships, it makes sense to not put all our faith in one predictor set; instead, for a particular target gene, a better approach is to consider a number of predictor sets with high CoDs. Considering each retained predictor set to be indicative of the tran-

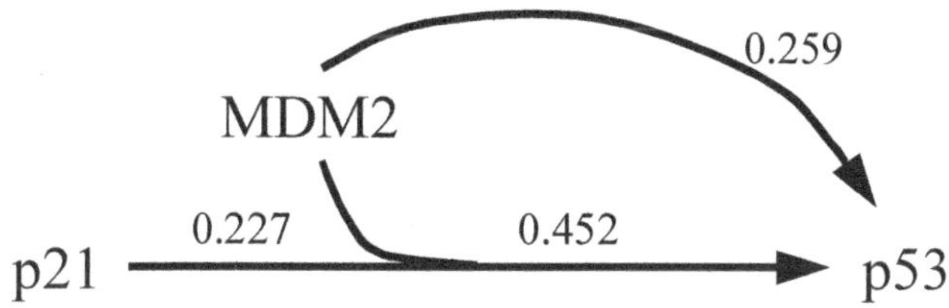

FIGURE 15.3 CoD diagram for p21 and MDM2 predicting p53.

scriptional activity of the target gene with a probability proportional to its CoD represents feature selection for gene prediction.

Having inferred inter-gene relationships in some manner, this information can be used to model the evolution of the gene activity profile over time. It is unlikely that the determinism of the Boolean-network model will be concordant with the data. One could pick the predictor set with the highest measure of predictability, but as remarked previously in the case of the CoD, there are usually a number of almost equally performing predictor sets, and for them we will have only estimates from the data. By associating several predictor sets with each target gene, it is not possible to obtain with certainty the transcriptional status of the target gene at the next time point; however, one can compute the probability that the target gene will be transcriptionally active at time $t + 1$ based on the gene activity profile at time t. The time evolution of the gene activity profile then defines a stochastic dynamical system. Since the gene activity profile at a particular time point depends only on the profile at the immediately preceding time point, the dynamical system is Markovian. Such systems can be studied in the established framework of Markov Chains and Markov Decision Processes. These ideas are mathematically formalized in *probabilistic Boolean networks* (*PBN*s) [104, 108]. In a PBN, the transcriptional activity of each gene at a given time point is a Boolean function of the transcriptional activity of the elements of its predictor sets at the previous time point. The choice of Boolean function and associated predictor set can vary randomly from one time point to another. For instance, when using the CoD, the choice of Boolean function and predictor set can depend on CoD-based selection probabilities associated with the different predictor sets. This kind of probabilistic generalization of a Boolean network, in which the Boolean function is randomly selected at each time point, defines an *instantaneously random PBN*.

Instead of simply assigning Boolean functions at each time point, one can take the perspective that the data come from distinct sources, each representing a *context* of the cell. From this viewpoint, the data derive from a family of deterministic networks and, were we able to separate the samples according to the contexts from which they have been derived, then there would in fact be CoDs with value 1, indicating deterministic biochemical activity for the wiring of a particular constituent network. Under this perspective, the only reason that it is not possible to find predictor sets with CoD equal (or very close to) 1 is because they represent averages across the various cellular contexts. This perspective results in the view that a PBN is a collection of Boolean networks in which one constituent network governs gene activity for a random period of time before another randomly chosen constituent network takes over, possibly in response to some random event, such as an external stimulus or genes not included in the model network. Since the latter is not part of the model, network switching is random. This model defines a *context-sensitive PBN*. The probabilistic nature of the constituent choice reflects the fact that the system is open, not closed, the idea being that network changes result from the genes responding to *latent* variables external to the model network. The context-sensitive model reduces to the instantaneously random model by having network switching at every time point.

Much of the theory and application of PBNs applies directly to the more general case which need not possess binary quantization and which are also called PBNs, owing to the multi-valued logical nature of functional relations for finite quantization. A particularly important case is ternary quantization, where expression levels take on the values +1 (up-regulated), −1 (down-regulated), and 0 (invariant).

A PBN is composed of a set of n genes, $x_1, x_2, \ldots, x_n$, each taking values in a finite set V (containing d values), and a set of vector-valued *network functions*, $\mathbf{f}_1, \mathbf{f}_2, \ldots \mathbf{f}_r$, governing the state transitions of the genes. To every node x_i, there corresponds a set

$$F_i = \{f_j^{(i)}\}_{j=1,\ldots,l(i)}, \tag{15.5}$$

where each $f_j^{(i)}$ is a possible function, called a *predictor*, determining the value of gene x_i and $l(i)$ is the number of possible functions assigned to gene x_i. Each network function is of the form $\mathbf{f}_k = (f_{k_1}^{(1)}, f_{k_2}^{(2)}, \ldots, f_{k_n}^{(n)})$, for $k = 1, \ldots, r$, $1 \leq k_i \leq l(i)$ and where $f_{k_i}^{(i)} \in F_i (i = 1, 2, \ldots, n)$. Each vector function $\mathbf{f}_k : \{0,1\}^n \to \{0,1\}^n$ acts as a transition function (mapping) representing a possible realization of the entire PBN. Thus, given the value of all genes, $(x_1, \ldots, x_n)$, $\mathbf{f}_k(x_1, x_2, \ldots, x_n) = (x_1', x_2', \ldots, x_n')$ gives us the state of the genes after one step of the network given by the realization $\mathbf{f}_k$.

The choice of which network function $\mathbf{f}_j$ to apply is governed by a selection procedure. At each time point a random decision is made as to whether to switch the network function for the next transition, with a probability q of a change being a system parameter. If a decision is made to change the network function, then a new function is chosen from among $\mathbf{f}_1$, $\mathbf{f}_2$,..., $\mathbf{f}_r$, with the probability of choosing $\mathbf{f}_k$ being the selection probability c_k.

Now, let $\mathbf{F}=(f^{(1)}, f^{(2)}, \ldots, f^{(n)})$ be a random vector taking values in $F_1 \times F_2 \ldots \times F_n$. That is, $\mathbf{F}$ can take on all possible realizations of the PBN. Then, the probability that predictor $f_j^{(i)}$ is used to predict gene i $(1 \leq j \leq l(i))$ is equal to

$$c_j^{(i)} = P\{f^{(i)} = f_j^{(i)}\} = \sum_{k:f_{k_i}^{(i)}=f_j^{(i)}} P\{\mathbf{F} = \mathbf{f}_k\}. \tag{15.6}$$

Since $c_j^{(i)}$ are probabilities , they must satisfy

$$\sum_{j=1}^{l(i)} c_j^{(i)} = 1. \tag{15.7}$$

It is not necessary that the selection of Boolean functions composing a specific network be independent. This means that it is not necessarily the case that

$$P\{f^{(i)} = f_j^{(i)}, f^{(l)} = f_k^{(l)}\} = P\{f^{(i)} = f_j^{(i)}\}.P\{f^{(l)} = f_k^{(l)}\}. \tag{15.8}$$

A PBN is said to be independent if the random variables $f^{(1)}, f^{(2)}, \ldots, f^{(n)}$ are independent. In the dependent case, product expansions such as the one given in Eq. 15.8, as well as ones involving more functions, require conditional probabilities. If the PBN is independent, then there are $L = \prod_{i=1}^{n} l(i)$ realizations (constituent Boolean networks). Moreover, for an independent PBN, if the kth network is obtained by selecting $f_{i_r}^{(i)}$ for gene i , $i = 1, 2, \ldots, n$, $1 \leq i_r \leq l(i)$, then the selection probability c_k is given by $c_k = \prod_{i=1}^{n} c_{i_r}^{(i)}$.

A PBN with perturbation can be defined by there being a probability p of any gene changing its value uniformly randomly to another value in V at any instant of time. Whereas a network switch corresponds to a change in a latent variable causing a structural change in the functions governing the network, for instance, in the case of a gene outside the network model that participates in the regulation of a gene in the model, a random perturbation corresponds to a transient value flip that leaves the network wiring unchanged, as in the case of activation or inactivation owing to external stimuli such as mutagens, heat stress, etc. [105].

The state space S of the network together with the set of network functions, in conjunction with transitions between the states and network functions, determine a Markov chain, the states of the Markov chain being of the form $(\mathbf{x}^i, \mathbf{f}_j)$. If there is random perturbation, then the Markov chain is ergodic, meaning that it has the possibility of reaching any state from another state and that its stationary distribution becomes a steady-state distribution. In the special case when $q = 1$, a network function is randomly chosen at each time point and the Markov chain consists only of the PBN states.

For a PBN, characterization of its long-run behavior is described via the Markov chain it defines. In particular, an instantaneously random PBN has equivalence classes of communicating states analogous to the basins of attraction for Boolean networks, and if there is perturbation, which we will always suppose, then the Markov chain is ergodic, which then guarantees the existence of a global steady-state distribution. In general, whether the PBN is instantaneously random or context-sensitive, by definition its attractors consist of the attractors of its constituent Boolean networks. Two events can remove a network from an attractor cycle C: (1) a perturbation can send it to a different state, and assuming the constituent network remains unchanged and there are no further perturbations for a sufficient time, then it will return to C if the perturbation leaves it in the basin of C or it will transition to a different attractor cycle of the same constituent network if the perturbation sends it to a different basin; (2) a network switch will put it in the basin of an attractor cycle for the new constituent network and it will transition to the attractor cycle for that basin so long as the constituent network remains unchanged and there are no further perturbations for a sufficient time. Whereas the attractor cycles of a Boolean network are mutually disjoint, the attractor cycles of a PBN can intersect because different cycles can correspond to different constituent Boolean networks. Assuming that the switching and perturbation probabilities are very small, a PBN spends most of its time in its attractors. The probabilities of PBN attractors have been analytically characterized [109].

15.4 Network Inference

For genetic regulatory networks to be of practical benefit, there must be methods to design them based on experimental data. We confront three impediments: (1) model complexity, (2) limited data, and (3) lack of appropriate time-course data to model dynamics. Numerous approaches to the *network inference problem* have been proposed in the literature, many based on gene-expression microarray data. Here, we briefly outline some of the proposed methods for PBNs and the rationale behind each of them (there also having been substantial study of inferring Boolean networks [108, 110]).

As first proposed, the inference of the PBN is carried out using the CoD [104]. For each gene in the network, a number of high CoD predictor sets are found and these high-CoD predictor sets determine the evolution of the activity status of that particular gene. Furthermore, the selection probability of each predictor set for a target gene is assumed to be the ratio of the CoD of that predictor set to the sum of the CoDs of all predictor sets used for that target gene. This approach makes intuitive sense since it is reasonable to assign the selection probability of each predictor set in a PBN to be proportional to its predictive worth as quantified by the CoD.

A second approach to PBN construction uses mutual information clustering and reversible-jump Markov-chain-Monte-Carlo predictor design [111]. First, mutual-information-minimization clustering is used to determine the number of possible parent gene sets and the input sets of gene variables corresponding to each gene. Thereafter, each (predictor) function from the possible parent gene sets to each target gene is modeled by a simple neural network consisting of a linear term and a nonlinear term, and a reversible-jump Markov-chain-Monte-Carlo (MCMC) technique is used to calculate the model order and the parameters. Finally, the selection probability for each predictor set is calculated using the ratio of the CoDs.

In most expression studies, there is some degree of previous knowledge regarding genes that play a role in the phenotypes of interest, for instance, p53 in unregulated proliferation. To take advantage of this knowledge, and to obtain networks relating to genes of interest, it has been proposed to construct networks in the context of directed graphs by starting with a seed consisting of one or more genes believed to participate in a meaningful subnetwork [112]. Given the seed, a network is grown by iteratively adjoining new genes that are sufficiently interactive with genes in the growing network in a manner that enhances subnetwork autonomy. The proposed algorithm has been applied using both the CoD and the Boolean-function influence [104], which measures interaction between genes. The algorithm has the benefit of producing a collection of small tightly knit autonomous subnetworks as opposed to one massive network with a large number of genes. Such small subnetworks are more amenable to modeling and simulation studies, and when properly seeded are more likely to capture a small set of genes that may be maintaining a specific core regulatory mechanism.

Figure 15.4 shows a melanoma network grown from four seed genes, WNT5A, RTN1, S100B and SNCA. The CoD has been used as the measure of gene interaction and the boxes denote the seed genes while the ellipses are the genes added by the algorithm. The solid lines represent strong connections (connection strengths exceeding 0.3) while the dotted lines represent weak connections (connection strengths between 0.2 and 0.3). This network reveals some very interesting insights that are highly consistent with prior biological knowledge derived from earlier gene expression studies using melanoma cell lines [113, 114]. For instance, it is known that the WNT5A gene product has the capability to drive aspects of cell motility and invasiveness. That being the case, it is to be expected that genes playing a part in either mediating extra-cellular matrix remodeling/interaction, such as MMP3 (matrix metalloproteinase 3), SERPINB2 (serine (or cysteine) proteinase inhibitor), and MCAM (melanoma adhesion molecule), or cellular movement, such as MYLK (myosin light polypeptide kinase), would share regulatory information with WNT5A. On the other hand, it is not known how WNT5A regulation is coupled to other genes playing a part in melanoma cell proliferation, such as MAP2K1 (mitogen-activated protein kinase kinase 1), and the regulation of apoptosis, such as CASP3 (cysteine aspartate protease 3) and BIRC1 (baculoviral IAP repeat-containing 1). Nevertheless, it is quite possible that high-level coordination of these activities exists through either specific circuitry, or as a consequence of differing extra-cellular interactions that arise from metastatic cell movement.

A key issue in network design arises because much of the currently available gene-expression data comes to us from *steady-state* phenotypic behavior and does not capture any temporal history. Consequently, the process of inferring a PBN, which is a dynamical system, from steady-state data is a severely *ill-posed inverse problem.* Steady-state behavior constrains the dynamical behavior of the network but does not determine it and, therefore, building a dynamical model from steady-state data is a kind of overfitting. It is for this reason that a designed network should be viewed as providing a regulatory structure that is consistent with the observed steady-state behavior. Also, it is possible that several networks may emerge as candidates for explaining the steady-state data. Under the assumption that we are sampling from the steady-state, a key criterion for checking the validity of a designed network is that much of its steady state mass lies in the states observed in the sample data because it is expected that the data states consist mostly of attractor states [115].

A number of recent papers have focused on network inference keeping in mind that most of the data states correspond to steady-state behavior. In one of these, a fully Bayesian approach has been proposed that emphasizes network topology [116]. The method computes the possible parent sets of each gene, the corresponding predictors and the associated probabilities based on a neural-network model, using a reversible jump MCMC technique; and an MCMC method is employed to search the network configurations to find those with the highest Bayesian scores to construct the PBNs. This method has been applied to a melanoma cell line data set. The steady-state distribution of the resulting model contains attractors that are either identical or very similar to the states observed in the data, and many of the attractors are singletons, which mimics the biological propensity to stably occupy a given state. Furthermore, the connectivity rules for the most optimally generated networks constituting the PBN were found to be remarkably similar, as would be expected for a network operating on a distributed basis, with strong interactions between the components.

If we consider network inference from the general perspective of an ill-posed inverse problem, then one can formalize inference by postulating criteria that constitute a solution space in which a designed network must lie. For this we propose two kinds of criteria [117]:

- *Constraint criteria* are composed of restrictions on the form of the network, such as biological and complexity constraints.

- *Operational criteria* are composed of relations that must be satisfied between the model and the data.

Examples of constraint criteria include limits on connectivity and attractor cycles. One example of an operational criterion is some degree of concordance between sample and model CoDs, and another is the requirement that data states are attractor states in the model. The inverse problem may still be ill-posed with such criteria, but all solutions in the resulting space can be considered satisfactory relative to the requirements imposed by the criteria. This kind of approach has been implemented by finding constituent Boolean networks satisfying constraints such as limited attractor structure, transient time, and connectivity [118].

Let us close by noting that, in addition to the ongoing effort to infer PBNs, there has been a continuing effort to infer Bayesian and dynamic Bayesian networks (DBNs) [88, 119, 120]. A *Bayesian network* is essentially a compact graphical representation of a joint probability distribution [121, 122, 123]. This representation takes the form of a directed acyclic graph in which the nodes of the graph represent random variables and the directed edges, or lack thereof, represent conditional dependencies, or independencies. The network also includes conditional probability distributions for each of the random variables. In the case of genetic networks, the values of the nodes can correspond to gene-expression levels or other measurable events, including external conditions. There is a precisely characterized relation between certain DBNs and PBNs in the sense that they can represent the same joint distribution over their corresponding variables [124]. PBNs are more specific in the sense that the mapping between PBNs and DBNs is many-to-one, so that a DBN does not specify a specific PBN.

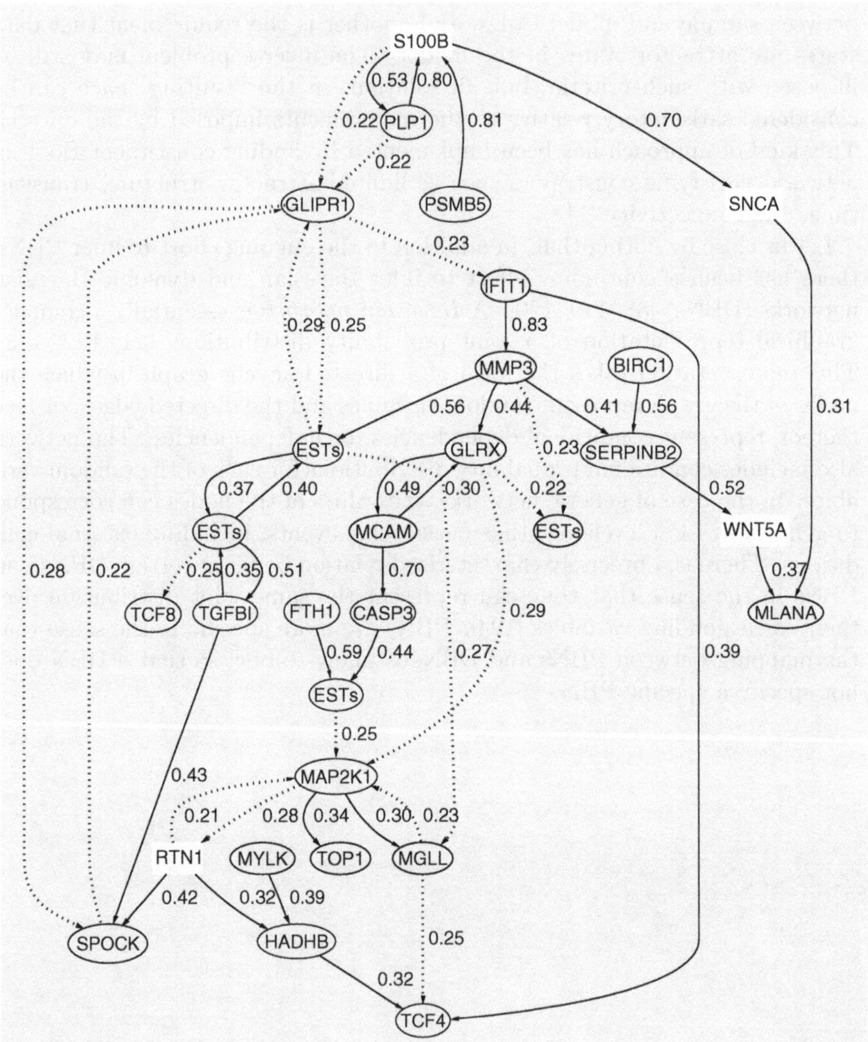

FIGURE 15.4 Melanoma network grown from four seed genes, WNT5A, RTN1, S100B and SNCA.

16

Intervention

From a translational perspective, the ultimate objective of genetic regulatory network modeling is to use the network to design different approaches for affecting network dynamics in such a way as to avoid undesirable phenotypes, for instance, cancer. The pragmatic manifestation of this goal is the development of therapies based on the disruption or mitigation of aberrant gene function contributing to the pathology of a disease. Mitigation would be accomplished by the use of drugs to act on the gene products. Engineering therapeutic tools involves synthesizing nonlinear dynamical networks, analyzing these networks to characterize gene regulation, and developing intervention strategies to modify dynamical behavior. For instance, changes in network connectivity or functional relationships among the genes in a network, via mutations or re-arrangements, can lead to steady-state behavior associated with tumorigenesis, and this is likely to lead to a cancerous phenotype unless corrective therapeutic intervention is applied.

To date, intervention studies using PBNs have used three different approaches: (i) resetting the state of the PBN, as necessary, to a more desirable initial state and letting the network evolve from there [125]; (ii) changing the steady-state (long-run) behavior of the network by minimally altering its rule-based structure [126]; and (iii) manipulating external (control) variables that alter the transition probabilities of the network and can, therefore, be used to desirably affect its dynamic evolution [127].

In this chapter we present the intervention results obtained using the first two approaches. The results obtained using the third approach and their variants will be discussed in Chapter 17. Given a PBN, the transition from one state to the next takes place in accordance with certain transition probabilities and their dynamics, and hence intervention, can be studied in the context of homogeneous Markov chains with finite state spaces. The chapter is organized as follows: Section 16.1 summarizes PBN notation that is essential for the subsequent development; Section 16.2 discusses intervention limited to a one-time flipping of the expression status of a single gene; and Section 16.3 considers intervention to alter the steady-state behavior of the network.

16.1 PBN Notation

To characterize the Markov chain associated with an instantaneously random PBN, we first focus on Boolean networks, for which the state vector $x(k)$ at any time step k is essentially an n-digit binary number whose decimal equivalent is given by

$$y(k) = \sum_{j=1}^{n} 2^{n-j} x_j(k). \tag{16.1}$$

As $x(k)$ ranges from 00...0 to $11\ldots1$, $y(k)$ takes on all values from 0 to $2^n - 1$. To be consistent with the development in [104], we define

$$z(k) = 1 + y(k). \tag{16.2}$$

As $x(k)$ ranges from 00...0 to $11\ldots1$, $z(k)$ takes on all values from 1 to 2^n. The mapping from $x(k)$ to $z(k)$ is one-to-one and onto, and hence invertible. Thus, instead of the binary representation $x(k)$ for the state vector, we can equivalently work with the decimal representation $z(k)$ for which the state space is $S = \{1, 2, 3, \cdots, 2^n\}$. Furthermore, each $z(k)$ can be uniquely represented by a basis vector $w(k) \in R^{2^n}$, where $w(k) = e_{z(k)}$, e.g. if $z(k)$=1, then $w(k)$ = [1, 0, 0,.....]. Then, as discussed in [104], the evolution of the vector $w(k)$ proceeds according to the difference equation

$$w(k+1) = w(k)A \tag{16.3}$$

where A is a $2^n \times 2^n$ matrix having only one non-zero entry (equal to one) in each row. Equation 16.3 derived for a deterministic Boolean network has a stochastic counterpart for a PBN. To arrive at this counterpart, let $w(k)$ denote the probability distribution vector for a PBN at time k, i.e. $w_i(k) = Pr\{z(k) = i\}$. Then it can be shown [104] that $w(k)$ evolves according to the equation

$$w(k+1) = w(k)A \tag{16.4}$$

where A is the stochastic matrix of transition probabilities. This completes our limited discussion of PBNs, as Markov chains.

As with the majority of the literature, in this section we have focused on binary quantization; nevertheless, as pointed out in the last chapter, most of the theory and application carry over to any finite quantization in a fairly obvious fashion; indeed, it is in the ternary setting that we will consider the application of external control in Chapter 17.

16.2 Intervention by Flipping the Status of a Single Gene

Recognizing that a key goal of PBN modeling is the discovery of possible intervention targets (genes) by which the network can be "persuaded" to transition into a desired state or set of states, in this section, we consider the effects of intervention by deliberately affecting a particular gene in an instantaneously random PBN. Whereas in Boolean networks attractors are hypothesized to correspond to functional cellular states [106], in PBNs this role is played by irreducible subchains. Absent the possibility of perturbation ($p = 0$), a PBN is unable to escape from an irreducible subchain, implying that the cellular state cannot be altered. If p is positive, then the Markov chain is ergodic and there is a chance that the current cellular state may switch to another cellular state by means of a random gene perturbation. Clearly, flipping the values of certain genes is more likely to achieve the desired result than flipping the values of some other genes. Our goal is to discover which genes are the best potential "lever points," to borrow the terminology from [106], in the sense of having the greatest possible impact on desired network behavior so that we can intervene with them by changing their value (1 or 0) as needed. In addition, we wish to be able to intervene with as few genes as possible in order to achieve our goals. To motivate the discussion, let us illustrate the idea with an example.

Example 16.1

[104]. Suppose we are given a PBN consisting of three genes x_1, x_2, x_3. There are two functions$f_1^{(1)}, f_2^{(1)}$ associated with x_1, one function associated with x_2 and two functions $f_1^{(3)}, f_2^{(3)}$ associated with x_3. These functions are given by the truth table in Table 16.1. This truth table results in four possible Boolean networks $N_1 =(f_1^{(1)}, f_1^{(2)}, f_1^{(3)})$, $N_2=(f_1^{(1)}, f_1^{(2)}, f_2^{(3)})$, $N_3= (f_2^{(1)}, f_1^{(2)}, f_1^{(3)})$ and $N_4= (f_2^{(1)}, f_1^{(2)}, f_2^{(3)})$ possessing the probabilities $c_1 = 0.3$, $c_2 = 0.3$, $c_3 = 0.2$ and $c_4 = 0.2$, respectively. The state diagram of the Markov chain corresponding to this PBN is shown in Fig. 16.1. Suppose that we are currently in state (111) and wish to eventually transition to state (000). The question is, with which of the three genes, x_1, x_2, or x_3, should we intervene such that the probability is greatest that we will end up in (000). By direct inspection of the diagram in Fig. 16.1, we see that if we make $x_1 = 0$, then with probability 0.2 we will transition into (000), whereas if we make $x_2 = 0$ or $x_3 = 0$, then it will be impossible for us to end up in (000) and with probability 1 we will eventually return to (111). In other words, the network is resistant to perturbations of the second or third genes and will eventually maintain the same state. Thus, the answer to our question in this rather simple example is that only by intervening with gene x_1 do we have a chance of achieving our goal. To answer such questions in general, we need to develop

several tools.

TABLE 16.1 Truth Table [104].

$x_1x_2x_3$	$f_1^{(1)}$	$f_2^{(1)}$	$f_1^{(2)}$	$f_1^{(3)}$	$f_2^{(3)}$
000	0	0	0	0	0
001	1	1	1	0	0
010	1	1	1	0	0
011	1	0	0	1	0
100	0	0	1	0	0
101	1	1	1	1	0
110	1	1	0	1	0
111	1	1	1	1	1
$c_j^{(i)}$	0.6	0.4	1	0.5	0.5

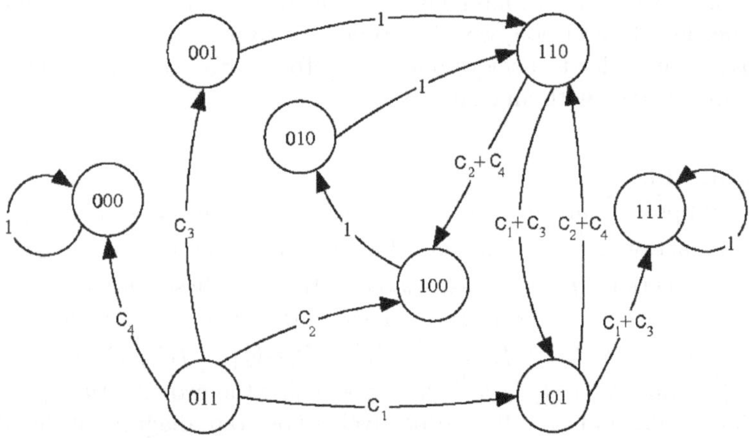

FIGURE 16.1 State Transition Diagram [104].

□

Assume there is independent random perturbation with $p > 0$, so that the Markov chain is ergodic and every state will eventually be visited. The question of intervention can be posed in the sense of reaching a desired state as soon as possible. For instance, in the example considered above, if p is very small and we are in state (111), then it will be a long time until we reach (000) and setting $x_1 = 0$ is much more likely to get us there faster. Hence, we are interested in the probability $F_k(x, y)$ that, starting in state x, the first time the PBN will reach some given state y will be at time k. This is known as the *first passage time* from state x to state y. For $k = 1$, $F_k(x, y) = A(x, y)$, which is just the transition probability from x to y. For $k \geq 2$, it can be shown [128] that

$$F_k(x,y) = \sum_{z \in S - \{y\}} A(x,z) F_{k-1}(z,y). \tag{16.5}$$

We can examine our results by considering

$$H_{K_0}(x,y) = \sum_{k=1}^{K_0} F_k(x,y) \tag{16.6}$$

which is the probability that the network, starting in state x, will visit state y before time K_0. (Note that the events {the first passage time from x to y will be at time k} are disjoint for different values of k.) As a special case, when $K_0 = \infty$, $H_{K_0}(x, y)$ is the probability that the chain ever visits state y, starting at state x, which is equal to 1 since the Markov chain is ergodic.

A related measure of interest is the *mean first passage time* from state x to state y, defined as

$$M(x,y) = \sum_k k F_k(x,y). \tag{16.7}$$

$M(x, y)$ is the average time it will take to get from state x to state y.

Example 16.2

[125] Continuing with the PBN of Example 16.1, the entries of the matrix A can be computed directly using the results of [125]. Supposing $p = 0.01$, the steady-state distribution is given by [0.0752, 0.0028, 0.0371, 0.0076, 0.0367, 0.0424, 0.0672, 0.7310], where the leftmost element corresponds to (000) and the rightmost to (111). The PBN spends much more time in state (111) than in any other state. Let our starting state x be (111) and the destination state y be (000), as before. Should we intervene with gene x_1, x_2, or x_3? Using first-passage time, we compute $F_k((011), (000))$, $F_k((101), (000))$, and $F_k((110), (000))$. Figure 16.2 shows the plots of $H_{K_0}(x, y)$ for $K_0 = 1, 2, \ldots, 20$ and for the three states of interest, namely, (011), (101), and (110). The plots indicate that starting at state (011), the network is much more likely to enter state (000) sooner than by starting at states (110) or (101). For instance, during the first 20 steps, there is almost a 0.25 probability of entering (000) starting at (011), whereas starting at (110) or (101),there is only a 0.05 probability. Thus, we should intervene with gene x_1 rather than with x_2 or x_3. Were we to base intervention on mean first passage time (Eq. 16.7), then the best gene for intervention would be the one possessing the smallest mean first passage time to the destination state. For this example, the mean first passage times corresponding to the perturbations of genes x_1, x_2, and x_3 are 337.51, 424.14,

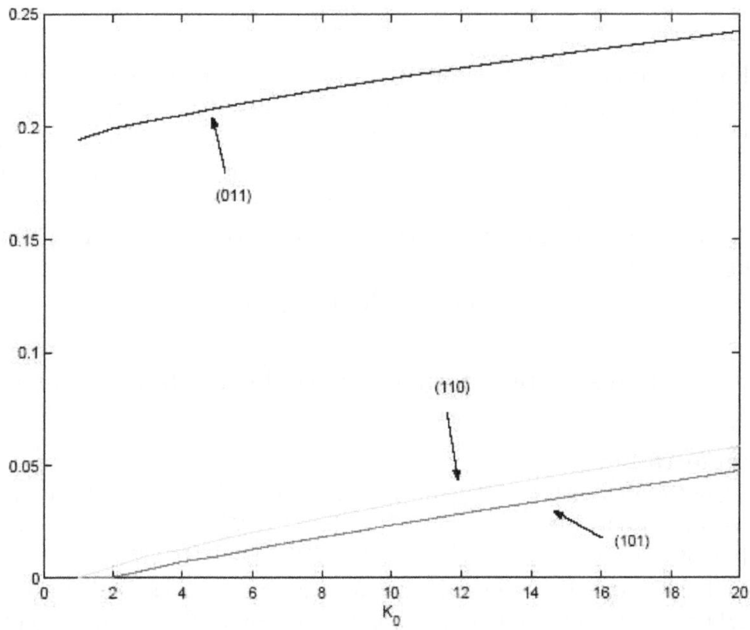

FIGURE 16.2 $H_{K_0}(x^{(i)}, y)$ for $K_0 = 1, \ldots, 20$, for starting states (011), (101), and (110), corresponding to perturbations of first, second, and third genes, respectively [125].

and 419.20, respectively. Since the first one is the smallest, this again supports the conclusion that gene x_1 is the best candidate for intervention.

□

To summarize the results of this section, given an initial state x, we generate different states $x^{(i)} = x \oplus e_i$, $i = 1, 2, \ldots, n$, where e_i is the unit binary vector with a 1 in the ith coordinate, by perturbing each of the n genes, and compute $H_{K_0}(x^{(i)}, y)$ for some desired destination state y and constant K_0. Then, the best gene for intervention is the one for which $H_{K_0}(x^{(i)}, y)$ is maximum; that is, given a fixed K_0, the optimal gene $x_{i_{opt}}$ satisfies

$$i_{opt} = \arg\max_i H_{k_0}(x^{(i)}, y).$$

Alternatively, by minimizing the mean first passage times, the optimal gene satisfies

$$i_{opt} = \arg\min_i M(x^{(i)}, y).$$

16.3 Intervention to Alter the Steady-State Behavior

The type of intervention described in the last section can be useful for modulating the dynamics of the network but it does not alter the underlying network structure. Accordingly, the stationary distribution remains unchanged. However, an imbalance between certain sets of states can be caused by mutations of the "wiring" of certain genes, thereby permanently altering the state-transition structure and, consequently, the long-run behavior of the network [106]. Therefore, it is prudent to develop a methodology for altering the steady-state probabilities of certain states or sets of states with minimal modifications to the rule-based structure. The motivation is that these states may represent different phenotypes or cellular functional states, such as cell invasion and quiescence, and we would like to decrease the probability that the whole network will end up in an undesirable set of states and increase the probability that it will end up in a desirable set of states. One way to accomplish this is by altering some Boolean functions (predictors) in the PBN. An additional goal is to alter as few functions as possible. In [126], formal methods and algorithms have been developed for addressing such a problem. Here we briefly discuss the results.

Consider an instantaneously random PBN with perturbation and two sets of states $A, B \subseteq \{0,1\}^n$. Since the Markov chain is ergodic, each state $x \subseteq \{0,1\}^n$ has a positive stationary probability $\pi(x)$. Thus, we can define $\pi(A) = \sum_{x \in A} \pi(x)$, and $\pi(B)$ similarly. Suppose that we are interested in altering the stationary probabilities of these two sets of states in such a way that the stationary probability of A is decreased and the stationary probability of B

is increased by λ, $0 < \lambda < 1$. As already mentioned above, these two states may represent two different cellular functional states or phenotypes. In order to achieve this, suppose we alter function $f_{j_0}^{(i_0)}$ by replacing it with a new function $g_{j_0}^{(i_0)}$. The probability $c_{j_0}^{(i_0)}$ corresponding to $g_{j_0}^{(i_0)}$ must remain the same as for $f_{j_0}^{(i_0)}$, since $c_1^{(i)} + c_2^{(i)} + \ldots + c_{l(i)}^{(i)} = 1$. Thus, we have a new PBN whose stationary distribution we can denote by μ. Letting $\mu(A)$ and $\mu(B)$ be the stationary probabilities of A and B under the altered PBN model, we pose the following optimization problem:

Given sets A and B, predictor functions $f_j^{(i)}$ together with their selection probabilities $c_j^{(i)}$, i =1, 2,. . . ,n, j =1, 2,. . . ,$l(i)$, and $\lambda \in (0, 1)$, select i_0 and j_0, and a function $g_{j_0}^{(i_0)}$ to replace $f_{j_0}^{(i_0)}$, such that

$$\varepsilon(\pi(A) - \lambda, \mu(A)) \tag{16.8}$$

and

$$\varepsilon(\pi(B) + \lambda, \mu(B)) \tag{16.9}$$

are minimum among all i,j, $g_j^{(i)}$, where $\varepsilon(a, b)$ is some error function, such as the absolute error $\varepsilon(a, b) = |a - b|$. An additional constraint can be that $g_{j_0}^{(i_0)}$ has no more essential variables than $f_{j_0}^{(i_0)}$. In this scenario, we are only allowing the alteration of one predictor function. More generally, we can pre-select a number of predictor functions that we are willing to alter.

Example 16.3

[126]. For the PBN of Example 16.1, Fig. 16.1 shows the state transition diagram assuming no perturbation (p =0). From the figure we see that there are two absorbing states, (000) and (111). For the sake of this example, suppose (111) corresponds to cell invasion (and rapid proliferation) and (000) corresponds to quiescence. Now assume perturbation probability $p = 0.01$. A simple analysis based on the probability transition matrix shows that the stationary probabilities of states (000) and (111) are 0.0752 and 0.7310, respectively. Thus, in the long run, the network will be in quiescence only 7% of the time and will be in proliferation 73% of the time. Suppose we wish to alter this imbalance and require the stationary probabilities to be approximately 0.4 for both (000) and (111). The other six states will then be visited only 20% of the time. In the framework of the above optimization problem, $A = \{(111)\}$, $B = \{(000)\}$, $\pi(A) = 0.7310$, $\pi(B) = 0.0752$, $\mu(A) = \mu(B) = 0.4$, and $\lambda = 0.3279$. Finally, suppose we are allowed to change only one predictor function. In Table 16.1, this corresponds to changing only one column, while keeping the selection probabilities $c_j^{(i)}$ unchanged. Thus, there are five possible columns (predictors) and 256 possibilities for each. The $5 \times 256 = 1280$ possible alterations have been generated and the stationary probabilities $\mu(000)$ and$\mu(111)$ have been computed for each (see Fig. 16.3). The optimal values of $\mu(000)$ and

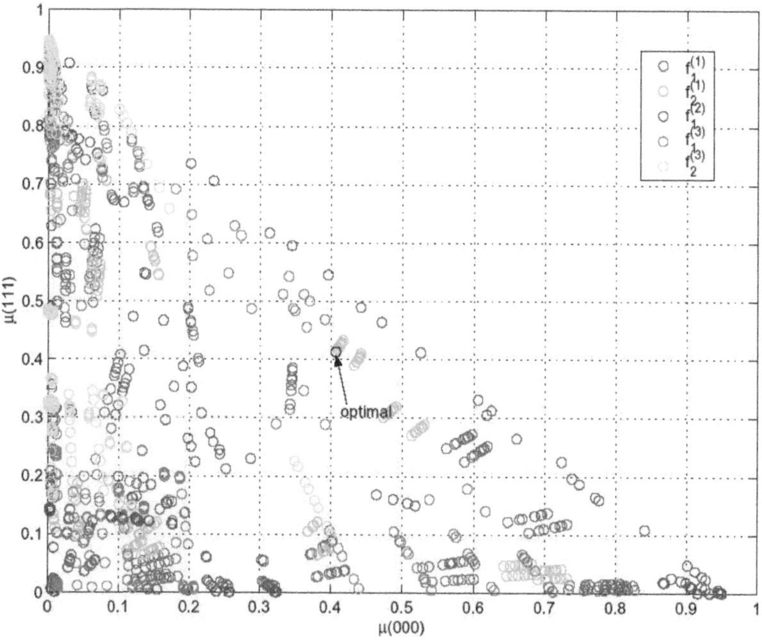

FIGURE 16.3 Each circle represents one of the 1280 possible alterations to the predictors. The x-axis is $\mu(000)$ and the y-axis is $\mu(111)$. The optimal choice is shown with an arrow, as it comes closest to 0.4 for both stationary probabilities. The colors of the circles represent the predictor that is altered (See color figure following page 146) [126].

$\mu(111)$ for the error function $\varepsilon(a,b) = |a - b|$ are indicated by an arrow. The objective function to be minimized is $|\mu(000) - 0.4| + |\mu(111) - 0.4|$, which corresponds to the sum of the two objective functions in Eqs. 16.8 and 16.9. The colors of the circles represent which predictor is altered. For example, red denotes that predictor $f_1^{(1)}$ is altered. The optimal predictor is the one that alters $f_2^{(1)}$ for gene 1 (column 2 in the truth tables) and the truth table of the optimal predictor is $(00010101)^T$. This predictor achieves the stationary probabilities $\mu(000) = 0.4068$ and $\mu(111) = 0.4128$. The structure of the plot in Fig. 16.3 reveals an interesting phenomenon: the two stationary probabilities exhibit regularities, forming clusters of points arranged in a linear fashion, with different directions. In fact, this phenomenon has been observed in numerous examples. It appears that the alterations of different predictors tend to occupy different parts of the space, implying that for a given predictor, there is a certain "range of action" that can be achieved by manipulating it. This suggests that a brute-force search for the optimal predictor alteration may possibly be avoided by following a number of search directions simultaneously, with the more promising ones being explored further. This, in turn suggests the use of genetic algorithms for optimization [129]. In fact, genetic algorithms have been used to solve the optimal structural intervention problem posed here and the resulting savings in computational effort have been remarkable [126]. Nonetheless, this remains an essentially brute force procedure and better approaches need to be developed.

□

17

External Intervention Based on Optimal Control Theory

Probabilistic Boolean networks can be used for studying the dynamic behavior of gene regulatory networks. Once a probability distribution vector has been specified for the initial state, the probability distribution vector evolves according to Eq. 16.4. From this perspective PBNs are *descriptive* in nature. There is no mechanism for controlling the evolution of the probability distribution vector. For treatment or intervention purposes, we are interested in working with PBNs in a *prescriptive* fashion, where the transition probabilities of the associated Markov chain depend on certain external variables, whose values can be chosen to make the probability distribution vector evolve in some desirable manner.

The use of such external variables makes sense from a biological perspective. For instance, in the case of diseases like cancer, external treatment inputs such as radiation, chemotherapy, etc. may be employed to move the state probability distribution vector away from one associated with uncontrolled cell proliferation or markedly reduced apoptosis. The variables could also include genes that serve as external master-regulators for all the genes in the network. To be consistent with the binary nature of the expression status of individual genes in a PBN, we will assume that these variables (*control inputs*) can take on only the binary values 0 or 1. The values of the individual control inputs can be changed from one time step to another in an effort to make the network behave in a desirable fashion.

In this chapter, we present the results obtained to date on intervention in PBNs using external control variables that alter the transition probabilities of the network and can, therefore, be used to desirably affect its dynamic evolution. Since its initial introduction in [127] this control-theoretic approach has subsequently been extended in several directions. First, the optimal intervention algorithm has been modified to accommodate the case where the entire state vector, or gene activity profile (*GAP*), is not available for measurement [130]. Second, whereas the original control-theoretic approach has been developed in the framework of instantaneously random PBNs, the intervention results have been extended to *context-sensitive* PBNs [131]. Third, in [132], control algorithms have been developed for a family of genetic regulatory networks as opposed to a single network. Finally, in [133], the earlier finite horizon results have been extended to the infinite horizon case in an effort to alter the steady-state behaviour of the genetic regulatory network.

The chapter is organized as follows. Section 17.1 formulates the intervention problem in probabilistic Boolean networks as an optimal control problem that is then solved using the standard approach of *dynamic programming.* Section 17.2 extends the results of section 17.1 to the imperfect information case. Section 17.3 extends the results to context-sensitive PBNs while section 17.4 develops intervention results for a family of Boolean networks. Section 17.5 extends the results of section 17.1 to the infinite horizon case, and finally section 17.6 contains some concluding remarks.

17.1 Finite-Horizon-Control

Suppose that a PBN with n genes has m control inputs $u_1, u_2, \ldots, u_m$. Then at any given time step k, the row vector

$$u(k) = [u_1(k), u_2(k), ..., u_m(k)] \tag{17.1}$$

describes the complete status of all the control inputs. $u(k)$ can take on all binary values from $00\cdots0$ to $11\cdots1$. Letting

$$v(k) = 1 + \sum_{i=1}^{m} 2^{m-i} u_i(k) \tag{17.2}$$

as $u(k)$ takes on binary values from $00\cdots0$ to $11\cdots1$, the variable $v(k)$ranges from 1 to 2^m. We can equivalently use $v(k)$ as an indicator of the complete control input status of the PBN at time step k and define the control space $U = \{1, 2, 3, \cdots, 2^m\}$.

We now proceed to derive the counterpart of Eq. 16.4 in the last chapter for a PBN subject to auxiliary controls. Let v^* be any integer between 1 and 2^m and suppose that $v(k) = v^*$. The same reasoning as in the derivation of Eq. 16.4 can be used to compute the corresponding A matrix, which will now depend on v^* and can be denoted by $A(v^*)$. Furthermore, the evolution of the probability distribution vector at time k will take place according to the equation

$$w(k+1) = w(k)A(v^*). \tag{17.3}$$

Since the choice of v^* is arbitrary, the one-step evolution of the probability distribution vector in the case of a PBN with control inputs takes place according to the equation

$$w(k+1) = w(k)A(v(k)). \tag{17.4}$$

The system in Eq. 17.4 can be equivalently represented as a stationary discrete-time dynamic system

$$z(k+1) = f(z(k), v(k), d(k)),\ k = 0, 1, ..., \tag{17.5}$$

where for all k, the state* $z(k)$ is an element of S, the control input $v(k)$ is an element of U, the disturbance $d(k)$ is an element of a space D and $f : S \times U \times D \mapsto S$. The disturbance $d(k)$ is manifested in terms of change of network based on the network transition probability q or change of state due to perturbation probability p. $d(k)$ is independent of prior disturbances $d(0), d(1)....d(k-1)$. In this chapter, we will interchangeably use either representation (17.4) or (17.5) depending on their suitability for a particular context or a particular derivation.

The transition probability matrix in Eq. 17.4 is a function of all the control inputs $u_1(k)$, $u_2(k)$,..., $u_m(k)$. Consequently, the evolution of the probability distribution vector of the PBN with control now depends not only on the initial distribution vector but also on the values of the control inputs at different time steps. Furthermore, intuitively it appears that it may be possible to make the states of the network evolve in a desirable fashion by appropriately choosing the control input at each time step. We next proceed to formalize these ideas.

Equation 17.4 is referred to in the control literature as a *controlled Markov chain* or a *Markov decision process* [134]. Markov chains of this type occur in many real life applications, the most notable example being the control of queues. Given a controlled Markov chain, the objective is to find a sequence of control inputs, usually referred to as a *control strategy*, so that an appropriate cost function is minimized over the entire class of allowable control strategies. To arrive at a meaningful solution, the cost function must capture the costs and the benefits of using any control. The design of a "good" cost function is application dependent and is likely to require considerable expert knowledge. We next outline a procedure that we believe would enable us to arrive at a reasonable cost function for determining the course of therapeutic intervention using PBNs.

In the case of diseases like cancer, treatment is typically applied over a finite time horizon. For instance, in the case of radiation treatment, the patient may be treated with radiation over a fixed interval of time following which the treatment is suspended for some time as the effects are evaluated. After that, the treatment may be applied again but the important point to note is that the treatment window at each stage is usually finite. We consider a finite-horizon problem, where the control is applied only over a finite number of steps.

Suppose that the number of steps over which the control input is to be applied has been *a priori* determined to be M and we are interested in controlling the behavior of the PBN over the interval $k = 0, 1, 2, \ldots, M-1$. Suppose at time step k, the state of the PBN is given by $z(k)$ and the corresponding control input is $v(k)$. Then we can define a cost $C_k(z(k), v(k))$ as being the cost of applying the control input $v(k)$ when the state is $z(k)$. With

*In the rest of this chapter, we will be referring to $z(k)$ as the state of the probabilistic Boolean network since, as discussed in section 16.1, $z(k)$ is equivalent to the actual state $x(k)$.

this definition, the expected cost of control over the entire treatment horizon becomes

$$E[\sum_{k=0}^{M-1} C_k(z(k), v(k))|z(0)]. \quad (17.6)$$

Note that even if the network starts from a given (deterministic) initial state $z(0)$, the subsequent states will be random because of the stochastic nature of the evolution in Eq. 17.4. Consequently, the cost in Eq. 17.6 must be defined using expectation. Equation 17.6 provides one component of the finite-horizon cost, namely the cost of control. We next introduce the second component.

The net result of the control actions $v(0)$, $v(1)$,..., $v(M-1)$ is that the state of the PBN will transition according to Eq. 17.4 and will end up in some state $z(M)$. Owing to the probabilistic nature of the evolution, the terminal state $z(M)$ is a random variable that can possibly take on any of the values in S. Depending on the particular PBN and the control inputs used at each step, it is possible that, unless the PBN is ergodic, some of these states may never be reached because of non-communicating states in the resulting Markov chains; however, since the control strategy itself has not yet been determined, it would be difficult, if not impossible, to identify and exclude such states from further consideration. Instead, we assume that all 2^n terminal states are reachable and assign a penalty, or terminal cost, $C_M(z(M))$ to each of them. We next consider penalty assignment.

First, consider the PBN with all controls set to zero i.e. $v(k) \equiv 1$ for all k. Then, divide the states into different categories depending on how desirable or undesirable they are and assign higher terminal costs to the undesirable states. For instance, a state associated with rapid cell proliferation leading to cancer should be associated with a high terminal penalty, while a state associated with normal behavior should be assigned a low terminal penalty. For the purposes of this section, we will assume that the assignment of terminal penalties has been carried out and we have at our disposal a terminal penalty $C_M(z(M))$ that is a function of the terminal state. Thus, we have arrived at the second component of our cost function. Once again, note that the quantity $C_M(z(M))$ is a random variable and so we must take its expectation while defining the cost function to be minimized. In view of Eq. 17.6, the finite-horizon cost to be minimized is given by

$$E[\sum_{k=0}^{M-1} C_k(z(k), v(k)) + C_M(z(M))|z(0)]. \quad (17.7)$$

To proceed further, let us assume that at time k the control input $v(k)$ is a function of the current state $z(k)$, namely,

$$v(k) = \mu_k(z(k)) \quad (17.8)$$

where $\mu_k : S \rightarrow U$. The *optimal control problem* can now be stated:

Given an initial state $z(0)$, find a control law $\pi = \{\mu_0, \mu_1, ..., \mu_{M-1}\}$that minimizes the cost functional

$$J_\pi(z(0)) = E[\sum_{k=0}^{M-1} C_k(z(k), \mu_k(z(k))) + C_M(z(M))|z(0)] \tag{17.9}$$

subject to the constraint

$$\Pr\{z(k+1) = j|z(k) = i, v(k) = v\} = a_{ij}(v) \tag{17.10}$$

where $a_{ij}(v)$ is the ith row, j^{th} column entry of the matrix $A(v)$.

17.1.1 Solution Using Dynamic Programming

Optimal control problems of the type described by Eqs. 17.9 and 17.10 can be solved using the technique of *dynamic programming.* This technique pioneered by Bellman in the 1960s, is based on the so-called *principle of optimality.* This principle is a simple but powerful concept and can be explained as follows. Consider an optimization problem where we are interested in optimizing a performance index over a finite number, M, of steps. At each step, a decision is made and the objective is to devise a strategy or *sequence* of M decisions that is optimal in the sense that the cumulative performance index over all the M steps is optimized. In general, such an optimal strategy may not exist. However, when such an optimal strategy does exist, the principle of optimality asserts: if one searches for an optimal strategy over a subset of the original number of steps, then this new optimal strategy will be given by the overall optimal strategy, restricted to the steps being considered. Although intuitively obvious, the principle of optimality can have far reaching consequences. For instance, it can be used to obtain the following proposition [134].

PROPOSITION 17.1
Let $J^*(z(0))$ be the optimal value of the cost functional in Eq. 17.9. Then

$$J^*(z(0)) = J_0(z(0)),$$

where the function J_0 is given by the last step of the following dynamic programming algorithm, which proceeds backward in time from time step $M-1$ to time step 0:

$$J_M(z(M)) = C_M(z(M)) \tag{17.11}$$

$$J_k(z(k)) = \min_{v(k)\in U} E\{C_k(z(k), v(k)) + J_{k+1}[z(k+1)]\} \tag{17.12}$$

for $k = 0, 1, \ldots, M-1$. Furthermore, if $v^*(k) = \mu_k^*(z(k))$ minimizes the right hand side of (17.12) for each $z(k)$ and k, then the control law $\pi^* = \{\mu_0^*, \mu_1^*, \ldots, \mu_{M-1}^*\}$ is optimal.

Note that the expectation on the right hand side of Eq. 17.12 is conditioned on $z(k)$ and $v(k)$. Hence, in view of Eq. 17.10, it follows that

$$E[J_{k+1}(z(k+1))|z(k), v(k)] = \sum_{j=1}^{2^n} a_{z(k),j}(v(k))J_{k+1}(j).$$

Thus the dynamic programming solution to Eqs. 17.9 and 17.10 is given by

$$J_M(z(M)) = C_M(z(M)) \tag{17.13}$$

$$J_k(z(k)) = \min_{v(k)\in U} \{C_k(z(k), v(k)) + \sum_{j=1}^{2^n} a_{z(k),j}(v(k))J_{k+1}(j)\} \tag{17.14}$$

for $k = 0, 1, \ldots, M-1$.

We next present two extensive examples to show optimal control design using the dynamic programming approach. The first is contrived for illustrative purposes only, while the second is realistic and based on actual gene expression data.

17.1.2 A Simple Illustrative Example

We consider an example of a PBN with control and work through the details to show how Eqs. 17.13 and 17.14 can be used to arrive at an optimal control strategy. The example is adapted from the one used in sections 16.2 and 16.3 and involves the truth table in Table 16.1, which corresponds to an uncontrolled PBN. To introduce control, let us assume that x_1 in that table is now going to be a control input whose value can be externally switched between 0 and 1 and the genes of the new PBN are x_2 and x_3. To be consistent with the notation introduced in this section, the variables x_1, x_2 and x_3 will be renamed; the earlier variable x_1 now becomes u_1 while the earlier variables x_2 and x_3 become x_1 and x_2 respectively. With this change, we have the truth table shown in Table 17.1, which also contains the values of the variables v and z corresponding to u_1 and x_1x_2, respectively. The values of $c_j^{(i)}$ in the table dictate that there are two possible networks, the first network N_1 corresponding to the choice of functions $(f_1^{(1)},f_1^{(2)})$ and the second network N_2 corresponding to the choice of functions $(f_1^{(1)},f_2^{(2)})$. The probabilities c_1 and c_2 associated with each of these networks is given by $c_1 = c_2 = 0.5$. We next proceed to compute the matrices $A(1)$ and $A(2)$ corresponding to the two possible values for v. From Table 17.1, it is clear that when $v=1$, the following transitions are associated with the network N_1 and occur with probability c_1:

$$z=1 \rightarrow z=1,\ z=2 \rightarrow z=3,\ z=3 \rightarrow z=3,\ z=4 \rightarrow z=2 \tag{17.15}$$

The corresponding transitions associated with network N_2 that occur with probability c_2 are given by:

$$z=1 \rightarrow z=1,\ z=2 \rightarrow z=3,\ z=3 \rightarrow z=3,\ z=4 \rightarrow z=1 \tag{17.16}$$

TABLE 17.1 Truth Table for the example of this section [127] (With kind permission of Springer Science and Business Media).

u_1	v	x_1	x_2	z	$f_1^{(1)}$	$f_1^{(2)}$	$f_2^{(2)}$
0	1	0	0	1	0	0	0
0	1	0	1	2	1	0	0
0	1	1	0	3	1	0	0
0	1	1	1	4	0	1	0
1	2	0	0	1	1	0	0
1	2	0	1	2	1	1	0
1	2	1	0	3	0	1	0
1	2	1	1	4	1	1	1
$c_j^{(i)}$					1	0.5	0.5

In view of Eqs. 17.15 and 17.16, the matrices $A(1)$ and $A(2)$ are given by

$$A(1) = \begin{bmatrix} 1 & 0 & 0 & 0 \\ 0 & 0 & 1 & 0 \\ 0 & 0 & 1 & 0 \\ c_2 & c_1 & 0 & 0 \end{bmatrix} \tag{17.17}$$

$$A(2) = \begin{bmatrix} 0 & 0 & 1 & 0 \\ 0 & 0 & c_2 & c_1 \\ c_2 & c_1 & 0 & 0 \\ 0 & 0 & 0 & 1 \end{bmatrix}. \tag{17.18}$$

In this example, $n=2$ so that the variable z can take on any one of the four values 1, 2, 3, or 4. Since $m=1$, the control variable v can take on any one of the two values 1 or 2. Suppose that the control action is to be carried out over five steps, so that $M=5$. Moreover, assume that the terminal penalties are given by

$$C_5(1) = 0,\ C_5(2) = 1,\ C_5(3) = 2,\ C_5(4) = 3 \tag{17.19}$$

Note that the choices of M and the values of the terminal penalties are completely arbitrary; in a real-world example, this information would be obtained from biologists. The current choice of terminal penalties indicates that the most desirable terminal state is 1 and the least desirable terminal state is 4. For the optimization problem of Eqs. 17.9 and 17.10, we need to define the function $C_k(z(k), v(k))$. For the sake of simplicity, let us define

$$C_k(z(k), v(k)) = \sum_{i=1}^{m} u_i(k) = u_1(k) \tag{17.20}$$

where $v(k)$ and $u_i(k)$, i= 1, 2,..., m, are related by Eq. 17.2. The cost $C_k(z(k), v(k))$ captures the cost of applying the input $u_1(k)$ at the kth step. The optimization problem of Eqs. 17.9 and 17.10 can now be posed using the quantities defined in Eqs. 17.17, 17.18, 17.19, 17.20. The dynamic programming algorithm resulting from Eqs. 17.13 and 17.14 becomes

$$J_5(z(5)) = C_5(z(5)) \tag{17.21}$$

$$J_k(z(k)) = \min_{v(k)\in\{1,2\}} [u_1(k) + \sum_{j=1}^{4} a_{z(k),j}(v(k))J_{k+1}(j)], \quad k = 0, 1, 2, 3, 4 \tag{17.22}$$

We proceed backwards step by step from k= 4 to obtain a solution to Eqs. 17.21 and 17.22. The resulting optimal control strategy for this finite horizon problem is:

$$\mu_0^*(z(0)) = \mu_1^*(z(1)) = \mu_2^*(z(2)) = \mu_3^*(z(3)) = 1 \text{ for all } z(0), z(1), z(2), z(3) \tag{17.23}$$

$$\mu_4^*(z(4)) = \begin{cases} 2 \text{ if } z(4) = 3 \\ 1 \text{ otherwise.} \end{cases} \tag{17.24}$$

Thus, the control input is applied only in the last time step, provided the state z of the system at that time step is equal to 3; otherwise, the optimal control strategy is to not apply any control at all. Let us now consider a few different initial states $z(0)$ and see whether this optimal control strategy makes intuitive sense.

Case 1. $z(0) = 1$: According to Eqs. 17.23 and 17.24, the optimal control strategy in this case is no control. Note from Eq. 17.19 that the evolution of the PBN is starting from the most desirable terminal state. Furthermore, from Eq. 17.17, it is clear that in the absence of any control, the state of the network remains at this position. Hence, the control strategy arrived at is, indeed, optimal and the value of the optimal cost is 0.

Case 2. $z(0) = 4$: In this case, from Eq. 17.19, it is clear that the evolution of the PBN is starting from the most undesirable terminal state. Moreover, from Eq. 17.18, note that if the control input were kept turned ON over the entire control horizon, then the state would continue to remain in this most undesirable position during the entire control duration. Such a control strategy cannot be optimal since not only does the network end up in the most undesirable terminal state but also the maximum possible control cost is incurred over the entire time horizon.

To get a more concrete feel for the optimal control strategy, let us focus on the cases where the PBN degenerates into a standard (deterministic) Boolean network. There are two cases to consider:

1. $c_2 = 1$, $c_1 = 0$: In this case, from Eq. 17.17 we have

$$A(1) = \begin{bmatrix} 1\,0\,0\,0 \\ 0\,0\,1\,0 \\ 0\,0\,1\,0 \\ 1\,0\,0\,0 \end{bmatrix}. \tag{17.25}$$

   Clearly, if no control is employed then, starting from $z(0) = 4$, the network will reach the state $z(1) = 1$ in one step and stay there forever. Thus, this no-control strategy is optimal and the optimal cost is 0.

2. $c_2 = 0$, $c_1 = 1$: In this case, from Eqs. 17.17 and 17.18 we have

$$A(1) = \begin{bmatrix} 1\,0\,0\,0 \\ 0\,0\,1\,0 \\ 0\,0\,1\,0 \\ 0\,1\,0\,0 \end{bmatrix}, \; A(2) = \begin{bmatrix} 0\,0\,1\,0 \\ 0\,0\,0\,1 \\ 0\,1\,0\,0 \\ 0\,0\,0\,1 \end{bmatrix} \tag{17.26}$$

   From Eq. 17.23 the optimal control strategy is no control over the first four time steps. From Eq. 17.26 it follows that, with $z(0) = 4$, we will have $z(1) = 2$, $z(2) = 3$, $z(3) = 3$ and $z(4) = 3$. Then at the last time step, the control input is turned ON and from Eq. 17.26 the resulting state is $z(5) = 2$. The optimal cost is 2 (the sum of the terminal cost and the cost of control).

17.1.3 Melanoma Example

We now apply the methodology of this section to derive an optimal intervention strategy for a particular gene regulatory network. The network chosen as an example of how control might be applied is one developed from data collected in a study of metastatic melanoma [113]. In this expression profiling study, the abundance of messenger RNA for the gene WNT5A was found to be a highly discriminating difference between cells with properties typically associated with high metastatic competence versus those with low metastatic competence. These findings were validated and expanded in a second study [114]. In this study, experimentally increasing the levels of the Wnt5a protein secreted by a melanoma cell line via genetic engineering methods directly altered the metastatic competence of that cell as measured by the standard in vitro assays for metastasis. A further finding of interest in the current study was that an intervention that blocked the Wnt5a protein from activating its receptor, the use of an antibody that binds Wnt5a protein, could substantially reduce Wnt5a's ability to induce a metastatic phenotype. This suggests a study of control based on interventions that alter the contribution of the WNT5A gene's action to biological regulation, since the available data suggest that disruption of this influence could reduce the chance of a melanoma metastasizing, a desirable outcome.

The methods for choosing the genes involved in a small local network that includes the activity of the WNT5A gene and the rules of interaction have been described in [115]. The WNT5A network was obtained by studying the predictive relationship between 587 genes. The expression status of each gene was quantized to one of three possible levels: -1 (down-regulated), 0 (unchanged) and 1 (upregulated). In this case, the gene activity profile at any time step is ternary, not binary; nonetheless, the PBN formulation and the associated control strategy can be developed exactly as described, the only difference being that now, for an n-gene network, we will have 3^n states instead of 2^n states. In this context, it is appropriate to point out that to apply the control algorithm of this section, it is not necessary to actually construct a PBN; all that is required are the transition probabilities between the different states under the different controls.

A ternary network with 587 genes will have 3^{587} states, which is an intractably large number to use either for modeling or for control. Consequently, the number of genes was reduced to the ten most significant ones and the resulting multivariate relationships, using the best three-gene predictor for each gene, are shown in Fig. 17.1. These relationships were developed using the CoD technique (see Chapter 15) applied to the gene-expression patterns across 31 different conditions and prior biological knowledge (a detailed description being given in [115]).

Because it is biologically known that WNT5A ceasing to be down-regulated is strongly predictive of the onset of metastasis, the control objective for this 10-gene network is to externally down-regulate theWNT5A gene. Controlling the 10-gene network using dynamic programming would require designing a control algorithm for a system with 3^{10} (59,049) states. Although there is nothing conceptually difficult about doing this, it was beyond the limits of our software at that time. Accordingly, we further narrowed down the number of genes in the network to 7 by using CoD analysis on the 31 samples. The resulting genes, along with their multivariate relationships, are shown in Fig. 17.2. For each gene in this network, we determined its two best two-gene predictors and their corresponding CoDs. Using the procedure sketched in Chapter 15 and elaborated upon in [104], the CoD information for each of the predictors was then used to determine the $3^7 \times 3^7$ matrix of transition probabilities for the Markov chain corresponding to the dynamic evolution of the gene-activity profile of the seven-gene network.

The optimal control problem can now be completely specified by choosing (i) the treatment/intervention window, (ii) the terminal penalty and (iii) the types of controls and the costs associated with them. For the treatment window, we arbitrarily chose a window of length 5, i.e. control inputs would be applied only at time steps 0, 1, 2, 3 and 4. The terminal penalty at time step 5 was chosen as follows. Since our objective was to ensure that WNT5A was down regulated, we assigned a penalty of 0 to all states for which WNT5A equaled -1, a penalty of 3 to all states for which WNT5A equaled 0 and a penalty of 6 to all states for which WNT5A equaled 1. Here the choice of the

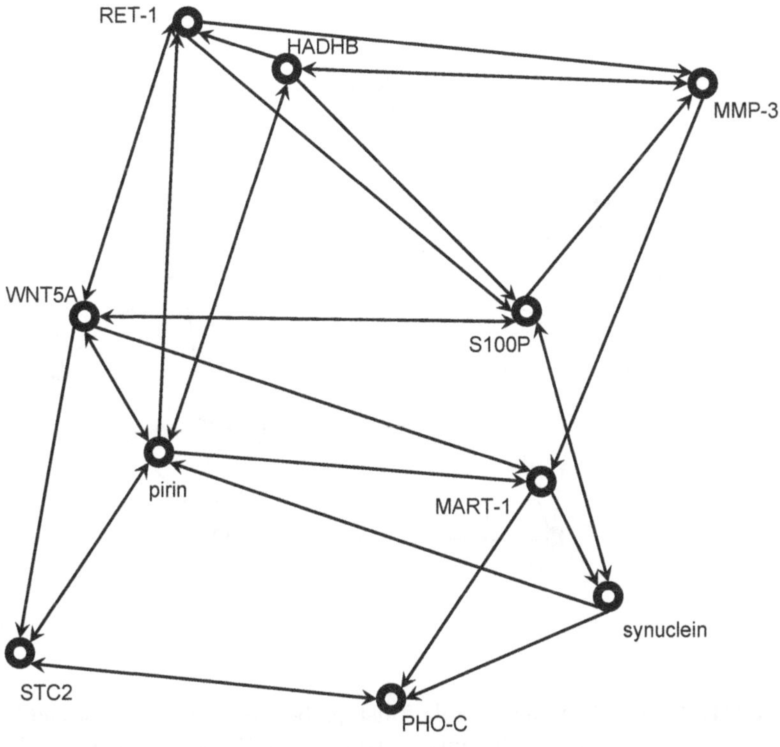

FIGURE 17.1 Multivariate relationship between the genes of the 10-gene WNT5A network [115].

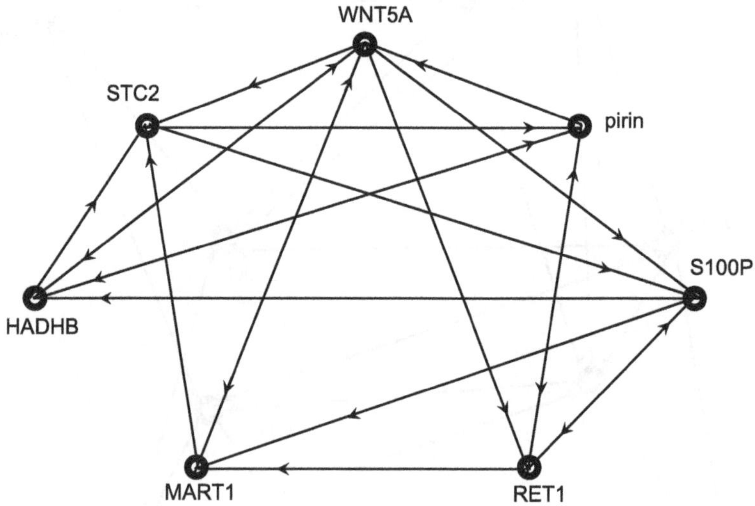

FIGURE 17.2 Multivariate relationships between the genes of the 7-gene WNT5A network [127]. (With kind permission of Springer Science and Business Media)

numbers 3 and 6 was arbitrary but it did reflect our attempt to capture the intuitive notion that states where WNT5A equaled 1 were less desirable than those where WNT5A equaled 0. Two types of possible controls were used and we discuss the two cases separately.

Case 1. WNT5A controlled directly: In this case, the control action at any given time step is to force WNT5A equal to −1, if necessary, and let the network evolve from there. Biologically such a control could be implemented by using a WNT5A inhibitory protein. In this case, the control variable is binary with 0 indicating that the expression status of WNT5A has not been *forcibly* altered while 1 indicates that such a forcible alteration has taken place. Of course, whether at a given time step such intervention takes place or not is decided by the solution to the resulting dynamic programming algorithm and the actual state of the network immediately prior to the intervention. With this kind of intervention strategy, it seems reasonable to incur a control cost at a given time step if and only if the expression status of WNT5A has to be forcibly changed at that time step. Once again, we arbitrarily assigned a cost of 1 to each such forcible change and solved for the optimal control using dynamic programming. The net result was a set of optimal control inputs for each of the 2187 (3^7) states at each of the five time points. Using these control inputs, we studied the evolution of the state probability distribution vector with and without control. For every possible initial state, our simulations indicated that at every time step from 1 to 5, the probability of WNT5A being equal to −1 was higher with control than that without control. Furthermore, with control, WNT5A always reached −1 at the final time point ($k = 5$). Thus, we concluded that the optimal control strategy of this section was successful in achieving the desired control objective. In this context, it is significant to point out that if the network started from the initial state STC2 = −1, HADHB = 0, MART-1 = 0, RET-1 = 0, S100P = −1, pirin = 1, WNT5A = 1 and if no control was used, then it quickly transitioned to a *bad absorbing* state (absorbing state with WNT5A = 1). With optimal control, however, this did not happen.

Case 2. WNT5A controlled through pirin: In this case, the control objective was the same as in Case 1, namely to keep WNT5A down-regulated. The only difference was that this time we used another gene, pirin, to achieve this control. The treatment window and the terminal penalties were kept exactly the same as before. The control action consisted of either forcing pirin to −1 (corresponding to a control input of 1) or letting it remain wherever it was (corresponding to a control input of 0). As before, at any step, a control cost of 1 was incurred if and only if pirin was forcibly reset to −1 at that time step. Having chosen these design parameters, we implemented the dynamic programming algorithm with pirin as the control. Using the resulting optimal controls, we studied the evolution of the state probability distribution vector with and without control. For every possible initial state, our simulations indicated that, at the final state, the probability of WNT5A being equal to −1 was higher with control than that without control. In this case, however, there was no definite ordering of probabilities between the controlled and uncontrolled cases at the intermediate time points. Moreover, the probability of WNT5A being equal to -1 at the final time point was not, in general, equal to 1. This was not surprising given that in this case we were trying to control the expression status of WNT5A using another gene and the control horizon of length 5 simply might not have been adequate for achieving the desired objective with such a high probability. Nevertheless, even in this case, if the network started from the state corresponding to STC2 $=-1$, HADHB $= 0$, MART-1 $= 0$, RET-1 $= 0$, S100P $= -1$, pirin $= 1$, WNT5A $= 1$ and evolved under optimal control, then the probability of WNT5A $= -1$ at the final time point equaled 0.673521. This was quite good in view of the fact that the same probability would have been equal to 0 in the absence of any control action.

17.2 External Intervention in the Imperfect Information Case

The control law that emerged from the solution of the dynamic programming problem of Eqs. 17.13 and 17.14 took the form of a state feedback †

$$v_k = \mu_k(z_k), \; k = 0, 1, 2, \ldots, M-1. \tag{17.27}$$

When the state vector z_k of the PBN is not available for measurement, such a control law cannot be implemented. In that case, we will assume that when the PBN is in the state z_k, it emits q_0 *measurable* outputs, each of which could take on the value 0 or 1. Thus, the output status of the PBN at any time k can be captured by a q_0-digit binary number or, alternatively, by its decimal equivalent plus one, which we shall call θ_k. As the outputs range over all possible binary values, θ_k takes on all values from 1 to 2^{q_0}.

The design of the optimal control in this case can make use of only the signals available to the controller. In other words, at time k, the controller tries to design the control input v_k using all the available signals, $\theta_0, \theta_1, \ldots, \theta_k, v_0, v_1, \ldots, v_{k-1}$. Although the state z_k evolves according to Eq. 17.10 and is not available for measurement, we assume that the output θ_k at time k is probabilistically related to the state z_k at time k and the input v_{k-1} through the known conditional probability measure $\Pr_{\theta_k}(.|z_k, v_{k-1})$ defined by

$$\Pr\{\theta_k = j | z_k = i, v_{k-1} = v\} = r_{ij}^v. \tag{17.28}$$

†In the rest of this chapter, we will be denoting $w(k)$, $z(k)$, $v(k)$, $d(k)$ by w_k, z_k, v_k, d_k respectively, mainly for the purpose of simplifying the notation.

The total information available for control at time k is given by $I_k = [\theta_0, v_0, \theta_1, v_1, \ldots, v_{k-1}, \theta_k]^T$. I_k can be generated recursively using the equation

$$I_{k+1} = [I_k^T, v_k, \theta_{k+1}]^T, \; I_0 = \theta_0. \tag{17.29}$$

Since the state z_k is not available, it seems reasonable to replace the state feedback control of Eq. 17.27 by the information feedback control

$$v_k = \mu_k(I_k), \; k = 0, 1, 2, \ldots, M-1 \tag{17.30}$$

and search for the optimal μ_k over the space of all functions μ_k mapping the space of information vectors I_k into the control space U. Thus the counterpart to the optimization problem of Eqs. 17.9 and 17.10 for this case becomes [130, 134]

$$\min_{\mu_0,\mu_1,\ldots,\mu_{M-1}} E_{z_0,d_0,d_1,\ldots,d_{M-1},\theta_0,\theta_1,\ldots,\theta_{M-1}}\{\sum_{k=0}^{M-1} C_k(z_k, \mu_k(I_k), d_k) + C_M(z_M)\} \tag{17.31}$$

subject to

$$z_{k+1} = d_k, \tag{17.32}$$

$$Pr\{d_k = j | z_k = i, v_k\} = a_{ij}(v_k), \tag{17.33}$$

$$I_{k+1} = [I_k^T, v_k, \theta_{k+1}]^T, \; I_0 = \theta_0 \tag{17.34}$$

The dynamic programming algorithm for the above problem is given by [130, 134]

$$J_{M-1}(I_{M-1}) = \min_{v_{m-1} \in U}\{E_{z_{M-1},d_{M-1}}[C_M(d_{M-1}) + C_{M-1}(z_{M-1}, v_{M-1}, d_{M-1})|I_{M-1}, v_{M-1}]\} \tag{17.35}$$

$$J_k(I_k) = \min_{v_k \in U}\{E_{\theta_{k+1},z_k,d_k}[\{C_k(z_k, v_k, d_k) + J_{k+1}([I_k^T, v_k, \theta_{k+1}]^T)\} |I_k, v_k]\} \tag{17.36}$$

for $k = 0, 1, 2, \ldots, M-2$, and the optimal control input is obtained from the values minimizing the right-hand side of Eqs. 17.35 and 17.36. Using this algorithm, we will ultimately arrive at $J_0(I_0) = J_0(\theta_0)$. The optimal cost J^* can be obtained by taking the expectation of this quantity with respect to θ_0, i.e.

$$J^* = E_{\theta_0}[J_0(\theta_0)] \tag{17.37}$$

17.2.1 Melanoma Example

Consider a seven-gene network which is a slight variation of the one considered in Section 17.1.3. Since implementing the imperfect information based control is computationally more intensive compared to the perfect information case, we have developed a binary seven-gene network using CoD analysis on the same experimental data. The resulting genes, along with their multivariate relationship, are shown in Fig. 17.2. For each gene in this network, we have determined their two best two-gene predictors and their corresponding CoDs. Using the procedure discussed in Chapter 15 and elaborated upon in [104], the CoD information for each of the predictors is used to determine the $2^7 \times 2^7$ matrix of transition probabilities for the Markov chain corresponding to the dynamic evolution of the GAP of the seven-gene network. The transition probability matrix $A(v(k))$, the probability distribution of the observations given the current state and the immediately prior control r_{ij}^v, and the initial state probability distribution vector together constitute the data needed for the optimal control problem with imperfect state information. In our construction, the vector r_{ij}^v does not depend on the prior control input v and probabilistically relates the observation to the current state of the network. This relationship is shown in Fig. 17.3 and it closely mimics the behavior of a gene MMP-3 that appears in the 10-gene network of Fig. 17.1 but does not appear in the 7-gene network of Fig. 17.2.

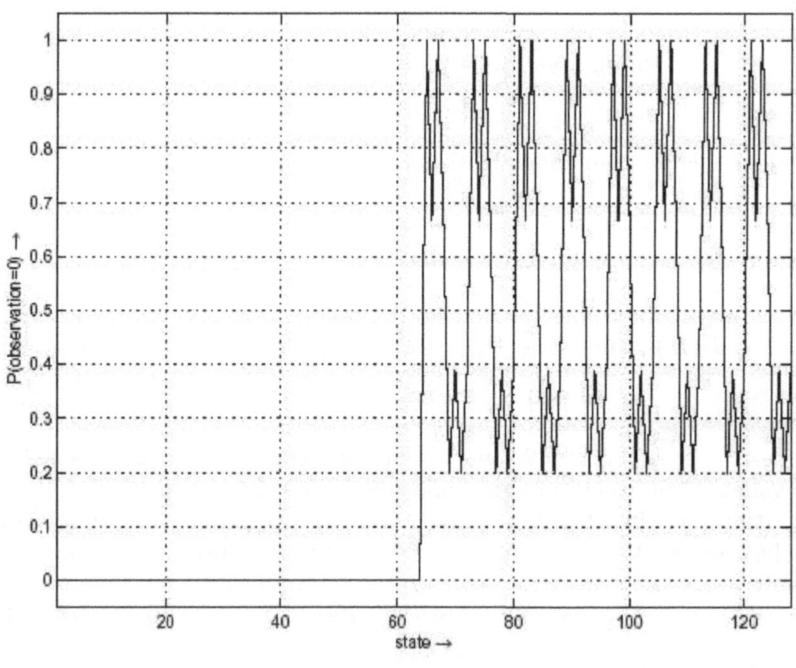

FIGURE 17.3 Probability (observed variable =0) versus current state [130].

The optimal control problem is completely specified by choosing (i) the treatment/intervention window, (ii) the terminal penalty and (iii) the types of controls and the costs associated with them. For the treatment window, we arbitrarily choose a window of length 5, i.e. time steps 0, 1, 2, 3 and 4. Based upon the same reasoning as in the full-information case, the terminal penalty at time step 5 was chosen as 0 for all states for which WNT5A equals 0 and 3 for all states for which WNT5A equals 1. We now discuss two possible types of control actions for various initial state probability distributions.

Case 1. WNT5A controlled directly: In this case, the control action at any given time step is to force WNT5A equal to 0, if necessary, and let the network evolve from there. The control variable is binary with 1 and 0 indicating intervention and no intervention, respectively. The one-step cost of control is taken to be the value of the control variable. Whether at a given time step intervention takes place is decided by the solution to the resulting dynamic programming algorithm depending on the initial distribution and the subsequent total information vector I_k. Note that unlike the perfect information scenario considered in the last section, we are now not in a position to determine if forcible alteration of the state takes place or not. Consequently, it is reasonable to expect that WNT5A inhibition may be used, even when not absolutely necessary, thereby contributing to a possible increase in the total optimal expected cost, compared to the perfect information case. We recursively used Eqs. 17.35 and 17.36 to calculate the optimal controls for certain initial state probability distributions. The net result, in each case, was a tree of control actions corresponding to each control action and subsequent observation. Starting with P_{data}, the distribution of states in the 31 point data set, we found the optimal expected cost based on imperfect information to be 0.4079. The corresponding optimal cost using full state observation as in the last section was 0.3226. The expected cost incurred by not using any control was 0.9677. We computed these quantities for a few different cases of initial state distributions. The relevant quantities are tabulated in Table 17.2.

TABLE 17.2 Expected costs for various initial state distributions [130].

Initial distribution	*Control using observation*	*Full state*	*No control*
$P_{sample-data}$	0.4079	0.3226	0.9677
$[\frac{1}{128}, \frac{1}{128}, ...]$	0.7068	0.3395	0.9990
$[0, \frac{1}{64}, 0, \frac{1}{64}, ...]$	0.7296	0.3395	0.9990
$[\frac{1}{64}, 0, \frac{1}{64}, 0, ...]$	0.5692	0.3395	0.9990

We also calculated the optimal expected costs when the initial state was deterministic. These values for all the 128 possible initial states are shown in Fig. 17.4. As expected, the optimal cost for control with imperfect informa-

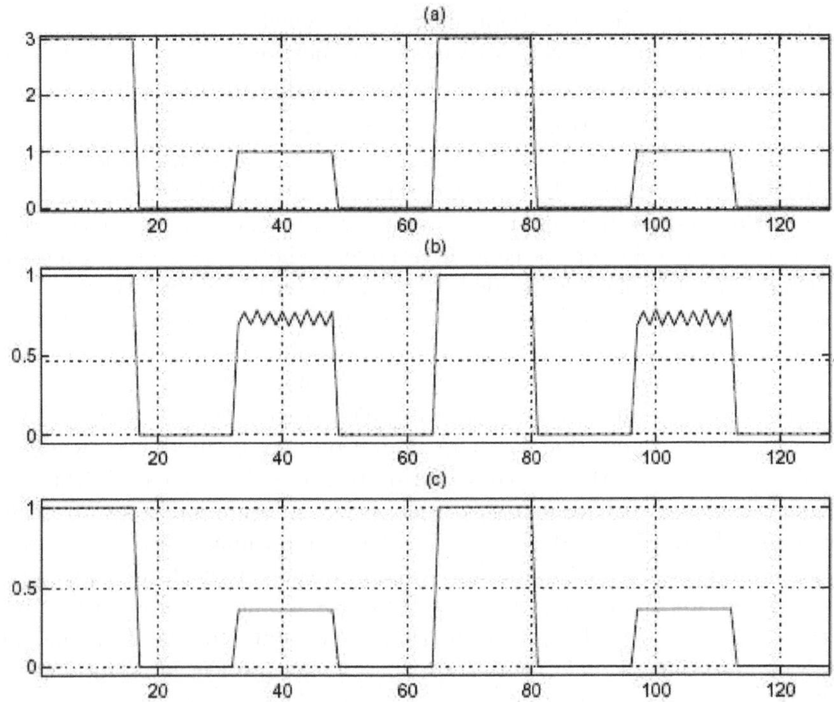

FIGURE 17.4 Optimal expected cost versus initial states **(a)** uncontrolled, **(b)** control using imperfect information, **(c)** control using full state information [130].

tion was higher than that for control with perfect state information. The cost function, however, is a somewhat subjective quantity chosen by us to mathematically capture the underlying biological objective. A more natural way to look at the performance of the control scheme would be to examine the probability of WNT5A being equal to 0 at the final time step, i.e. at $k=5$. This quantity has been computed for each (deterministic) initial state for both the uncontrolled and imperfect-information-based controlled cases. These plots are shown in Fig. 17.5. From this figure, it is clear that the control strategy for each initial state is increasing the probability for WNT5A being equal to 0 at the terminal time point relative to the corresponding probability in the uncontrolled case. This is a desirable outcome achieved by using control.

Case 2. WNT5A controlled through pirin: In this case, the control objec-

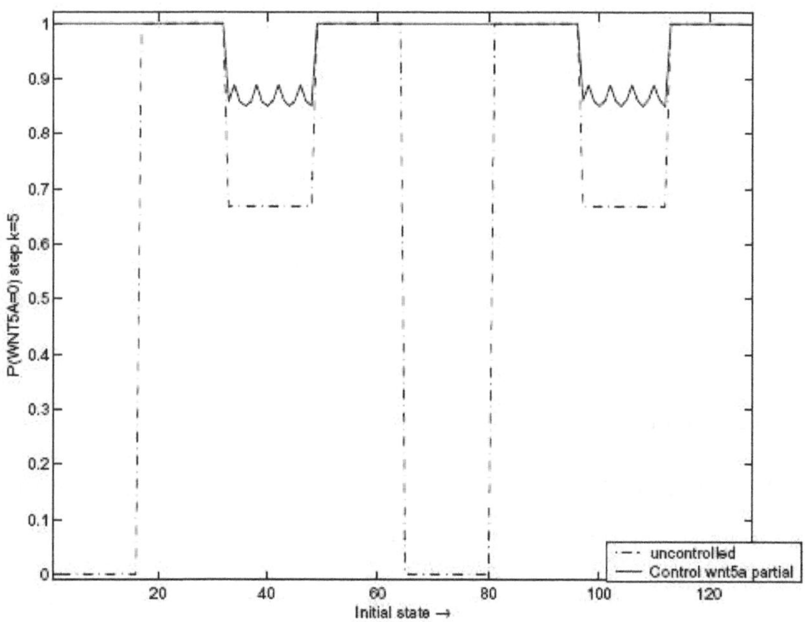

FIGURE 17.5 Probability of WNT5A= 0 at the terminal time point versus the initial state for the uncontrolled and imperfect-information-based controlled cases [130].

tive was the same as in Case 1, namely to keep WNT5A at 0; however, we now used pirin to achieve control. The treatment window and the terminal penalties were kept exactly the same as before. The control action consisted of either using a pirin inhibitor (corresponding to a control input of 1) or not employing such an inhibitor (corresponding to a control input of 0). The one-step cost of control was taken to be equal to the value of the control variable. As before, at any step, whether such intervention took place or not was decided by the solution to the resulting dynamic programming algorithm. Having chosen these design parameters, we implemented the algorithm with pirin as the control. We found that using pirin as a control was totally ineffective. The expected cost with pirin as the control was found to be the same as the one obtained in Table 17.2 with no control. Even with full state feedback, we still found that pirin was similarly ineffective (data not shown). This was in stark contrast to the results in the last section where we had demonstrated the feasibility of doing full state feedback control of WNT5A through pirin. It is possible that going from a ternary setup in the last section to the binary setup here may have drastically reduced our ability to control WNT5A through pirin. This suggests that the standard control theoretic notions of *controllability* and *observability* [135] may have to be revisited in the context of genetic regulatory networks to enable us to decide which genes can be used as effective controls and which ones can be used as meaningful observations.

17.3 External Intervention in the Context-Sensitive Case

This section extends the results of Section 17.1 to *context-sensitive* PBNs with perturbation. The intervention results from Section 17.1 carry over to this case and the only difference is that the entries of the transition probability matrix have to be derived differently. Since there are n genes, the probability of there being a random perturbation at any time point is $1-(1-p)^n$. For a context-sensitive PBN, the state z_k at time k could be originating from any one of the L possible networks. To keep track of the network emitting a particular state, let us redefine the states by incorporating the network number inside the state label. Since we have L different Boolean networks forming the PBN, the total number of states becomes 2^nL and we label these states as $S_1, S_2, ..., S_{2^nL}$, where for each r= 1, 2, ..., L, states $S_{2^n(r-1)+1}, S_{2^n(r-1)+2}, ..., S_{2^nr}$ belong to network r. Equivalently, $S_{2^n(r-1)+i}$ corresponds to z_{r_i} , where z_{r_i} is the decimal representation of the ith state in the network r. Denote the redefined state at time k by w_k[‡]. We need to derive the transition probability expressions for the uncontrolled and controlled cases. First, we treat the uncontrolled case.

[‡]This w_k should not be confused with the probability distribution vector w_k of section 17.1.

In a context-sensitive PBN with perturbation, one of the following four mutually exclusive events occurs at each time point k: (1) the current network function is applied, the PBN transitions accordingly, and the network function remains the same for the next transition; (2) the current network function is applied, the PBN transitions accordingly, and a new network function is selected for the next transition; (3) there is a random perturbation and the network function remains the same for the next transition; or (4) there is a random perturbation and a new network function is selected for the next transition. Assuming that the individual genes perturb independently, and letting mod(v, w) denote the remainder left over when v is divided by w, we consider two cases for determining the transition probability of going from state a to state b:

Case 1: $[(a-1)/2^n]=[(b-1)/2^n]$, meaning $2^n(r-1)+1 \leq a,b \leq 2^n r$ for the same r. This corresponds to the events (1) and (3) above and the transition probabilities are given by

$$\Pr(w_{k+1}=b|w_k=a) = (1-q)(1-p)^n\zeta_{r,a,b}+(1-q)(1-p)^{n-h}p^h s(h) \quad (17.38)$$

where h is the Hamming distance between $\text{mod}(a-1\ ,\ 2^n)$ and $\text{mod}(b-1,\ 2^n)$, i.e. the number of genes which differ between the two states,

$$\zeta_{r,a,b} = \begin{cases} 1, & \text{if } a \text{ transitions to } b \text{ in a single step in network } r, \\ 0, & \text{otherwise} \end{cases}$$

and

$$s(h) = \begin{cases} 0, & \text{if } h = 0, \\ 1, & \text{otherwise.} \end{cases}$$

The first term in Eq. 17.38 corresponds to event (1) above, where $1-q$ is the probability that the network selection does not change, $(1-p)^n$ is the probability that none of the n genes undergoes a perturbation, we assume that network selection and random gene perturbation are independent events, and $\zeta_{r,a,b}=1$ if that particular transition is possible in the rth Boolean network. The second term corresponds to event (3), where h genes have to be perturbed to go from state a to state b.

Case 2: $2^n(r_1-1)+1 \leq a \leq 2^n r_1$ and $2^n(r_2-1)+1 \leq b \leq 2^n r_2$, where $r_1 \neq r_2$. This corresponds to the events (2) and (4) above and the transition probabilities are given by

$$Pr(w_{k+1}=b|w_k=a) = qc_{r_2}\left(\sum_{i=1,i\neq r_1}^{L} c_i\right)^{-1}[(1-p)^n\zeta_{r_1,a,b} + (1-p)^{n-h}p^h s(h)] \quad (17.39)$$

If we define

$$g(a,b) = \begin{cases} 1, & \text{if } [(a-1)/2^n] - [(b-1)/2^n] = 0 \\ 0, & \text{otherwise} \end{cases}$$

then a unified transition probability expression encompassing the two cases is given by

$$\begin{aligned}
Pr(w_{k+1} = b|w_k = a) = {} & [(1-q)(1-p)^n \zeta_{r,a,b} + (1-q)(1-p)^{n-h}p^h s(h)]g(a,b) \\
& + [qc_{r_2} \left(\sum_{i=1, i \neq r_1}^{L} c_i \right)^{-1} [(1-p)^n \zeta_{r_1,a,b} \\
& + (1-p)^{n-h} p^h s(h)]] \times [1 - g(a,b)] \qquad (17.40)
\end{aligned}$$

By letting a and b range over all integers from 1 to $2^n L$ and using Eq. 17.40, we can determine all the entries of the $2^n L \times 2^n L$ matrix of transition probabilities.

In practice, it will likely be impossible to detect the Boolean network from which the current gene activity profile is being emitted. In most cases, we will have knowledge only of the states of the genes. To handle such situations, we can derive an expression for the transition probability from state s_2 to state s_1, where these states run from 1 to 2^n and reflect only the expression status of the n-gene state vector:

$$\begin{aligned}
\Pr[z_{k+1} = s_1 | z_k = s_2] &= \sum_{i=1}^{L} \Pr[z_{k+1} = s_1, s_2 \text{ belongs to network } i | z_k = s_2] \\
&= \sum_{i=1}^{L} \Pr[z_{k+1} = s_1 | z_k = s_2, s_2 \text{ belongs to network } i] \\
&\quad \times Pr[s_2 \text{ belongs to network } i] \\
&= \sum_{i=1}^{L} \Pr[z_{k+1} = s_1 | w_k = s_2 + 2^n(i-1)].c_i \\
&= \sum_{i=1}^{L} \sum_{j=1}^{L} c_i. \Pr[w_{k+1} = s_1 + 2^n(j-1) | w_k = s_2 + 2^n(i-1)] \qquad (17.41)
\end{aligned}$$

Note that here state s_1 is equivalent to the distinct states s_1, $s_1 + 2^n$,..., $s_1 + (L-1)2^n$ in the previous $2^n L$ formulation. Similarly, s_2 here is equivalent to s_2, $s_2 + 2^n$,..., $s_2 + (L-1)2^n$ in the earlier formulation. By letting s_1 and s_2 range from 1 to 2^n and using Eq. 17.41, we can derive the $2^n \times 2^n$ transition probability matrix A corresponding to the context-sensitive PBN.

If a control action is applied, then the transition probability expressions will change. Suppose our control action consists of forcibly altering the value of a single gene, g, from 0 to 1 or from 1 to 0. Thus, m=1. Then the new transition probabilities with control, denoted by $Prc1$, are given by

$$\begin{aligned}
Prc1(w_{k+1} = b|w_k = a) = {} & Pr(w_{k+1} = b|w_k = a + 2^{n-g})func(a) \\
& + Pr(w_{k+1} = b|w_k = a - 2^{n-g})(1 - func(a)) \quad (17.42)
\end{aligned}$$

where

$$func(a) = \begin{cases} 1, \text{ if state of gene } g \text{ is 0 for } a, \\ 0, \text{ if state of gene } g \text{ is 1 for } a, \end{cases}$$

and the transition probabilities, Pr, without control are given by Eq. 17.40. Here, a and b range over 1 through 2^nL. As before we can reduce the dimension of the state space by replacing the ws in Eq. 17.42 by zs and using Eq. 17.41 to determine the transition probabilities without the control action:

$$\begin{aligned} Prc1(z_{k+1} = b|z_k = a) &= Pr(z_{k+1} = b|z_k = a + 2^{n-g})func(a) \\ &+ Pr(z_{k+1} = b|z_k = a - 2^{n-g})(1 - func(a)) \end{aligned} \quad (17.43)$$

By letting a and b vary over 1 to 2^n and making use of Eq. 17.43, we can determine the $2^n \times 2^n$ matrix $A(v_k)$ of control-dependent transition probabilities.

From this point onwards, the formulation and solution of the control problem is exactly the same as in Section 17.1. To avoid unnecessary repetition, we proceed directly to the melanoma example considered in the two previous sections.

17.3.1 Melanoma Example

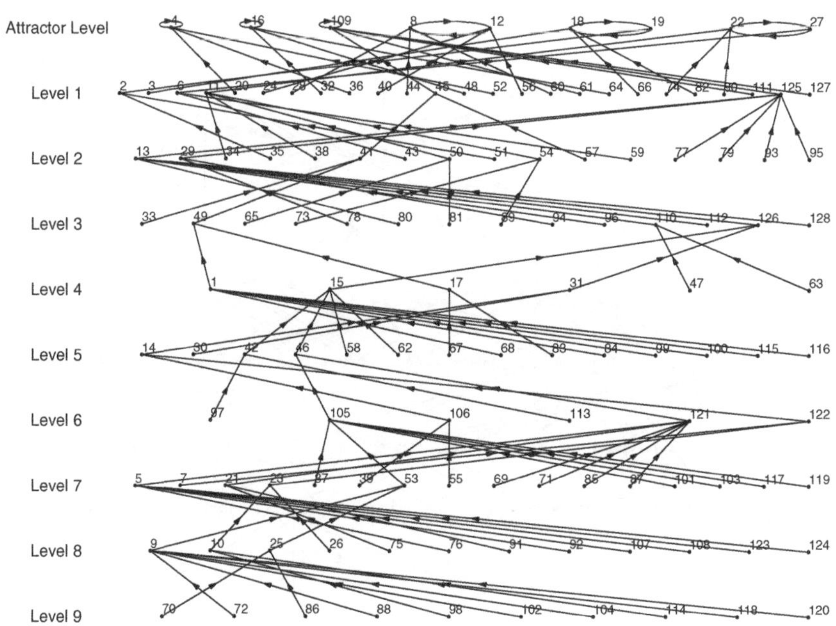

FIGURE 17.6 Network 1 [131].

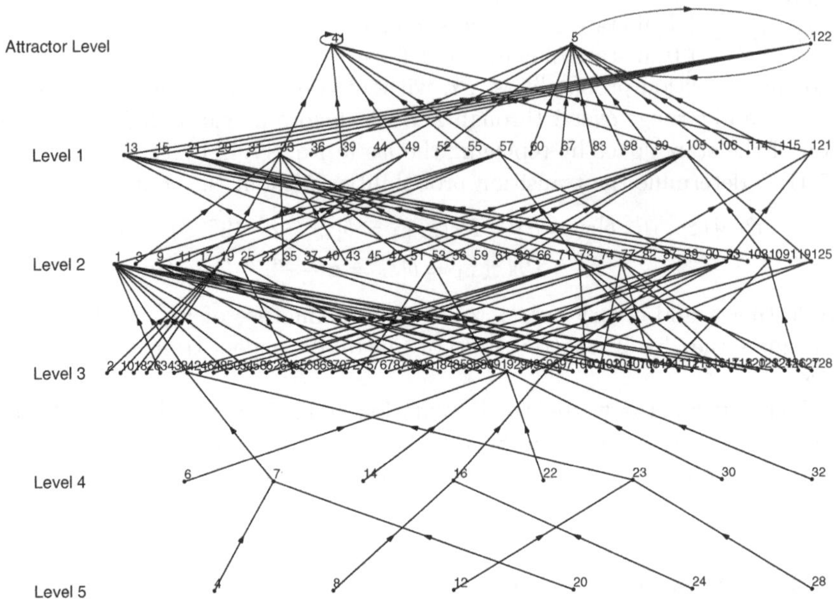

FIGURE 17.7 Network 2 [131].

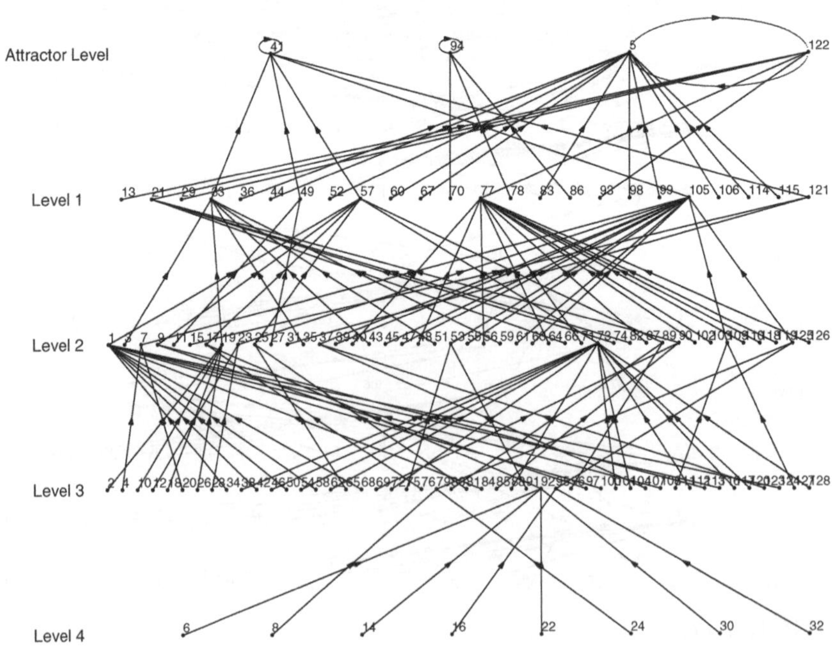

FIGURE 17.8 Network 3 [131].

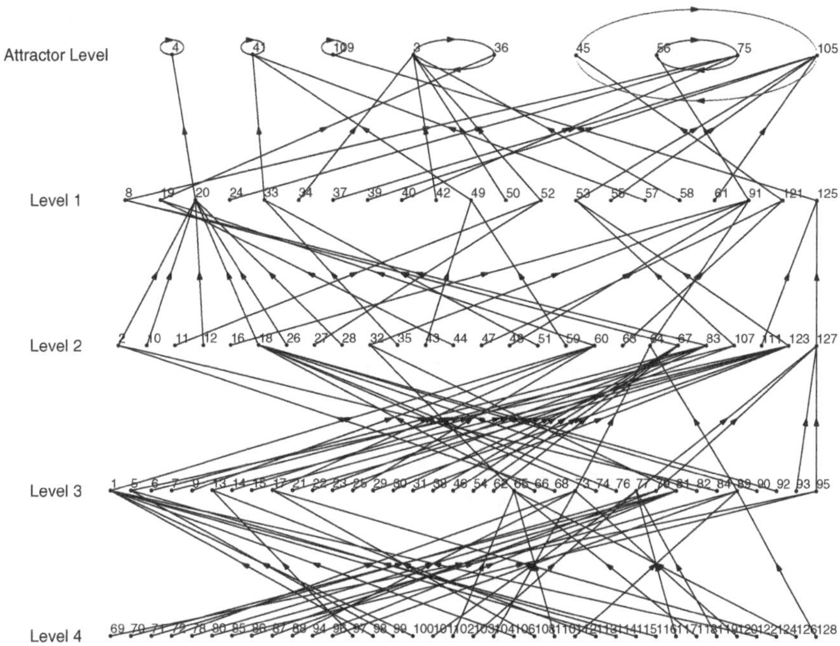

FIGURE 17.9 Network 4 [131].

We considered a 7-gene network with genes WNT5A, pirin, S100P, RET1, MART1, HADHB and STC2. Although derived from the same data, this network was designed based on *steady-state considerations* and, therefore, differs from the PBNs considered in sections 17.1 and 17.2. Carrying out the new design can be justified by the fact that most microarray-based gene-expression studies do not involve controlled time series experimental data; rather, it is assumed that data result from sampling from the steady state. Consequently, to obtain the PBN here, we used a Bayesian connectivity-based approach [116] to construct four highly probable Boolean networks that were used as the constituent Boolean networks in the PBN, with their selection probabilities based on their Bayesian scores. The four generated Boolean networks are shown in Figs. 17.6 through 17.9, where the states are labeled from 1 to $128 = 2^7$. Each constituent network was assumed to be derived from steady-state gene-expression data, and the attractor states and the level sets are shown in the figures. Observe that in each of these networks the state enters an attractor cycle in a small number of steps (at most nine), which is consistent with what is expected in real networks [116]. The control strategy of this section was applied to the designed PBN with pirin chosen as the control gene and $p = q = 0.01$. Fig. 17.10 shows the expected cost for a finite horizon problem of length 5 originating from each of the 128 states. In these simulations, the problem formulation for 2^n states was used. The cost of control was assumed to be 0.5 and the states were assigned a terminal penalty of 5 if WNT5A was 1 and 0 if WNT5A was 0. The control objective was to down-regulate the WNT5A gene. From Fig. 17.10, it is clear that the expected cost with control was much lower than that without control, which agreed with our objective.

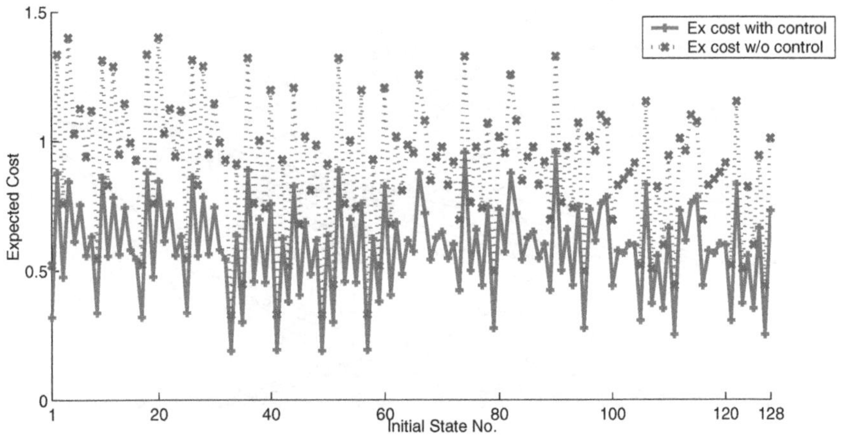

FIGURE 17.10 Expected cost for a finite horizon problem of length 5 originating from the different initial states [131].

17.4 External Intervention for a Family of Boolean Networks

The results of this section are motivated by the fact that most gene expression data used for PBN design are likely to come from the phenotype observed at steady-state. For instance, the gene expression data for cancer genomics studies are usually obtained from tumor biopsies. Given a data set consisting of gene-expression measurements, PBN design constitutes an ill-posed inverse problem that is treated by using a design algorithm to generate a solution. Inference can be formalized by postulating criteria that constitute a solution space for the inverse problem. As discussed in Chapter 15, the criteria come in two forms: (1) the *constraint criteria* are composed of restrictions on the form of the network, and (2) the *operational criteria* are composed of relations that must be satisfied between the model and the data. The solution space consists of all PBNs that satisfy the two sets of criteria. Recognizing that

PBNs are composed of Boolean networks, and since it is difficult to infer the probabilistic structure among the constituent Boolean networks from the steady-state data typically used for design, a more general view may be taken in which the inverse problem is restricted to determining a solution space of Boolean networks and then finding networks in that space [118]. Without a probabilistic structure among the Boolean networks, we have a family of Boolean networks satisfying both the constraint and operational criteria. If desired, one can then go further and construct a PBN by using networks from the family, or one can simply treat the family as a collection of solutions to the Boolean-network inverse problem.

In [132], we derived a control algorithm that can be applied to the second situation, i.e. to a family of Boolean networks. This is accomplished by minimizing a composite cost function that is a weighted average cost over the entire family. Ideally, the weighting for each member of the family at any time point would be proportional to the instantaneous probability of a particular network being the governing network. Although these instantaneous probabilities are not known, we adaptively estimate them from the available data and the estimate is used to implement the control algorithm.

To motivate the development here, we first revisit some of the characteristics of the optimal control solution presented in Section 17.1 for a single network. More specifically, let us focus attention on Eqs. 17.13 and 17.14. Eq. 17.14 states that the optimal cost to go from state $z(k)$ at the kth time step is the sum of the cost of the optimal control action at state $z(k)$ and the expected value of the cost to go at the $(k+1)$th time step. Since there is no control action in the terminal time step, Eq. 17.13 simply formalizes the fact that the cost to go at the terminal time step equals the penalty associated with the terminal state.

If a family of BNs is designed whose attractors match the data, assuming the family is not too small we have the expectation that the underlying biological phenomena are closely modeled by at least some of the BNs in the family. In the absence of perfect knowledge as to which BNs are capable of better representing the underlying phenomena, we develop a control policy that optimizes a composite cost function over the entire family of BNs.

Towards this end, let $\mathcal{N}$ be a set of L Boolean networks $N_1, N_2, \ldots, N_L$ possessing identical sets of singleton attractors, all sharing the same state space S and the same control space U. Associated with each network is an initial probability of it representing the underlying phenomenon. Since this information is not available, we will adaptively estimate these probabilities as more transitions are observed. For each network N_l, $l = 1, 2, \cdots, L$ define:

- $a^l_{ij}(v)$ to be the ith row, jth column entry of the matrix $A^l(v)$ of the network N_l;

- $C^l_k(i, v)$ to be the cost of applying the control v at the kth time step in state i in network N_l;

- $C_M^l(i)$ to be the terminal cost associated with state i in network N_l.

We define the *belief vector* $\pi_k = [\pi_k^1, \pi_k^2, \ldots, \pi_k^L]$, where π_k^l is the probability of network N_l being the underlying network at the kth time step. π_k is the probability distribution vector for the family of networks at the kth time step. Since π_k is unknown, we will make an initial guess for it and update it as more information becomes available. The use of this vector is inspired by the *information vector* in [136].

Suppose i is the current state at step k, π is the current estimate of the belief vector, and upon application of control v we observe state j at the next time step. Then the new belief vector is $\pi' = T(\pi, i|j, v)$, where the transformation T can be obtained by use of Bayes' theorem and the theorem of total probability,

$$\pi' = [\cdots, \frac{a_{ij}^l(v).\pi_k^l}{\sum_{s\in\mathcal{N}} a_{ij}^s(v).\pi_k^s}, \cdots] \tag{17.44}$$

We will now make use of this belief vector to set up the optimal control solution over a family of Boolean networks. Suppose we are given an initial belief vector π_0 and an initial state z_0. The initial belief vector is based on our prior knowledge of the system. It could be a function of likelihood or Bayesian scores of networks, or it could be uniform to reflect no prior knowledge. Our objective is to find controls $v_0, v_1, \cdots, v_k, \cdots, v_{M-1}$ to minimize the expectation of the cost-to-go function over all networks in $\mathcal{N}$. The cost to go at the kth time step $(0 \leq k < M)$ is a function of the current state z_k and the updated belief vector π_k. Motivated by (17.14) for the single PBN case, we define the average optimal cost-to-go function by

$$\begin{aligned} J_k(\pi_k, i) = \min_{v\in U}[\sum_{l\in\{1,2,\ldots,L\}} \pi_k^l \{C_k^l(i,v) + \\ \sum_{j\in S} a_{ij}^l(v).J_{k+1}(T(\pi_k, i|j, v), j)\}] \end{aligned} \tag{17.45}$$

The inner summation is the expectation over all $j \in S$ of the cost to go at the $(k+1)$th step in the lth network on observing j. We then add to it the cost of control at the kth step and average over all the networks in the family. Finally we take the minimum over all control actions in U to obtain the optimal policy and the cost to go at the kth step.

The terminal cost for a state i is trivially defined to be the average terminal cost over the entire family:

$$J_M(\pi_M, i) = \sum_{l\in\{1,2,\ldots,L\}} \pi_M^l.C_M^l(i). \tag{17.46}$$

In the melanoma examples of the previous sections, terminal penalties were assigned to states based on the expression level of a certain key gene, namely

WNT5A; however, as discussed in [137], it may be more reasonable to assign terminal penalties based on the long-term prospective behavior of the system in the absence of control. For any Markov chain with closed classes the procedure is summarized as follows:

- Partition the states of the Markov chain into transient and persistent states.
- For singleton attractors, the penalty C_M^l is set according to the status of the *penalty gene(s)*. A penalty gene is a gene for which certain expression statuses are known to be undesirable, e.g. WNT5A for the melanoma example.
- For a closed class, the penalty is based on the fraction of time spent in states having a penalty gene in an undesirable profile.
- For a transient state j, the terminal penalty

$$C_M^l(j) = \sum_i Prob(z_\infty = i | z_k = j, \text{ Network}{=}N_l).C_M^l(i), \tag{17.47}$$

where the summation variable i corresponds to a closed class or a singleton attractor. $Prob(z_\infty = i|z_k = j,$ Network$= N_l)$ is the long-term probability of getting absorbed in i starting from state j, in network N_l.

Since the attractors are shared by each network in the family, the attractor states will have the same penalty across the different networks; however, penalties for non-attractor states will differ across networks, depending on the particular attractor in whose basin that non-attractor state may happen to lie in.

17.4.1 Melanoma Example

Here we apply the methodology of this section to the same melanoma data considered earlier. As before, a family of networks with 7 genes: PIRIN, S100P, RET1, MART1, HADHB, STC2 and WNT5A is constructed. Since all 31 data points correspond to steady-state behavior, they should be considered as attractors in the networks. However, out of the 31 samples only 18 were distinct. To reduce the number of attractors, we formed seven clusters from the data points and treated the cluster centers as attractors. These attractors are shown in Table 17.3. The first column is used to classify them into two categories, GOOD and BAD, depending on the status of the WNT5A gene.

Using the procedure of [118], we obtained four distinct BN's (N_1, N_2, N_3, N_4) with the same set of seven attractors. We assigned a penalty of 5 to all states in the basin of the undesirable attractors (WNT5A $= 1$) and 0 to all the other states. We used pirin as the control gene. A forcible alteration in the expression level of pirin was associated with $v = 2$ while $v = 1$ represented no

TABLE 17.3 Cluster Centers as attractors for the WNT5A network. The good attractors are the ones with the profile of WNT5A gene downregulated. Pirin is the most significant bit (MSB) and WNT5A is the least significant bit (LSB). [132]

	z	Gene Activity Profile x						
		PIRIN	S100P	RET1	MART1	HADHB	STC2	WNT5A
B	4	0	0	0	0	0	1	1
A	32	0	0	1	1	1	1	1
D	82	1	0	1	0	0	0	1
G	33	0	1	0	0	0	0	0
O	57	0	1	1	1	0	0	0
O	95	1	0	1	1	1	1	0
D	109	1	1	0	1	1	0	0

control. In a reasoning similar to the previous sections, a terminal penalty of 5 for bad states vs. 0 for good states and a control cost of 1 for intervention vs. 0 for no intervention was our attempt to capture the intuitive notions of the relative costs of ending up in desirable vs. undesirable state and the cost of intervention.

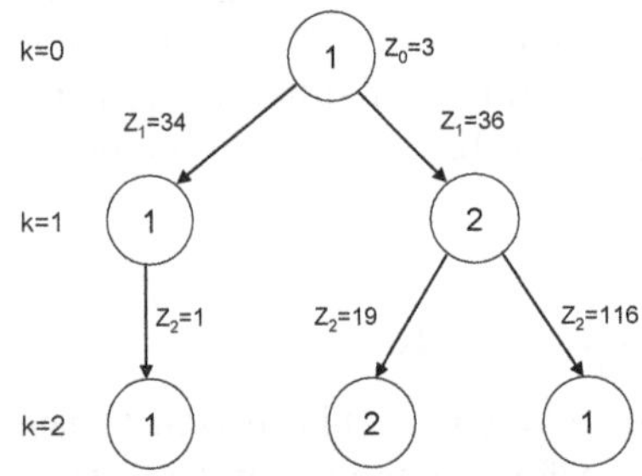

FIGURE 17.11 Policy tree for $M = 3$, initial state $z_0 = 3$ and initial belief vector $\pi_0 = [\frac{1}{4}, \frac{1}{4}, \frac{1}{4}, \frac{1}{4}]$ [132].

To present the results, we make use of pruned policy trees where the number inside each circle represents the optimal control action, and the arc following each circle corresponds to the next observed state which leads to the next optimal control action. A pruned policy tree for $M = 3$ with initial belief vector $\pi_0 = [\frac{1}{4}, \frac{1}{4}, \frac{1}{4}, \frac{1}{4}]$ and initial state $z_0 = 3$ is shown in Fig. 17.11.

For purposes of comparison, three different policies are considered side by side: Pol^{TR} being the optimal policy of this section; Pol^1, Pol^2, Pol^3, Pol^4

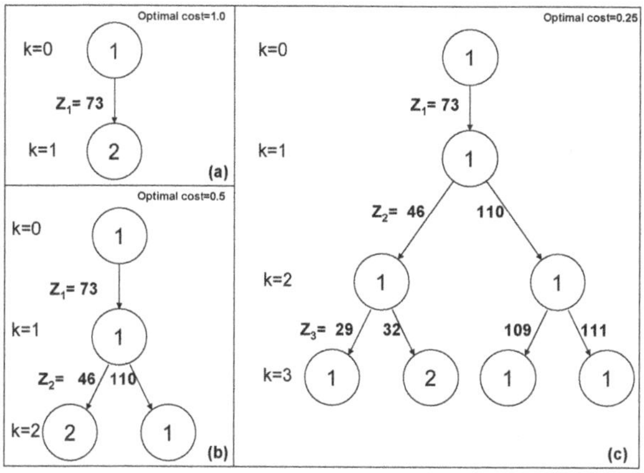

FIGURE 17.12 Policy trees and optimal costs, for initial state $z_0 = 93$, $\pi_0 = [\frac{1}{4}, \frac{1}{4}, \frac{1}{4}, \frac{1}{4}]$, $M = 2$ (a), $M = 3$ (b) and $M = 4$ (c) [132].

being the optimal policies tuned to the individual Boolean networks N_1, N_2, N_3 and N_4, respectively; and Pol^{SW} being the policy obtained for a PBN in which each Boolean network is assigned equal selection probability. The expected cost is 0.75 when we control using Pol^{TR}, 1.5 when using Pol^{SW}, and 1.75, 2.5, 1.5 and 1.75 when using Pol^1, Pol^2, Pol^3 and Pol^4, respectively. The expected uncontrolled cost is 2.5. For all horizons M and all initial states $z_0 = i \in S$, the method of this section is superior to the other methods considered. Out of the 128 states in the network, 89 states needed to be controlled in at least one of the 4 networks. In particular for $M = 5$, starting from such states Pol^{TR} was more effective than Pol^{SW} in reducing the cost by 0.1152 on average. In terms of absolute probabilities Pol^{TR} was able to take the system to a desirable attractor starting from all initial states and all networks with a probability 1.0, except for states $4, 36, 68, 100$ in network N_2, which are uncontrollable from pirin. For Pol^{SW}, states $4, 8, 24, 36, 68, 100$ are not taken to a desirable attractor in N_2. In the event of N_2 being the underlying network, starting from states $4, 36, 68, 100$, Pol^{TR} recognizes this and gives up promptly, while Pol^{SW} keeps on applying control, incurring extra costs, without any extra benefit.

Policy trees for initial state $z_0 = 93$, $\pi_0 = [\frac{1}{4}, \frac{1}{4}, \frac{1}{4}, \frac{1}{4}]$, and $M = 2, 3$ and 4 are shown in Fig. 17.12. The expected cost with $M = 2$ is 1.0 which can be further reduced to 0.25 if $M \geq 4$. This is reasonable because the algorithm has more time steps to identify and control the system.

17.5 External Intervention in the Infinite Horizon Case

The external control approaches presented so far have all focused on manipulating external (control) variables that affect the transition probabilities of a PBN to desirably affect its dynamic evolution over a finite time horizon. These short-term policies are not always effective in changing the steady-state behavior of the PBN, even though they can change the dynamical performance of the network for a small number of stages. Motivated by this, in [133] we considered intervention via external control variables in PBNs over an infinite length of time. We derived a control policy that does not change from one time step to the next because implementations of such stationary policies are often simple and stationary policies can be used to shift the steady-state distribution from undesirable states to desirable ones.

We first note that the problem formulation and results summarized in Section 17.1 for the finite horizon case serve to motivate the infinite horizon developments here. Consider the finite horizon cost function being minimized in Eq. 17.9 and suppose that the control horizon characterized by M is made larger and larger and in the limit we would like for it to tend to infinity. In trying to do so, we immediately encounter a number of potential obstacles that did not arise in the finite horizon case.

First, in the finite horizon case, since there is a terminal state which is being separately penalized, the cost per stage $C_k(z_k, v_k)$ is assumed to depend only on the control applied and the current state. In the infinite horizon problem, the control horizon is infinite and, therefore there is no terminal state or its associated terminal penalty. Consequently, for the infinite horizon case, the cost per stage should depend on the origin i, the destination j and the applied control input v. In other words, $C_k(i, v)$ of the finite horizon problem should now be replaced by $\tilde{C}(i, v, j)$ so that the per stage cost takes into account the origin, the destination and the control.[§]

Second, in the finite horizon problem, the summation in Eq. 17.9 is a finite one and so the quantity being minimized is finite. If we let the control horizon go to infinity, there is a possibility that the summation of the one stage costs may go to infinity (for all controls) leading to an ill-posed optimization problem. To make the optimization problem well posed, the cost considered in Eq. 17.9 has to be modified before letting the length M of the control horizon tend to infinity. Two such modifications have been extensively studied in the literature.

[§]Note that while finite horizon control problems in the literature allow for cost-per-stage functions that vary from one stage to another, infinite horizon control problems in the literature have typically been derived assuming that the same cost per stage function is used for all stages.

In the first case, we assume that the cost per stage $\tilde{C}(i, v, j)$ is *bounded* $\forall\, i, j \in S$ and $v \in U$ and a *discounting factor* $\alpha \in (0, 1)$ is introduced in the cost to make sure that the limit of the finite sums converges as the horizon length goes to infinity. More specifically, our objective is to find a policy $\pi = \{\mu_0, \mu_1......\}$, where $\mu_k : S \to U$, $k = 0, 1....$, that minimizes the cost function¶

$$J_\pi(z_0) = \lim_{M\to\infty} E\{\sum_{k=0}^{M-1} \alpha^k \tilde{C}(z_k, \mu_k(z_k), d_k)\}, \tag{17.48}$$

where the cost per stage $\tilde{C} : S \times U \times D \to \Re$ is given. This problem is referred to in the literature as the problem of *minimizing the total cost over an infinite number of stages with discounted and bounded cost per stage.* In the general formulation, the inclusion of α in the cost captures the fact that costs incurred at a later time are less significant. In the case of cancer treatment, $\alpha < 1$ signifies that the condition of the patient in the initial stages of treatment is more important than the condition at a later stage, or in other words, the reward for improving the condition of the patient in the present is more significant than the reward obtained from similar improvement at a later stage. This approach is reasonable if we keep in mind the expected life-span of the patient.

In the second case, one avoids the problem of a possibly infinite total cost by considering the *average cost per stage* which is defined by

$$J_\pi(z_0) = \lim_{M\to\infty} \frac{1}{M} E\{\sum_{k=0}^{M-1} \tilde{C}(z_k, \mu_k(z_k), d_k)\}. \tag{17.49}$$

In this formulation, a control policy $\pi = \{\mu_0, \mu_1, \cdots\}$ is chosen to minimize the above cost and the problem is referred to as the *average cost per stage problem.* Minimization of the total cost is feasible if $J_\pi(z_0)$ is finite for at least some admissible policies π and some admissible states z_0. If we consider no discounting, i.e. a *discount factor* of 1, and there is no zero-cost absorbing state (which is the case in context-sensitive PBNs with perturbation), then the total cost will frequently go to ∞. Hence the *average cost per stage* formulation is essential when we are interested in the condition of the patient in the long run and equal importance is given to the patient's condition in all stages.

For reasons already discussed, the cost per stage $\tilde{C}(z_k, v_k, d_k)$ depends on z_k, v_k and d_k. However, since in Eqns. 17.48 and 17.49, the cost is obtained only after taking the expectation with respect to the disturbances, it is possible to replace $\tilde{C}(z_k, v_k, d_k)$ by an equivalent cost per stage that does not depend on the disturbance d_k. This amounts to using the expected cost per stage in all calculations. More specifically, if $\tilde{C}(i, v, j)$ is the cost of using v at state i and

¶Note that a Markov chain can be modeled by $z_{k+1} = d_k$ [134]. Hence, the destination state is the same as the disturbance d_k.

moving to state j, we use as cost per stage the expected cost $C(i,v)$ given by [134]:

$$C(i,v) = \sum_{j=1}^{2^n} a_{ij}(v)\tilde{C}(i,v,j). \tag{17.50}$$

Now, the cost $\tilde{C}(i,v,j)$ of moving from state i to state j under control v may depend on the starting state i. However, in the case of PBNs, we have no obvious basis for assigning different costs based on different initial states. Accordingly, we assume that the penalty $\tilde{C}(i,v,j)$ is independent of the starting state i and its value is based on the control effort and the terminal state j. The penalty is high if the end state is a bad state regardless of the starting state, and vice-versa. Hence $\tilde{C}(i,v,j) = \tilde{C}(v,j)$ and Eq. 17.50 becomes

$$C(i,v) = \sum_{j=1}^{2^n} a_{ij}(v)\tilde{C}(v,j). \tag{17.51}$$

We next present the solution to the infinite horizon optimal control problem in the case where the performance index is *total cost with discounted and bounded cost per stage.* The solution for the case where the performance index is *average cost per stage* follows along similar lines and can be found in [133].

Let us denote by Π the set of all *admissible* policies π, i.e., the set of all sequences of functions $\pi = \mu_0, \mu_1, ...$ with $\mu_k(.) : S \to U, k = 0, 1, ...$ The optimal cost function J^* is defined by

$$J^*(z) = \min_{\pi \in \Pi} J_\pi(z), z \in S. \tag{17.52}$$

A *stationary policy* is an admissible policy of the form $\pi = \mu, \mu, ...,$ and its corresponding cost function is denoted by J_μ. We say that the stationary policy $\pi = \mu, \mu....$ is optimal if $J_\mu(z) = J^*(z)$ for all states z.

17.5.1 Optimal Control Solution

In this subsection, we solve the problem of minimizing the cost in Eq. 17.48 under the assumption that the cost per stage $\tilde{C}(i,v,d)$ is bounded, i.e. $\exists\, B > 0$ such that $\tilde{C}$ satisfies $|\tilde{C}(z,v,d)| \leq B$, for all $(z,v,d) \in S \times U \times D$. In the case of context-sensitive PBNs, this assumption holds since the expected cost, $C(i,v)$, for state i is given by Eq. (17.51), $\sum_{j=1}^{2^n} a_{ij}(v) = 1$, and $\tilde{C}(v,j)$ is bounded since the control and disturbance spaces are finite.

Observe that if we set $C_M(z_M) = 0 \;\forall\; z_M \in S$ and $C_k(z_k, v_k) = \alpha^k C(z_k, v_k)$ in the finite horizon problem of Eq. 17.9 and let $M \to \infty$, then we obtain the infinite horizon cost function considered in Eq. 17.48. Thus it seems reasonable that the finite horizon solution described by Eqs. 17.13 and 17.14 in Section 17.1 could provide a basis for arriving at the solution of the optimization problem in Eq. 17.52 where J_π is given by Eq. 17.48. A formal derivation of this connection is given in [134]. Here we simply state the result and present an intuitive justification for it.

Towards this end, note that Eq. 17.14 in the dynamic programming algorithm basically describes how the optimal cost J_{k+1} propagates backwards in time to the optimal cost J_k in the finite horizon problem of Eq. 17.9. For the cost function considered in Eq. 17.48, it is clear that the cost J_{k+1} must be discounted by the factor α while being propagated to the previous stage. Consequently, for the optimal control problem of this section, Eq. 17.14 will have to be replaced by

$$J_k(i) = \min_{v \in U} \left[C(i,v) + \alpha \sum_{j=1}^{2^n} a_{ij}(v) J_{k+1}(j) \right] . \qquad (17.53)$$

The above equation motivates the introduction of the following two mappings: For any cost function $J : S \to \Re$, define the mapping $TJ : S \to \Re$ by

$$(TJ)(i) = \min_{v \in U} [C(i,v) + \alpha \sum_{j=1}^{2^n} a_{ij}(v) J(j)], \ i \in S. \qquad (17.54)$$

Note that TJ is the optimal cost function for the one-stage (finite horizon) problem that has stage cost C and terminal cost αJ.

Similarly for any cost function $J : S \to \Re$ and control function $\mu : S \to C$, define the mapping $T_\mu J : S \to \Re$ by

$$(T_\mu J)(i) = C(i, \mu(i)) + \alpha \sum_{j=1}^{2^n} a_{ij}(\mu(i)) J(j), \ i \in S. \qquad (17.55)$$

$T_\mu J$ can be viewed as the cost function associated with the policy μ for the one-stage problem that has stage cost function C and terminal cost αJ. Since the mappings T and T_μ map functions $J : S \to \Re$ into new functions mapping S to $\Re$, one can define the composition of T with itself and T_μ with itself as follows:

$$(T^k J)(i) = (T(T^{k-1}J))(i), i \in S, \ k = 1, 2, \cdots , \qquad (17.56)$$

$$(T^0 J)(i) = J(i), \ i \in S, \qquad (17.57)$$

and

$$(T_\mu^k J)(i) = (T_\mu(T_\mu^{k-1}J))(i), i \in S, \ k = 1, 2, \cdots , \qquad (17.58)$$

$$(T_\mu^0 J)(i) = J(i), \ i \in S. \qquad (17.59)$$

The mappings T and T_μ play an important role in the solution of the optimal control problem of this section. Specifically, it can be shown that (i) the optimal cost function J^* is the unique fixed point of the map T; (ii) the iteration $J_{k+1} = TJ_k$ converges to J^* as $t \to \infty$; and (iii) the mapping T_μ can be used to characterize the conditions under which a given stationary policy μ

is optimal. These ideas are formalized in the following three theorems adapted from [134]. The proofs are available in [133].

THEOREM 17.1

Convergence of the discounted-cost algorithm: *For any bounded cost function $J : S \to \Re$, the optimal cost function J^* satisfies*

$$J^*(i) = \lim_{M\to\infty} (T^M J)(i), \text{for all } i \in S. \tag{17.60}$$

THEOREM 17.2

Bellman's Equation: *The optimal cost function J^* satisfies*

$$J^*(i) = \min_{v\in U}[C(i,v) + \alpha \sum_{j=1}^{2^n} a_{ij}(v)J^*(j))], \text{ for all } i \in S. \tag{17.61}$$

or, equivalently, $J^* = TJ^*$. Furthermore, J^* is the unique solution of this equation within the class of bounded functions.

THEOREM 17.3

Necessary and Sufficient Condition for Optimality: *A stationary policy μ is optimal if and only if $\mu(i)$ attains the minimum in Bellman's equation (17.61) for each $i \in S$; i.e.,*

$$TJ^* = T_\mu J^*. \tag{17.62}$$

The three theorems above provide the basis for coming up with computational algorithms for determining the optimal policy. Theorem 17.2 asserts that the optimal cost function satisfies Bellman's equation while Theorem 17.1 states that the optimal cost function can be iteratively determined by running the recursion

$$J_{k+1} = TJ_k, \ k = 0, 1, 2, \cdots \tag{17.63}$$

for any bounded initial cost function $J_0 : S \to \Re$. Since this iteration is guaranteed to converge to J^*, one can keep on running this iteration until some stopping criterion is reached. The resulting policy is a stationary one which, by Theorem 17.3, must be optimal. The iteration described in Eq. 17.63 above is referred to as the *value iteration* procedure since, at every stage we are iterating on the values of the cost function and the optimal policy simply falls out as a by product when the iteration converges to the optimal value of the cost function.

An alternative approach for solving the optimal control problem of this section is referred to as *policy iteration*. Before presenting this approach, we introduce the following matrix and vector notations.

$$J = \begin{pmatrix} J(1) \\ \cdot \\ \cdot \\ J(2^n) \end{pmatrix},$$

$$J_\mu = \begin{pmatrix} J_\mu(1) \\ \cdot \\ \cdot \\ J_\mu(2^n) \end{pmatrix},$$

$$TJ = \begin{pmatrix} (TJ)(1) \\ \cdot \\ \cdot \\ (TJ)(2^n) \end{pmatrix},$$

$$T_\mu J = \begin{pmatrix} (T_\mu J)(1) \\ \cdot \\ \cdot \\ (T_\mu J)(2^n) \end{pmatrix}.$$

The transition probability matrix corresponding to the stationary policy μ is represented as

$$A_\mu = \begin{pmatrix} a_{11}(\mu(1)) & ... & a_{1,2^n}(\mu(1)) \\ \cdot & \cdot & \cdot \\ \cdot & \cdot & \cdot \\ a_{2^n,1}(\mu(2^n)) & ... & a_{2^n,2^n}(\mu(2^n)) \end{pmatrix}$$

and C_μ represents the cost vector

$$C_\mu = \begin{pmatrix} C(1, \mu(1)) \\ \cdot \\ \cdot \\ C(2^n, \mu(2^n)) \end{pmatrix}.$$

Using the above notation, it is clear that for any stationary policy μ, Eq. 17.55 can be rewritten as

$$T_\mu J = C_\mu + \alpha A_\mu J.$$

Furthermore, it can be shown [133] that the cost J_μ corresponding to the policy μ satisfies

$$J_\mu = C_\mu + \alpha A_\mu J_\mu$$

or

$$[I - \alpha A_\mu] J_\mu = C_\mu. \tag{17.64}$$

Equation 17.64 above is a system of linear equations that can be solved to calculate the cost J_μ corresponding to a given stationary policy μ. In the policy iteration algorithm, one starts with a given stationary policy, evaluates the corresponding cost using Eq. 17.64 and tries to find a policy that yields a smaller cost. The process is terminated when we arrive at a fixed point of the mapping T. We next formally present the steps involved in the policy iteration algorithm.

Step 1: (initialization) An initial policy μ^0 is selected.

Step 2: (policy evaluation) Given a stationary policy μ^k, we compute the corresponding cost function J_{μ^k} from the linear system of equations

$$(I - \alpha A_{\mu^k})J_{\mu^k} = C_{\mu^k}. \tag{17.65}$$

A_{μ^k} is the probability transition matrix obtained using control policy μ^k.

Step 3: (policy improvement) An improved (in terms of the cost J) stationary policy μ^{k+1} satisfying $T_{\mu^{k+1}} J_{\mu^k} = T J_{\mu^k}$ is obtained.

The iterations are stopped if $J_{\mu^k} = T J_{\mu^k}$; otherwise, we return to Step 2 and repeat the process.

17.5.2 Melanoma Example

In this subsection, we describe the results of applying the infinite horizon control policy to a context-sensitive PBN derived from the same gene expression data as before. The network contains the seven genes WNT5A, pirin, S100P, RET1, MART1, HADHB and STC2. In this case, to obtain the PBN, we used the algorithms described in [118] to construct four highly probable Boolean networks (Fig. 17.13, Fig. 17.14, Fig. 17.15, Fig. 17.16) to use as the constituent Boolean networks in the PBN. Each constituent network was assumed to be derived from steady-state gene-expression data. The states were ordered as WNT5A, pirin, S100P, RET1, MART1, HADHB and STC2, with WNT5A as the most significant bit (MSB) and STC2 as the least significant bit (LSB).

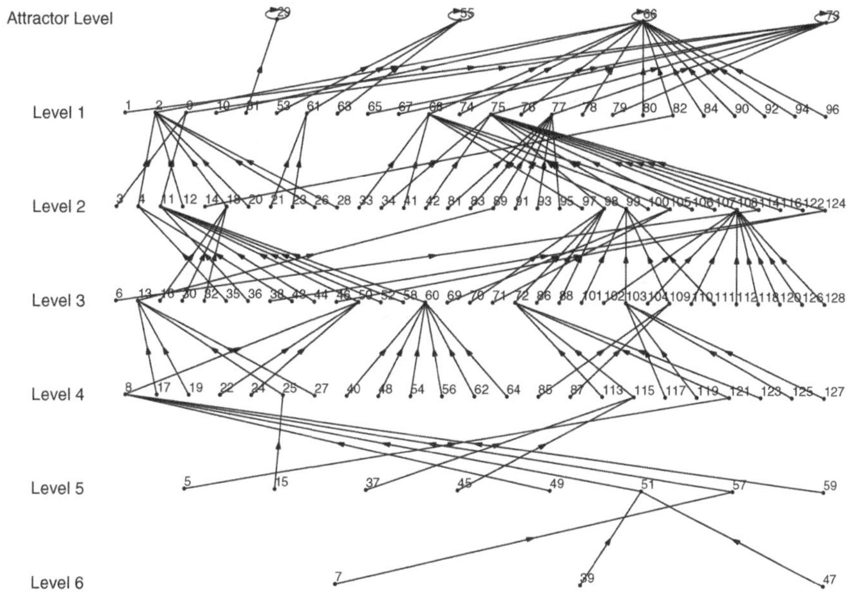

FIGURE 17.13 Network 1 [133] ©2006 IEEE.

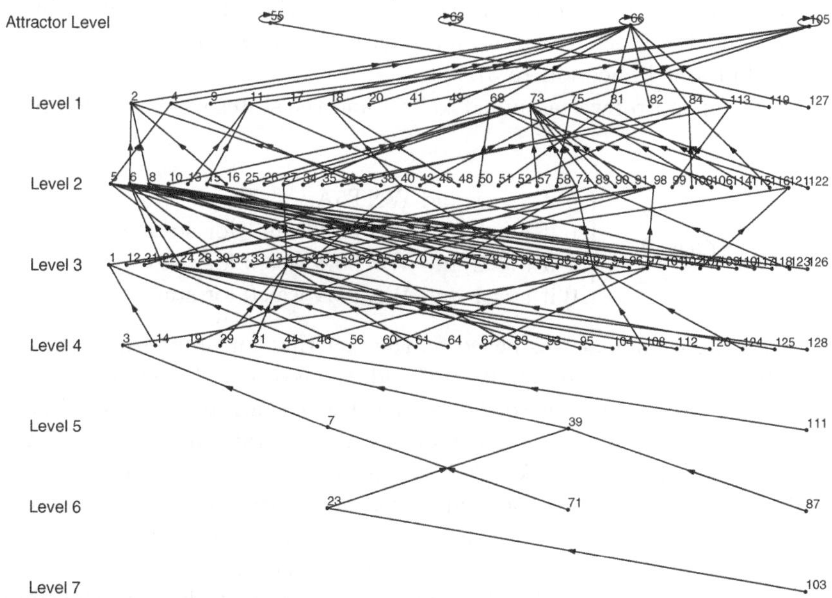

FIGURE 17.14 Network 2 [133] ©2006 IEEE.

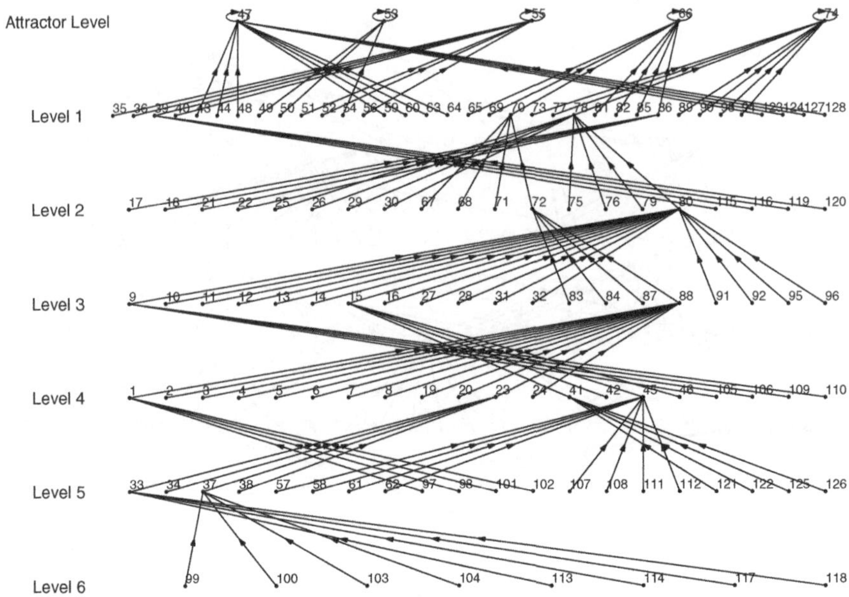

FIGURE 17.15 Network 3 [133] ©2006 IEEE.

The control strategy of this section was applied to the designed PBN with pirin chosen as the control gene ($v = 2$ signifying the state of pirin was reversed and $v = 1$ signifying no intervention) and $p = q = 0.01$ (see the p and q definitions in Section 15.3).

The cost of control was assumed to be 1 and the states were assigned penalties as follows:

$$\tilde{C}(v, j) = \begin{cases} 5 \text{ if } v = 1 \text{ and WNT5A was 1 for state } j \\ 6 \text{ if } v = 2 \text{ and WNT5A was 1 for state } j \\ 1 \text{ if } v = 2 \text{ and WNT5A was 0 for state } j \\ 0 \text{ if } v = 1 \text{ and WNT5A was 0 for state } j \end{cases}$$

The penalty assignment was based on the fact that for infinite-horizon problems, there is no terminal penalty; instead, the cost per stage C contains the penalties of each state. Since our objective was to down-regulate the WNT5A gene, a higher penalty was assigned for destination states having WNT5A up-regulated. Also for a given WNT5A status for the destination state, a higher penalty was assigned when the control was active versus when it was not.

Figure 17.17 shows the total cost for the discounted cost function with bounded cost per stage originating from each of the 128 states after the iterations had converged, with the discount factor α chosen to be 0.9. The control objective was to down-regulate the WNT5A gene. From Fig. 17.17, it is clear that the total cost with an optimal stationary policy was much lower than

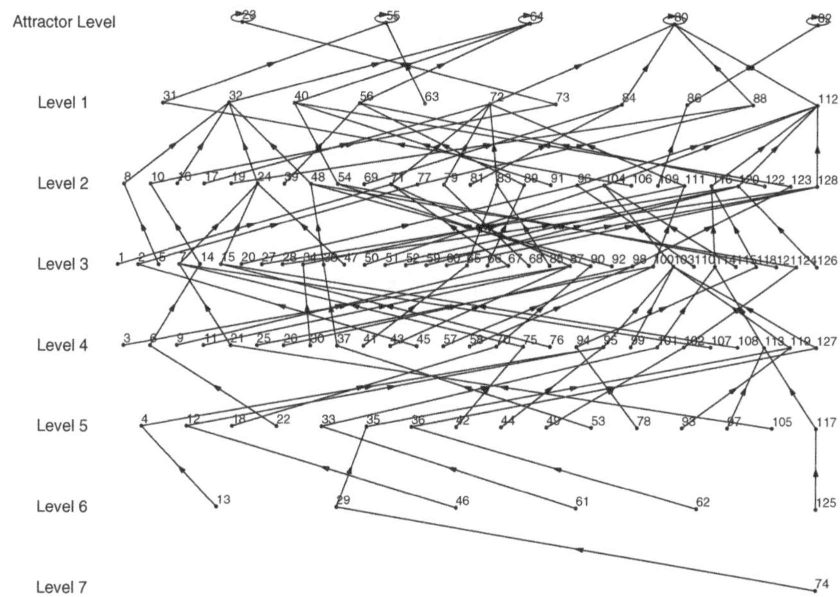

FIGURE 17.16 Network 4 [133] ©2006 IEEE.

that without control, which agreed with our objective. Fig. 17.18 shows the stationary policy obtained from the solution of the discounted cost problem. Fig. 17.19 shows the average total cost per state for each iteration. The stationary policy was obtained using value iteration and policy iteration. The starting policy for the policy iterations was $\mu = 0, 0, 0....$, i.e. no control, and hence the initial cost for the policy iteration was the same as the eventual total uncontrolled cost (Fig. 17.19). We should note that the policy iteration provided us the optimal policy in a smaller number of steps as compared to value iteration. Moreover, as the collection of stationary policies was finite (in this particular case, it was 2^{128}), the policy iteration was bound to give us an optimal stationary policy in a finite number of steps, whereas value iteration might converge in an infinite number of steps. On the other hand, the problem with policy iteration is solving the system of linear equations $(I - \alpha A_{\mu^k})J_{\mu^k} = C_{\mu^k}$, which becomes very complicated as the number of states increases.

Figure 17.20 shows the steady-state distributions of the PBN using the obtained stationary policy (Fig. 17.18) and Fig. 17.21 shows the original PBN steady state for comparison. We should note that the states from 1 to 64 have WNT5A =0 and hence are desirable states, as compared to states 65 to 128 that have WNT5A= 1 and hence are undesirable. The steady-state distribution Figures 17.20 and 17.21 show that the stationary policy enabled us to shift the probability mass from the bad states to states with lower

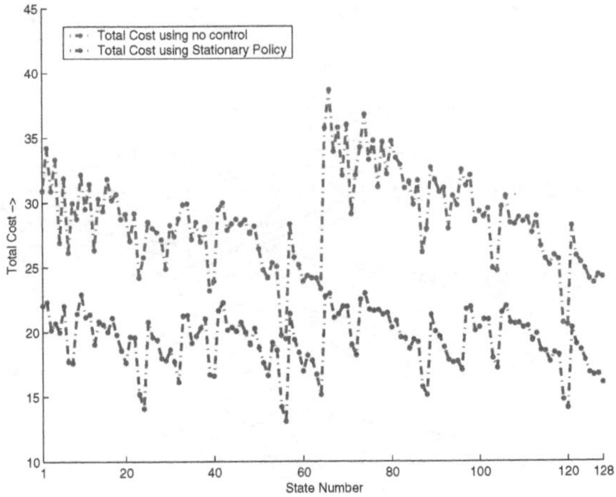

FIGURE 17.17 Total cost originating from different initial states (See color figure following page 146)[133] ©2006 IEEE.

metastatic competence. For example, state 66 (WNT5A is 1) has a high probability mass (0.15) in the original steady state but stationary control reduced its steady-state mass to 0.04. Similarly, the probability mass of state 64 (desirable state) was high when using the stationary policy. To numerically quantify the change, we multiplied the stationary distribution with the cost vector. For the original PBN, the cost vector was 0 for states 1 to 64 and 5 for states 65 to 128. For the stationary policy the cost vector was $\tilde{C}(\mu(z), z)$, $z \in [1, 2, 3, ...128]$. The value for the stationary policy using discounted cost formulation was 1.7465 as compared to 2.9830 for no control.

17.6 Concluding Remarks

In this chapter, we have discussed several approaches that have been recently developed for addressing the issue of external intervention in probabilistic Boolean networks. The results reported indicate that significant progress has been made in this area; however, numerous open issues remain and these will have to be successfully tackled before the methods suggested in this chapter find application in actual clinical practice. We next discuss some of the issues that we are aware of at the current time:

Methodical Assignment of Terminal Penalties

The formulation of the optimal control problem in Section 17.1 assumes that there is a terminal penalty associated with each state of the PBN; however, assignment of these terminal penalties for cancer therapy is by no means a

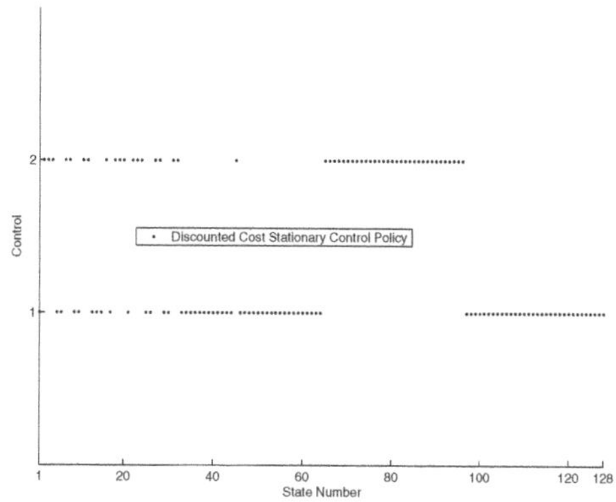

FIGURE 17.18 Stationary policy obtained using discounted cost formulation [133] ©2006 IEEE.

straightforward task. The reason is that while the intervention will be carried out only over a finite horizon, one would like to continue to enjoy the benefits in the steady state. For such purposes, the kind of terminal penalty used for the melanoma cell line study of Section 17.1 is inadequate since it fails to capture the steady-state behavior once the intervention has ceased. To remedy the situation, one could assign terminal penalties based on equivalence classes. The results of preliminary simulation studies in this regard [137] appear to be encouraging.

Choice of Control Input

In the case of the melanoma cell line study presented in Section 17.1, one of the genes in the PBN, namely pirin, has been used as a control input. The question is how to decide which gene to use. Of course, one consideration is to use genes for which inhibitors or enhancers are readily available. However, even if such a gene is chosen, how can we be certain that it is capable of controlling some other gene(s)? Although the answer is not clear at this stage, we do believe that the traditional control theoretic concept of *controllability* [135] may yield some useful insights. Another possibility is to use the concept of gene *influence* introduced in [104], an approach that was preliminarily explored in [131].

Robustness of the Control Strategies

The control algorithms presented in this chapter have all been analyzed assuming that the PBN model perfectly captures the actual behavior of the gene regulatory network. Since errors between the PBN model and the actual gene regulatory network are inevitable, the designed control algorithms will have to be *robust to modeling errors* if there is to be any hope of success upon actual implementation. Such robustness considerations have dominated the control literature for more than two decades now and we believe that some of the results obtained could be exploited in the context of application to genetic regulatory networks.

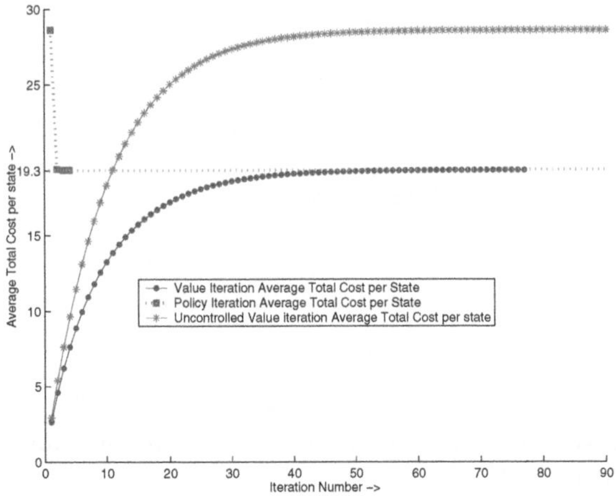

FIGURE 17.19 Average cost per state using discounted total cost (See color figure following page 146) [133] ©2006 IEEE.

The optimal control results presented in this chapter assume known transition probabilities and pertain to a problem of known length for the finite-horizon case. Their extension to the situation where the transition probabilities and the horizon length are unknown is a topic that merits further investigation. Finally, the results presented in this chapter correspond to the following stages in standard control design: modeling, controller design and verification of the performance of the designed controller *via computer simulations.* The designed controllers will have to be successfully implemented in practical studies, at least with cancer cell lines, to validate the use of engineering approaches in translational medicine. A considerable amount of future effort needs to be focused on this endeavor.

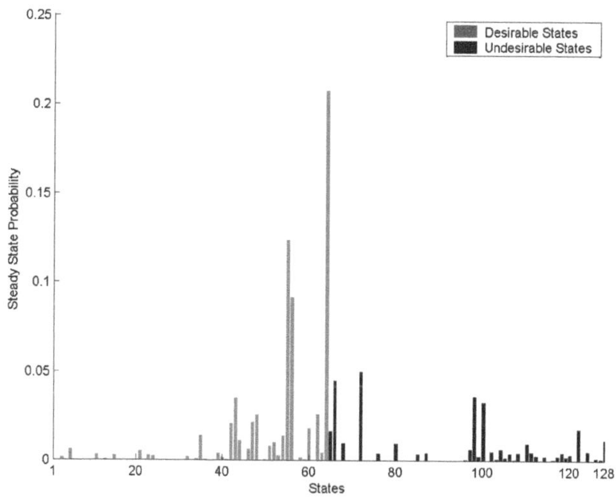

FIGURE 17.20 Steady state using discounted cost stationary policy (See color figure following page 146) [133] ©2006 IEEE.

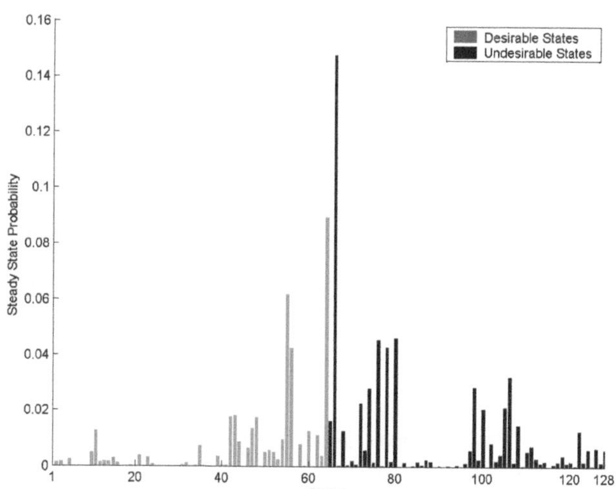

FIGURE 17.21 Original steady state (See color figure following page 146) [133] ©2006 IEEE.

References

[1] Alberts, B., Johnson, A., Lewis, J., Raff, M., Bray, D., Hopkin, K., Roberts, K., and Walter, P. *Essential Cell Biology.* Garland Science/ Taylor and Francis Group, 2nd edition, 2003.

[2] Murray, K. R., Granner, D. K., Mayes, P. A., and Rodwell, V. W. *Harpers Biochemistry.* McGraw-Hill Publishing Co, 25th edition, 1999.

[3] Schena, M., Shalon, D., Davis, R. W., and Brown, P. O. Quantitative monitoring of gene expression patterns with a complementary dna microarray. *Science*, 270:467 – 470, 1995.

[4] Chen, Y., Dougherty, E. R., and Bittner, M. L. Ratio-based decisions and the quantitative analysis of cdna microarray images. *Journal Of Biomedical Optics*, 2:364–374, 1997.

[5] Golub, T. R., Slonim, D. K., Tamayo, P., Huard, C., Gaasenbeek, M., Mesirov, J. P., Coller, H., Loh, M. L., Downing, J. R., Caligiuri, M. A., Bloomfield, C. D., and Lander, E. S. Molecular classification of cancer: Class discovery and class prediction by gene expression monitoring. *Science*, 286:531–537, 1999.

[6] Dougherty, E., Kim, S., and Chen, Y. Coefficient of determination in nonlinear signal processing. *Signal Process.*, 80:2219–2235, 2000.

[7] Schena, M. *Microarray Analysis.* Wiley, Hoboken, New Jersey, 2003.

[8] Blalock, E. M. *A Beginner's Guide to Microarrays.* Kluwer Academic Publishers, Boston, 2003.

[9] Hardiman, G. *Microarray Methods and Applications.* DNA Press Inc., Eagleville, PA, 2003.

[10] Zhang, W., Shmulevich, I., and Astola, J. *Microarray Quality Control.* Wiley, Hoboken, New Jersey, 2004.

[11] Kerr, M. K., Martin, M., and Churchill, G. A. Analysis of variance for gene expression microarray data. *Journal of Computational Biology*, 7:819–837, 2000.

[12] Kerr, M. K. and Churchill, G. Experimental design for gene expression microarrays. *Biostatistics*, 2:183–201, 2001.

[13] Lee, M. T., Kuo, F. C., Whitmore, G. A., and Sklar, J. Importance of replication in microarray gene expression studies: Statistical methods and evidence from repetitive cdna hybridizations. *Proc Natl Acad Sci USA*, 97:9834–9839, 2000.

[14] Yang, M. C. K., Ruan, Q. G., Yang, J. J., Eckenrode, S., Wu, S., McIndoe, R. A., and She, J. X. A statistical method for flagging weak spots improves normalization and ratio estimates in microarrays. *Physiological Genomics*, 7:45–53, 2001.

[15] Chen, Y., Kamat, V., Dougherty, E. R., Bittner, M. L., Meltzer, P. S., and Trent, J. M. Ratio statistics of gene expression levels and applications to microarray data analysis. *Bioinformatics*, 18:1207–1215, 2002.

[16] Ben-Dor, A., Bruhn, L., Friedman, N., Nachman, I., Schummer, M., and Yakhini, Z. Tissue classification with gene expression profiles. *Journal of Computational Biology*, 7:559–583, 2000.

[17] Wernisch, L., Kendall, S. L., Soneji, S., Wietzorrek, A., Parish, T., Hinds, J., Butcher, P. D., and Stoker, N. G. Analysis of whole-genome microarray replicates using mixed models. *Bioinformatics*, 19:53–61, 2003.

[18] Tu, Y., Stolovitzky, G., and Klein, U. Quantitative noise analysis for gene expression microarray experiments. *Proc Natl Acad Sci USA*, 99:14031–14036, 2002.

[19] Quackenbush, J. Microarray data normalization and transformation. *Nature Genetics*, 32:496–501, 2002.

[20] Bilban, M., Buehler, L. K., Head, S., Desoye, G., and Quaranta, V. Normalizing dna microarray data. *Curr Issues Mol Biol.*, 4:57–64, 2002.

[21] Hua, J., Balagurunathan, Y., Chen, Y., Lowey, J., Bittner, M., Xiong, Z., Suh, E., and Dougherty, E. R. Normalization benefits microarray-based classification. *EURASIP Journal on Bioinformatics and Systems Biology*, 2006.

[22] Tseng, G. C., Oh, M. K., Rohlin, L., Liao, J., and Wong, W. H. Issues in cdna microarray analysis: quality filtering, channel normalization, models of variations and assessment of gene effects. *Nucleic Acids Res.*, 29:2549–2557, 2001.

[23] Yang, Y. H., Dudoit, S., Luu, P., Lin, D. M., Peng, V., Ngai, J., and Speed, T. P. Normalization for cdna microarray data: a robust composite method addressing single and multiple slide systematic variation. *Nucleic Acids Res.*, 30:e15, 2002.

[24] Hughes, T. R., Marton, M. J., Jones, A. R., Roberts, C. J., Stoughton, R., Armour, C. D., Bennett, H. A., Coffey, E., Dai, H., He, Y. D., Kidd,

M. J., King, A. M., Meyer, M. R., Slade, D., Lum, P. Y., Stepaniants, S. B., Shoemaker, D. D., Gachotte, D., Chakraburtty, K., Simon, J., Bard, M., and Friend, S. H. Functional discovery via a compendium of expression profiles. *Cell*, 102:109–126, 2000.

[25] Nadon, R., Shi, P., Skandalis, A., Woody, E., Hubschle, H., Susko, E., Rghei, N., Ramm, P., and Brock. Statistical inference methods for gene expression arrays. *Proceedings of SPIE, BIOS2001, Microarrays, Optical Technologies and Informatics*, pages 46–55, 2001.

[26] Li, C. and Wong, W. H. Model-based analysis of oligonucleotide arrays: expression index computation and outlier detection. *Proc Natl Acad Sci USA*, 98:31–36, 2001.

[27] Hubbell, E. Multiplex sequencing by hybridization. *Journal of Computational Biology*, 8:141–149, 2001.

[28] Naef, F., Hacker, C., Patil, N., and Magnasco, M. Characterization of the expression ratio noise structure in high-density oligonucleotide arrays. *Genome Biol.*, 3:RESEARCH 0018, 2002.

[29] Irizarry, R. A. Exploration, normalization, and summaries of high density oligonucleotide array probe level data. *Biostatistics*, 4:249–264, 2003.

[30] Hedenfalk, I., Duggan, D., Chen, Y., Radmacher, M., Bittner, M., Simon, R., Meltzer, P., Gusterson, B., Esteller, M., Raffeld, M., Yakhini, Z., Ben-Dor, A., Dougherty, E. R., Kononen, J., Bubendorf, L., Fehrle, W., Pittaluga, S., Gruvberger, S., Loman, N., Johannsson, O., Olsson, H., Wilfond, B., Sauter, G., Kallioniemi, O. P., Borg, A., and Trent, J. Gene expression profiles in hereditary breast cancer. *New England Journal of Medicine*, 334:539–548, 2001.

[31] van de Vijver, M. J., He, Y. D., van't Veer, L. J., Dai, H., Hart, A. A., Voskuil, D. W., Schreiber, G. J., Peterse, J. L., Roberts, C., Marton, M. J., Parrish, M., Atsma, D., Witteveen, A., Glas, A., Delahaye, L., van der Velde, T., Bartelink, H., Rodenhuis, S., Rutgers, E. T., Friend, S. H., and Bernards, R. A gene-expression signature as a predictor of survival in breast cancer. *N. Engl. J. Med.*, 347:1999–2009, 2002.

[32] Dougherty, E. R. Small sample issues for microarray-based classification. *Comparative and Functional Genomics*, 2:28–34, 2001.

[33] Duda, R., Hart, P. E., and Stork, D. G. *Pattern Classification.* Wiley, New York, 2001.

[34] Cover, T. M. and Hart, P. E. Nearest neighbor pattern classification. *IEEE Transactions on Information Theory*, 13:21–27, 1967.

[35] Stone, C. Nonparametric regression and its applications (with discussion). *Ann. Statist.*, 5:595–645, 1977.

[36] Wald, P. W. and Kronmal, R. A. Discriminant functions when covariances are unequal and sample sizes moderate. *Biometrics*, 33:479–484, 1977.

[37] Vapnik, V. *Statistical Learning Theory.* Wiley, 1998.

[38] Titterington, D. M. General structure of regularization procedures in image reconstruction. *Astron. Astrophys.*, 144:381–387, 1985.

[39] Friedman, J. H. Regularized discriminant analysis. *J. Amer. Statist. Assoc.*, 84:165 – 175, 1989.

[40] Sietsma, J. and Dow, R. J. F. Neural net pruning - why and how. *IEEE International Conference on Neural Networks*, I:325–333, 1988.

[41] Matsuoka, A. Skeletal growth of a spongiose radiolarian dictyocoryre truncatum in laboratory culture. *Marine Micropaleontology*, 19:287–297, 1992.

[42] Bishop, C. M. *Neural Networks for Pattern Recognition.* Oxford University Press, 1995.

[43] Holmstraom, L. and Koistinen, P. Using additive noise in back-propagation training. *IEEE Transactions on Neural Networks*, 3:24–38, 1992.

[44] Skurichina, M., Kuncheva, L., and Duin, R. Bagging and boosting for the nearest mean classifier: Effects of sample size on diversity and accuracy. *Proc. Int. Workshop on Multiple Classifier Systems, Calgiari, Italy*, pages 62–71, 2002.

[45] Hua, J., Lowey, J., Xiong, Z., and Dougherty, E. R. Noise-injected neural networks show promise for use on small-sample expression data. *BMC Bioinformatics*, 7:274–287, 2006.

[46] Kim, S., Dougherty, E. R., Barrera, J., Chen, Y., Bittner, M. L., and Trent, J. Strong feature sets from small samples. *Journal of Computational Biology*, 9:127–146, 2002.

[47] Hughes, G. F. On the mean accuracy of statistical pattern recognizers. *IEEE Trans. Information Theory*, 14:55–63, 1968.

[48] Jain, A. and Chandrasekaran, B. Dimensionality and sample size considerations in pattern recognition practice. *Handbook of Statistics*, 2:835–855, 1982. in Krishnaiah, P. R., Kanal, L. N., eds.

[49] Cover, T. and Campenhout, J. M. V. On the possible orderings in the measurement selection problem. *IEEE Transactions on Systems, Man and Cybernetics*, 7:657–661, 1977.

[50] Sima, C., Attoor, S., Braga-Neto, U., Lowey, J., Suh, E., and Dougherty, E. R. Impact of error estimation on feature-selection algorithms. *Pattern Recognition*, 38:2472–2482, 2005.

[51] Pudil, P., Novovicova, J., and Kittler, J. Floating search methods in feature selection. *Pattern Recog. Lett.*, 15:1119–1125, 1994.

[52] Sima, C., Braga-Neto, U., and Dougherty, E. R. Superior feature-set ranking for small samples using bolstered error estimation. *Bioinformatics*, 21:1046–1054, 2005.

[53] Jain, A. and Zongker, D. Feature selection-evaluation, application, and small sample performance. *IEEE Trans. Pattern Anal. Machine Intell.,*, 19:153–158, 1997.

[54] Kudo, M. and Sklansky, J. Comparison of algorithms that select features for pattern classifiers. *Pattern Recog.*, 33:25–41, 2000.

[55] Hua, J., Xiong, Z., and Dougherty, E. Determination of the optimal number of features for quadratic discriminant analysis via the normal approximation to the discriminant distribution. *Pattern Recog.*, 38:403–421, 2004.

[56] Dougherty, E. R. Feature-selection overfitting with small-sample classifier design. *IEEE Intelligent Systems Magazine*, pages 64–66, 2005.

[57] Sima, C. and Dougherty, E. R. What should be expected from feature selection in small-sample settings. to appear in Bioinformatics.

[58] Zhou, X. and Mao, K. Z. The ties problem resulting from counting-based error estimators and its impact on gene selection algorithms. to appear in Bioinformatics.

[59] Devroye, L., Gyorfi, L., and Lugosi, G. *A Probabilistic Theory of Pattern Recognition.* Springer, Berlin, 1997.

[60] Efron, B. Bootstrap methods: another look at jackknife. *Ann. Statist.*, 7:1–26, 1979.

[61] Efron, B. Estimating the error rate of a prediction rule: Improvements on cross-validation. *Journal of the American Statistical Association*, 78:316–331, 1983.

[62] Braga-Neto, U. and Dougherty, E. Bolstered error estimation. *Pattern Recognition*, 6:1267–1281, 2004.

[63] Braga-Neto, U. M., Hashimoto, R., Dougherty, E. R., Nguyen, D. V., and Carroll, R. J. Is cross-validation better than resubstitution for ranking genes. *Bioinformatics*, 20:253–258, 2004.

[64] Braga-Neto, U. and Dougherty, E. Is cross-validation valid for small-sample microarray classification? *Bioinformatics*, 20:374–380, 2004.

[65] Braga-Neto, U. and Dougherty, E. R. Exact performance measures and distributions of error estimators for discrete classifiers. *Pattern Recognition*, 38:1799–1814, 2005.

[66] Dougherty, E. R. and Braga-Neto, U. Epistemology of computational biology: Mathematical models and experimental prediction as the basis of their validity. *Journal of Biological Systems*, 14:65–90, 2006.

[67] Dougherty, E. R. and Brun, M. A probabilistic theory of clustering. *Pattern Recognition*, 37:917–925, 2004.

[68] Kohonen, T. Self organized formation of topologically correct feature maps. *Cybernetics*, 43:59–69, 1982.

[69] Kohonen, T. *Self Organizing Maps.* Springer-Verlag, Berlin, 1995.

[70] Cottrell, M., Fort, J. C., and Pages, G. Theoretical aspects of the som algorithm. *Neurocomputing*, 21:119–138, 1998.

[71] Flexer, A. On the use of self-organizing maps for clustering and visualization. *Intelligent Data Analysis*, 5:373–384, 2001.

[72] Kobayashi, T., Yamaguchi, M., Kim, S., Morikawa, J., Ueno, S., Suh, E., Dougherty, E. R., Shmulevich, I., Shiku, H., and Zhang., W. Microarray reveals differences in both tumors and vascular specific gene expression in de novo cd5+ and cd5- diffuse large b-cell lymphomas. *Cancer Research*, 63:60–66, 2003.

[73] Dougherty, E. R., Barrera, J., Brun, M., Kim, S., Cesar, R. M., Chen, Y., Bittner, M., and Trent, J. M. Inference from clustering with application to gene-expression microarrays. *Journal of Computational Biology*, 9:105–126, 2002.

[74] Iyer, V. R., Eisen, M. B., Ross, D. T., Schuler, G., Moore, T., Lee, J. C. F., Trent, J. M., Staudt, L. M., Hudson, J., Boguski, M. S., Lashkari, D., Shalon, D., Botstein, D., and Brown, P. O. The transcriptional program in the response of human fibroblasts to serum. *Science*, 283:83–87, 1999.

[75] Rand, W. M. Objective criteria for the evaluation of clustering methods. *Journal of the American Statistical Association*, 66:846–850, 1971.

[76] Milligan, G. W. and Cooper, M. C. A study of the comparability of external criteria for hierarchical cluster analysis. *Multivariate Behavioral Research*, 21:441–458, 1986.

[77] Davies, D. L. and Bouldin, D. A cluster separation measure. *IEEE Trans. on Pattern Analysis and Machine Intelligence*, 1:224–227, 1979.

[78] Dunn, J. C. A fuzzy relative of the isodata process and its use in detecting compact well-separated clusters. *J. Cybernetics*, 3:32–57, 1973.

[79] Maulik, U. and Bandyopadhyay, S. Performance evaluation of some clustering algorithms and validity indices. *IEEE Transactions on Pattern Analysis and Machine Intelligence*, 24:1650–1654, 2002.

[80] Brun, M., Sima, C., Hua, J., Lowey, J., Carroll, B., Suh, E., and Dougherty, E. R. Model-based evaluation of clustering validation measures. to appear in Pattern Recognition.

[81] Endy, D. and Brent, R. Modelling cellular behaviour. *Nature*, 409:391–395, 2001.

[82] Ideker, T., Galitski, T., and Hood, L. A new approach to decoding life : Systems biology. *Annu. Rev. Genomics Hum. Genet*, 2:343–372, 2001.

[83] Hasty, J., McMillen, D., Issacs, F., and Collins, J. J. Computational studies of gene regulatory networks: in numero molecular biology. *Nat. Rev. Genet.*, 2:268–279, 2001.

[84] Karamouzis, M., Gorgoulis, V., and Papavassiliou, A. Transcription factors and neoplasia: vistas in novel drug design. *Clin Cancer Res.*, 8:949–961, 2002.

[85] Wolf, D. M. and Eckman, F. H. On the relationship between genomic regulatory element organization and gene regulatory dynamics. *J. Theor. Biol*, 195:167–186, 1998.

[86] Ideker, T., Thorsson, V., Ranish, J. A., Christmas, R., Buhler, J., Eng, J. K., Bumgarner, R., Goodlett, D. R., Aebersold, R., and Hood, L. Integrated genomic and proteomic analyses of a systematically perturbed metabolic network. *Science*, 292:929–934, 2001.

[87] Murphy, K. and Mian, S. Modelling gene expression data using dynamic bayesian networks, 1999. (Technical Report, Computer Science Division, University of California, Berkeley, CA).

[88] Friedman, N., Linial, M., Nachman, I., and Pe'er, D. Bayesian networks to analyze expression data. *Proceedings of the Fourth Annual International Conference on Computational Molecular Biology*, pages 127–135, 2000.

[89] Hartemink, A., Gifford, D., Jaakkola, T., and Young, R. Maximum likelihood estimation of optimal scaling factors for expression array normalization. *SPIE BIOS*, 2001.

[90] Weaver, D. C., Workman, C. T., and Stormo, G. D. Modeling regulatory networks with weight matrices. *Pac. Symp. Biocomputing*, 4:112–123, 1999.

[91] Wahde, M. and Hertz, J. Coarse-grained reverse engineering of genetic regulatory networks. *BioSystems*, 55:129–136, 2000.

[92] Mestl, T., Plahte, E., and Omholt, S. W. A mathematical framework for describing and analysing gene regulatory networks. *J Theor Biol.*, 176:291–300, 1995.

[93] De Jong, H. Modeling and simulation of genetic regulatory systems: A literature review. *Journal of Computational Biology*, 9:67–103, 2001.

[94] Goutsias, J. and Kim, S. A nonlinear discrete dynamical model for transcriptional regulation: Construction and properties. *Biophys. J.*, 86:1922–1945, 2004.

[95] Ivanov, I. and Dougherty, E. R. Modeling genetic regulatory networks: Continuous or discrete? *Journal of Biological Systems*, 14:219–229, 2006.

[96] Kauffman, S. Metabolic stability and epigenesis in randomly constructed genetic nets. *Theor. Biol.*, 22:437–467, 1969.

[97] Kauffman, S. Homeostasis and differentiation in random genetic control networks. *Nature*, 224:177–178, 1969.

[98] Glass, L. and Kauffman, S. The logical analysis of continuous, nonlinear biochemical control networks. *Journal of Theoretical Biology*, 39:103–129, 1973.

[99] Shmulevich, I. and Zhang, W. Binary analysis and optimization-based normalization of gene expression data. *Bioinformatics*, 18:555–565, 2002.

[100] Zhou, X., Wang, X., and Dougherty, E. R. Binarization of microarray data based on a mixture model. *Molecular Cancer Therapeutics*, 2:679–684, 2003.

[101] Pal, R., Datta, A., Fornace, A., Bittner, M., and Dougherty, E. Boolean relationships among genes responsive to ionizing radiation in the nci 60 acds. *Bioinformatics*, 21:1542–1549, 2005.

[102] Yuh, C.-H., Bolouri, H., and Davidson, E. H. Genomic cis-regulatory logic: Experimental and computational analysis of a sea urchin gene. *Science*, 279:1896 - 1902, 1998.

[103] Davidson, E. H., Rast, J. P., Oliveri, P., Ransick, A., Calestani, C., Yuh, C.-H., Minokawa, T., Amore, G., Hinman, V., Arenas-Mena, C., Otim, O., Brown, C. T., Livi, C. B., Lee, P. Y., Revilla, R., Rust, A. G., jun Pan, Z., Schilstra, M. J., Clarke, P. J. C., Arnone, M. I., Rowen, L., Cameron, R. A., McClay, D. R., Hood, L., and Bolouri, H. A genomic regulatory network for development. *Science*, 295:1669–1678, 2002.

[104] Shmulevich, I., Dougherty, E. R., Kim, S., and Zhang, W. Probabilistic boolean networks: A rule-based uncertainty model for gene regulatory networks. *Bioinformatics*, 18:261–274, 2002.

[105] Kauffman, S. *The Origins of Order: Self-Organization and Selection in Evolution.* Oxford Univ. Press, New York, 1993.

[106] Huang, S. Gene expression profiling, genetic networks, and cellular states: An integrating concept for tumorigenesis and drug discovery. *Molec. Med.*, 77:469–480, 1999.

[107] Kim, S., Dougherty, E. R., Bittner, M. L., Chen, Y., Sivakumar, K., Meltzer, P., and Trent, J. M. General nonlinear framework for the analysis of gene interaction via multivariate expression arrays. *Journal of Biomedical Optics*, 5:411–424, 2000.

[108] Shmulevich, I., Dougherty, E. R., and Zhang, W. From boolean to probabilistic boolean networks as models of genetic regulatory networks. *Proceedings of IEEE*, 90:1778–1792, 2002.

[109] Brun, M., Dougherty, E. R., and Shmulevich, I. Steady-state probabilities for attractors in probabilistic boolean networks. *Signal Processing*, 85:1993–2013, 2005.

[110] Lahdesmaki, H., Shmulevich, I., and Yli-Harja, O. On learning gene regulatory networks under the boolean network model. *Machine Learning*, 52:147–167, 2003.

[111] Zhou, X., Wang, X., and Dougherty, E. R. Construction of genomic networks using mutual-information clustering and reversible-jump markov-chain-monte-carlo predictor design. *Signal Processing*, 83:745–761, 2003.

[112] Hashimoto, R. F., Seungchan, K., Shmulevich, I., Zhang, W., Bittner, M. L., and Dougherty, E. R. Growing genetic regulatory networks from seed genes. *Bioinformatics*, 20:1241–1247, 2004.

[113] Bittner, M., Meltzer, P., Chen, Y., Jiang, Y., Seftor, E., Hendrix, M., Radmacher, M., Simon, R., Yakhini, Z., Ben-Dor, A., Sampas, N., Dougherty, E., Wang, E., Marincola, F., Gooden, C., Lueders, J., Glatfelter, A., Pollock, P., Carpten, J., Gillanders, E., Leja, D., Dietrich, K., Beaudry, C., Berens, M., Alberts, D., and Sondak, V. Molecular classification of cutaneous malignant melanoma by gene expression profiling. *Nature*, 406:536–540, 2000.

[114] Weeraratna, A. T., Jiang, Y., Hostetter, G., Rosenblatt, K., Duray, P., Bittner, M., and M., T. J. Wnt5a signalling directly affects cell motility and invasion of metastatic melanoma. *Cancer Cell*, 1:279–288, 2002.

[115] Kim, S., Li, H., Dougherty, E. R., Chao, N., Chen, Y., Bittner, M. L., and Suh, E. B. Can markov chain models mimic biological regulation ? *Biological Systems*, 10:337–357, 2002.

[116] Zhou, X., Wang, X., Pal, R., Ivanov, I., Bittner, M. L., and Dougherty, E. R. A bayesian connectivity-based approach to constructing probabilistic gene regulatory networks. *Bioinformatics*, 20:2918–2927, 2004.

[117] Dougherty, E. R. and Xiao, Y. Design of probabilistic boolean networks under the requirement of contextual data consistency. to appear in IEEE Transactions on Signal Processing.

[118] Pal, R., Ivanov, I., Datta, A., Bittner, M. L., and Dougherty, E. R. Generating boolean networks with a prescribed attractor structure. *Bioinformatics*, 21:4021–4025, 2005.

[119] Peer, D., Regev, A., Elidan, G., and Friedman, N. Inferring subnetworks from perturbed expression profiles. *Bioinformatics*, 17:215–224, 2001.

[120] Husmeier, D. Reverse engineering of genetic networks with bayesian networks. *Biochem. Soc. Trans.*, 31:1516–1518, 2003.

[121] Pearl, J. *Probabilistic Reasoning in Intelligent Systems: Networks of Plausible Inference.* Morgan Kaufman Publishers, San Mateo, CA, 1988.

[122] Lauritzen, S. L. *Graphical Models.* Oxford University Press, 1996.

[123] Jensen, F. V. *Bayesian Networks and Decision Graphs.* Springer-Verlag, 2001.

[124] Lahdesmaki, H., Yli-Harja, O., Zhang, W., and Shmulevich, I. Intrinsic dimensionality in gene expression analysis. *IEEE International Workshop on Genomic Signal Processing and Statistics*, 2005.

[125] Shmulevich, I., Dougherty, E. R., and Zhang, W. Gene perturbation and intervention in probabilistic boolean networks. *Bioinformatics*, 18:1319–1331, 2002.

[126] Shmulevich, I., Dougherty, E. R., and Zhang, W. Control of stationary behaviour in probabilistic boolean networks by means of structural intervention. *Biological Systems*, 10:431–446, 2002.

[127] Datta, A., Choudhary, A., Bittner, M. L., and Dougherty, E. R. External control in markovian genetic regulatory networks. *Machine Learning*, 52:169–191, 2003.

[128] Cinlar, E. *Introduction to Stochastic Processes.* Prentice Hall, Englewood Cliffs,NJ, 1997.

[129] Goldberg, D. *Genetic Algorithms in Search, Optimization, and Machine Learning.* Addison-Wesley, Reading, MA, 1989.

[130] Datta, A., Choudhary, A., Bittner, M. L., and Dougherty, E. R. External control in markovian genetic regulatory networks: The imperfect information case. *Bioinformatics*, 20:924–930, 2004.

[131] Pal, R., Datta, A., Bittner, M. L., and Dougherty, E. R. Intervention in context-sensitive probabilistic boolean networks. *Bioinformatics*, 21:1211–1218, 2005.

[132] Choudhary, A., Datta, A., Bittner, M. L., and Dougherty, E. R. Intervention in a family of boolean networks. *Bioinformatics*, 22:226–232, 2006.

[133] Pal, R., Datta, A., and Dougherty, E. R. Optimal infinite horizon control for probabilistic boolean networks. *IEEE Transactions on Signal Processing*, 54:2375– 2387, 2006.

[134] Bertsekas, D. P. *Dynamic Programming and Optimal Control.* Athena Scientific, 2nd edition, 2001.

[135] Kalman, R. Canonical structure of linear dynamical systems. *Proc. Natl Acad. Sci.*, pages 596–600, 1962.

[136] Smallwood, R. D. and Sondik, E. J. Optimal control of partially observable markov processes over a finite horizon. *Operations Research*, 21:1071–1088, 1973.

[137] Choudhary, A., Datta, A., Bittner, M., and Dougherty, E. Assignment of terminal penalties in controlling genetic regulatory networks. *American Control Conference*, pages 417–422, 2005.

Index

Aniruddha Datta is a Professor in the Department of Electrical and Computer Engineering at Texas A & M University, College Station, Texas. He holds a B. Tech degree in Electrical Engineering from the Indian Institute of Technology, Kharagpur, an M.S.E.E. degree from Southern Illinois University, Carbondale and an M.S. (Applied Mathematics) and Ph.D. degrees from the University of Southern California. He is the author of three books in the controls area and has authored more than eighty journal and conference papers. He is a Senior Member of IEEE, and has served as an Associate Editor of the IEEE Transactions on Automatic Control and the IEEE Transactions on Systems, Man and Cybernetics-Part B. His areas of interest include adaptive control, robust control, PID control and Genomic Signal Processing.

Edward R. Dougherty is a Professor in the Department of Electrical and Computer Engineering at Texas A&M University in College Station, TX, Director of the Genomic Signal Processing Laboratory at Texas A&M University, and Director of the Computational Biology Division of the Translational Genomics Research Institute in Phoenix, AZ. He holds a Ph.D. in mathematics from Rutgers University and an M.S. in Computer Science from Stevens Institute of Technology. He is author of twelve books, editor of five others, and author of more than one hundred and eighty journal papers. He is an SPIE fellow, is a recipient of the SPIE President's Award, and served as editor of the Journal of Electronic Imaging for six years. Prof. Dougherty has contributed extensively to the statistical design of nonlinear operators for image processing and the consequent application of pattern recognition theory to nonlinear image processing. His current research is focused in genomic signal processing, with the central goal being to model genomic regulatory mechanisms for the purposes of diagnosis and therapy.

For Product Safety Concerns and Information please contact our
EU representative GPSR@taylorandfrancis.com Taylor & Francis
Verlag GmbH, Kaufingerstraße 24, 80331 München, Germany

